COMMUNITY, TECHNOLOGY & TRADITION

Emma C. Wager

COMMUNITY, TECHNOLOGY & TRADITION

A Social Prehistory of the Great Orme Mine

© 2024 Emma Wager

This book is published under a Creative Commons Attribution 4.0 International License (CC BY 4.0). This license does not apply to content that is attributed to sources other than the copyright holder mentioned above. Under this license you are free to share and adapt this work, provided the makers are attributed.

For more information about our licenses, please visit https://www.sidestone.com/publishing/creative-commons.

Published by Sidestone Press, Leiden
www.sidestone.com

Imprint: Sidestone Press Academics
This book has been peer-reviewed. For more information see www.sidestone.com

Lay-out & cover design: Sidestone Press
Photograph cover: Emma Wager

ISBN 978-94-6427-090-7 (softcover)
ISBN 978-94-6427-091-4 (hardcover)
ISBN 978-94-6427-092-1 (PDF e-book)

DOI: 10.59641/0a901dm

Contents

Acknowledgements	7
1. Bronze Age mining as a social phenomenon	9
2. The Bronze Age mine: structure and development	25
3. The technology of mining: its sequence and content	69
4. Mining and the taskscape	99
5. Mining and the construction of community	127
6. A fundamentally social archaeology of Bronze Age mining	149
Bibliography	159
Appendices	175

Acknowledgements

The original PhD research from which this book stems was kindly funded by AHRB, with additional financial support from the British Federation of Women Graduates and the Historical Metallurgy Society. I am very grateful to the latter, and to the Cambrian Archaeological Association, for their recent generous financial support towards the illustration costs for this volume.

The ideas in this book draw on and have been made possible by the scholarship of others. Many individuals and organisations have generously shared advice, assistance and information, including but not limited to: Victoria Allnatt, Heather Beeton, Sue Bridgford, Clwyd-Powys Archaeological Trust, Sean Derby, Gary Duckers, Ralph Fyfe, Mel Giles, Adam Gwilt, Gwynedd Archaeological Trust, Emma Henderson, Bob Johnston, Jane Kenney, Peter Northover, Ben Roberts and Rick Schulting. Special thanks to Nick Jowett and Tony Hammond, Great Orme Mines Ltd, for permission to include previously unpublished radiocarbon dates and sample data from the mine. This work contains CPAT HER Charitable Trust data © copyright and database right (2023) and data derived from information held by the GAT HER Charitable Trust Database Right.

This book could not have been written without Andy Lewis's groundbreaking research on the Great Orme mine and his kind permission to reproduce previously unpublished site plans, photographs and data from his MPhil dissertation. I would particularly like to thank him for generously giving his time to answer my questions about the evidence.

Sophia Adams gave her own time to carry out Bayesian modelling specifically for this project. I am immensely grateful for her contribution and expertise and for her initial suggestion to include this analysis in the book. Peter Marshall kindly provided the original data from his own site-based Bayesian analysis of the radiocarbon dates from the Great Orme mine. Special thanks too to Michalis Catapotis for assisting with the analysis and to Derek Hamilton for responding to my questions about the significance of the results. Any errors in interpretation are entirely mine.

I am particularly grateful to Caroline Jackson and William O'Brien for their encouragement and continuing support for this project and for reading draft chapters. Their insightful comments, along with those of two referees, have greatly helped to improve the work. I owe a large debt of thanks to Barbara Ottaway and Mark Edmonds, whose supervision during my PhD inspired my enduring interest in the Great Orme mine, prehistoric mining and the study of past technologies. During that period, I was also fortunate enough to spend time studying the theory and practice of mining archaeology under the late Geoff David and the late David Jenkins. I remember their mentorship with gratitude.

The following individuals and archaeological organisations kindly helped with images and the permissions to reproduce illustrations: Maurice Czerewko, Andy Lewis, Gwynedd Archaeological Trust, Simon Timberlake, Giwrgos Peppas, Genevieve Tellier, Kate Waddington and William O'Brien. I wish to extend particular thanks to Kate Waddington for her helpful guidance on how to make distribution maps; to Michalis Andreikos for producing many of the drawings; and to Natassa Georgopoulou for enhancing the reproduction quality of the photographs and her support with the superb cover design. I am also grateful to Karsten Wentink and the team at Sidestone Press for their editorial support.

Writing online together with Susanna Harris and Nicki Whitehouse gave me much-needed motivation, advice and encouragement, brightening my Monday mornings. Thank you to Sophia Adams for inviting me to join the group. I am very grateful to the Early Mines Research Group (Simon Timberlake, Brenda Craddock, Phil Andrews, John Pickin and Anthony Gilmour) for reigniting my enthusiasm for fieldwork and for fascinating discussions about prehistoric mining and other matters. Lucy Price was there at the (second) beginning; thank you for encouraging me to get started and for staying the course. Gabriella Blandy provided the tools I needed to complete this project and I thank her from the bottom of my heart. Finally, this book is dedicated to Vassilis, who makes everything seem possible.

Author's Note

The manuscript for this book was in the final stages of publication when Williams, 2023, 'Boom and Bust in Bronze Age Britain. The Great Orme Copper Mine and European Trade' (Archaeopress), was published. Consequently, that work, which defines the temporal and spatial distribution of the products of prehistoric copper mining on the Great Orme in the corpus of contemporary metalwork, is not referenced in the discussion that follows.

1

Bronze Age mining as a social phenomenon

This book is about mining in Britain during the Bronze Age. It uses the Great Orme copper mine (SH77078308) in Wales, in the western British Isles, as its case study (Figure 1.1).

Miners seeking copper ore there during the second and early first millennia BC created one of the largest Bronze Age copper mines in Europe (O'Brien, 2015, p. 146). The metal produced from the ore extracted on the Great Orme has been traced to objects of copper and bronze that date from the end of the third to the early first millennia BC. Some of these items have been found as far afield as France, The Netherlands and Sweden (Williams and Le Carlier de Veslud, 2019, p. 1185, fig. 6). Both copper and bronze are thought to have played important roles in the lives of people during the Bronze Age (Webley, Adams and Brück, 2020, pp. 1–2). The sequence of tasks needed to make things from these materials had

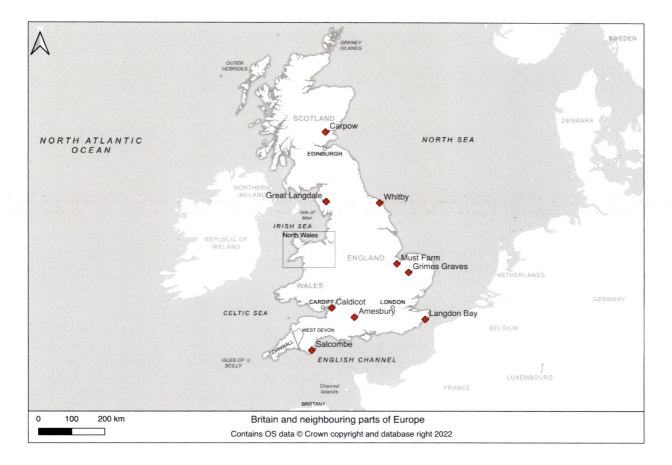

Figure 1.1: Britain and the neighbouring parts of Europe, showing the location of places mentioned in the text. For a larger-scale view of north Wales, see Figure 1.2.

five key stages, although the precise sequence and content of each stage could vary considerably. At its simplest, the first stage involved mining and processing copper ore to concentrate it for smelting. Secondly, the concentrated ore was smelted to produce copper metal. If bronze was the desired result, the smelted copper metal was then alloyed with tin, sometimes also with the addition of lead. In the third stage, the copper or bronze metal was melted and cast to create an artefact or blank. This was then followed by smithing and finishing to create the object's final form (stage four). During the final (fifth) stage, the artefact may have been repeatedly repaired and modified while in use, before being deposited (perhaps after having been intentionally destroyed) or remelted to recycle it into a new form (after Webley, Adams and Brück, 2020, pp. 19–20, fig. 2.1).

Although the initial stage of this production sequence is the primary focus of this book, the discussion will touch on evidence for all the subsequent stages, as together these provide an important practical and technological context for understanding routines of copper ore extraction and processing at the Great Orme mine. This remarkable site has much to tell us in general about mining for ore in the Bronze Age. Previous research at the Great Orme mine has already contributed significantly to our understanding of prehistoric mining technologies and how the practice of copper mining spread across Britain from the late third millennium BC onwards (e.g., Dutton, et al., 1994; Gale, 1995; Lewis, 1996; James, 2011; O'Brien, 2015; Timberlake and Marshall, 2019). This site has been identified as the location for "Britain's first mining boom", in which, over at least a few centuries (c.1600–1400 BC) of the Bronze Age life of the mine, this source was worked on an 'industrial' scale by full-time, specialist miners. This 'substantial workforce' was supported by a 'large and complex infrastructure'. The output from this operation then circulated outwards across southern and eastern Britain and into Europe as part of large-scale, long-distance networks in which ore or copper/bronze were traded in bulk for amber and other goods (Wood, 1992, p. 14; Wood and Campbell, 1995, p. 17; Williams and Le Carlier de Veslud, 2019, p. 1192).

While it is possible to agree with aspects of that model, this book aims to show how a strikingly different picture of the character, scale and organisation of mining at the Great Orme mine is possible from the available evidence. This chapter sets out the context for this research, beginning with the size and nature of the mining complex and its immediate landscape setting, followed by a detailed reassessment of the significance of the impressive physical extent of the surviving workings for the notion that mining took place there on a massive scale at any time during the Bronze Age. Some of the key trends in wider studies of Bronze Age settlement, land use and technology are then summarised, as these provide an important interpretative framework for the arguments presented in this book. The theoretical approach to understanding the evidence adopted here is briefly explained, with a short case study. Finally, the chapter closes by defining the book's chronological and geographic scope.

Introducing the Great Orme mining complex

The Great Orme headland is located at the tip of the Creuddyn Peninsula on the north coast of Wales, jutting into the Irish Sea midway between the island of Anglesey to the west and north-west England to the east (Figure 1.2).

The headland forms a long low promontory, approximately 3km long from west-north-west to east-south-east and 1.5km across, rising to a height of 207m. Its rolling central plateau is bound by sheer-sided crags and cliffs, which grade into more gently sloping vegetated scree slopes as they descend to the sea. The Creuddyn Peninsula is linked to the mainland by a series of low hills and ridges. Its south-western edge is formed by the estuary of the Afon Conwy, where it meets the sea at Conwy Bay after meandering northwards for 32km through the Conwy Valley. To the east, the low headland of the Little Orme (141m AOD) stands opposite the Great Orme at the end of Llandudno Bay (Gwyn and Thompson, 1999). Today, the Great Orme headland is joined to the Creuddyn Peninsula by a narrow strip of flat stabilised wind-blown sand where the modern resort town of Llandudno now stands. In the Bronze Age and up to the Medieval period, the Great Orme would have been almost an island, separated from the mainland by marshland (Smith, et al., 2012, p. 3) (Figure 1.3).

The visible surface traces of prehistoric mining on the Great Orme comprise an impressive series of open-cuts or 'trench workings' and a large hollow known as the 'Great Opencast'. Bronze Age miners used tools of bone, stone and probably also wood and bronze to extract secondary malachite-goethite copper ore from joints and fractures located in a narrow 'fault' zone c.100m wide in the dolomitised limestone rock in the eastern part of the headland. This activity also gave rise to an extensive complex of subterranean workings, with at least 6,000m of underground tunnels having so far been explored and/or surveyed and assigned to the Bronze Age (Lewis, 1994, p. 35, 1996, p. 151). The surface features and a segment of the underground workings have been painstakingly curated by Great Orme Mines Ltd and are now showcased to the public as a major tourist attraction.

The prehistoric mine is situated in a landscape littered with the residues of early/late modern copper mining. The earliest documented historic phase of copper mining on the Great Orme began in the 1690s and finally ceased in the 1880s. Mining for copper, together with stone quarrying, dominated activity on the headland for much of the 19th century AD (Williams, 1995, p. 6), producing abandoned workings and spoil tips that are

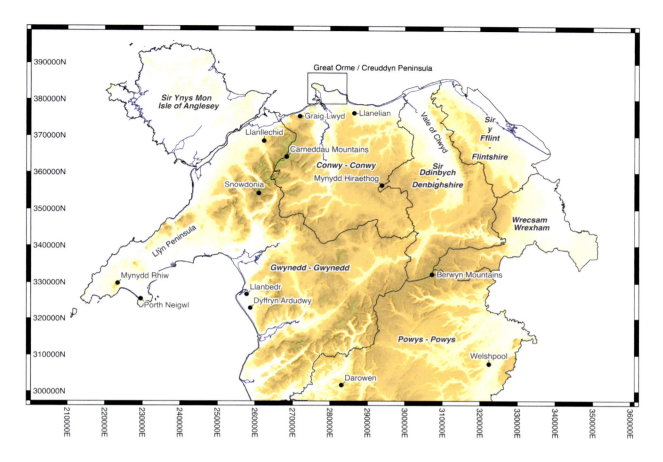

Figure 1.2: Topography and geographical extent of the north Wales study area, with the boundaries of the current administrative divisions (labelled in Welsh and English). For a larger-scale view of the Creuddyn Peninsula and the Great Orme, see Figures 1.4 and 1.5. Contains OS data © Crown copyright and database right 2022.

Figure 1.3: The Great Orme headland at the end of the Creuddyn Peninsula, facing west, as seen from the summit of the adjacent Little Orme headland. A – Creuddyn Peninsula, B – Carneddau Mountain Range and Snowdonia, C – Conwy Estuary, D – Conwy Bay, E – Llandudno, F – Llandudno Bay, G – Great Orme headland (photograph author's own).

today encountered all over the promontory. In places, the deposits of mainly 19th-century AD mining spoil are estimated to be at least 10–15m deep (Lewis, 1996, App. C, 3). Such material completely obscured the surface remains of prehistoric mining up to the late 1980s. The quantity and extent of these residues both at the mine and across the headland present a major challenge for research into the scale, chronology and routines of copper mining on the Great Orme during the Bronze Age, one which we will return to repeatedly in succeeding chapters. The surviving archaeology on the Creuddyn Peninsula, and the Great Orme in particular, is nonetheless plentiful. It encompasses not only the copper mine and associated metal production sites, but also caves and rock shelters, possible burial mounds and roundhouses, the megalithic chambered tomb of Llety'r Filiast, Pen y Dinas hillfort,

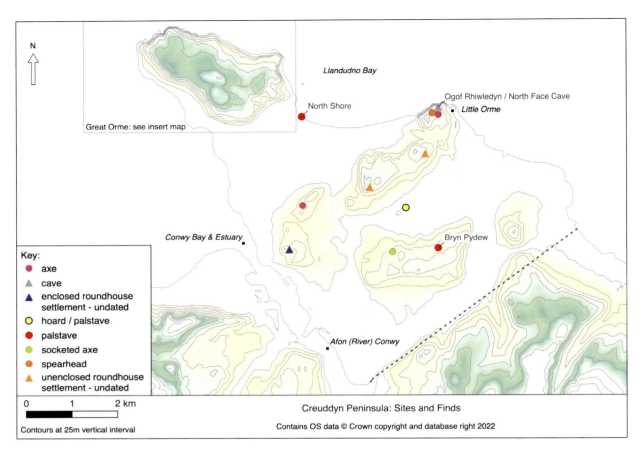

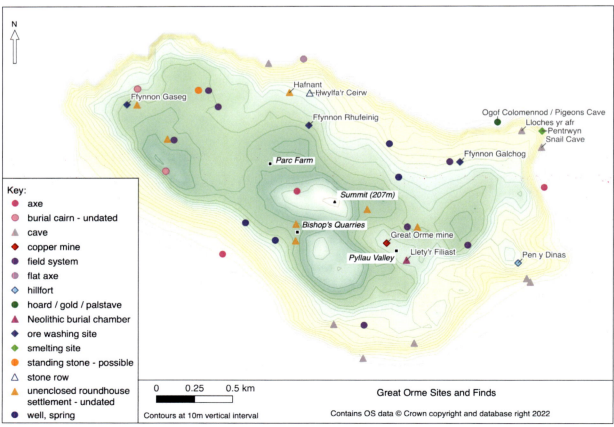

Figure 1.4: The Creuddyn Peninsula showing the distribution of roundhouse settlements, caves and hoards/isolated finds of copper/bronze metalwork. Dashed line indicates the approximate boundary between the Peninsula and the rest of north Wales.

Figure 1.5: The Great Orme headland showing the distribution of sites and finds discussed in the text.

a stone alignment (Hwylfa'r Ceirw), a possible standing stone, field systems, wells and springs and 'old' tracks and pathways (https://www.archwilio.org.uk/arch/ [Accessed: 23/03/2022]). A dozen copper and bronze artefacts – a spearhead and axes of various shapes, one of which was found with two gold objects – have also been recovered on the headland. Together, these sites and finds potentially offer a richly varied, local, multi-period context in which to study the prehistoric Great Orme mine, as we will see in Chapter Four (Figures 1.4, 1.5).

Reappraising the scale of mining at the Great Orme mine

The physical extent of the workings at the Great Orme mine and the monumental quantities of its spoil and finds are on a far larger scale than at any of the other known Bronze Age copper mines in Britain and Ireland (Timberlake, 2009, p. 272; O'Brien, 2015) (for the location of these sites, see Figure 3.3). Estimates for the total production of copper metal from this source during the Bronze Age life of the mine likewise dwarf those from these other sites. Output estimates for the Great Orme mine vary between 30 and 830 tonnes of copper metal in total (Lewis, 1996, p. 151; Timberlake, 2009, table 7.2; Williams and Le Carlier de Veslud, 2019, pp. 1188–9), with one suggestion of 1,800 tonnes (Randall, 1995, p. 1). It has been proposed that such high levels of production indicate a large workforce engaged in mining full time and supported by an extensive infrastructure (e.g., Wood, 1992, pp. 13–14; Wood and Campbell, 1995, p. 17; Williams and Le Carlier de Veslud, 2019, p.1189).

In comparison, the nearby Earlier Bronze Age copper mine of Mynydd Parys on Anglesey may have had a total output of only around 2 to 4 tonnes of copper metal. The largest known prehistoric copper mine in mid-Wales, working the Comet Lode at Copa Hill, Cwmystwyth, is likewise estimated to have produced no more than 1 or 2 tonnes of metal in total. Similar values are proposed for the mines at Alderley Edge, in Cheshire, north-west England, and Mount Gabriel in south-west Ireland (although it is estimated that the latter could have produced up to around 30 tonnes of metal [Jackson, 1979, 1980], a lower value closer to 1.5 tonnes is considered more probable) (O'Brien, 1994, table 12). The total estimated output from the Ecton mines, also in north-west England,

may have been even less than one tonne (500kg) of copper (Timberlake, 2009, table 7.2).

When the estimated scale of extraction at the Great Orme mine is considered in its broader European context, it can be seen that it has few other contemporary counterparts in Western Europe in terms of its complexity, extent and the abundance of finds (Lewis, 1996, p. 171). If the copper metal production value of 1,800 tonnes for the Great Orme mine is excluded as an outlier, then several larger mines – in terms of their total copper metal output – are known, specifically El Aramo in northern Spain, which was worked from the mid-third to the latter part of the second millennium BC (O'Brien, 2015, p. 96), and Saint-Véran in the French Alps, where the main mining phase occurred between c.2400–1700 BC. At Saint-Véran, 1,400 tonnes of copper are estimated to have been produced over the total life of the mine (O'Brien, 2015, 120). All these estimates are, of course, insignificant compared to those from the mines in the Mitterberg mining district of the Eastern Alps, Austria, where a total of approximately 20,000 tonnes of copper metal is calculated to have been produced between the 16th and 13th century BC (Pernicka, Lutz and Stöllner, 2016, p. 28).

Calculating the amount of metal produced from the extracted ore quantifies the qualitative impression of the scale of a mining operation at a source conveyed by, for example, the extent of the workings and the number of finds. Information about the following five variables is helpful for this calculation (adapted from O'Brien, 2015, pp. 269–70):

1. The size of the operation, as indicated by the dimensions of the prehistoric mine workings, including the size of the passage openings.
2. The total amount of rock removed, both barren material and mineralised ore, a variable referred to in modern mining as the 'Run of Mine' (RoM).
3. The percentage of copper-bearing rock in the RoM.
4. The percentage or grade of copper contained in the mineralised rock to constitute a viable ore.
5. The efficiency of the processing treatment (hand sorting, washing, smelting, and subsequent refining to ingot stage).

It is well recognised, however, that quantifying these parameters archaeologically is challenging, so that all estimates of prehistoric copper production must involve some informed guesswork (e.g., Budd, Pollard, *et al.*, 1992, p. 683; Lewis, 1996, p. 151; Timberlake, 2009, p. 300; O'Brien, 2015, p. 270; Williams and Le Carlier de Veslud, 2019, p. 1188). The proposed minimum and maximum metal output figures for the Great Orme mine differ significantly, due to the assumptions employed by each researcher (e.g., Randall, 1995, p. 1; Lewis, 1996, p. 151; Williams and Le Carlier de Veslud, 2019, pp. 1188–9).

In contrast, the metal output estimates for the mines in mid-Wales, England and on Anglesey, which were all produced by a single researcher, are similar in magnitude (Timberlake, 2009, table 7.2).

For the Great Orme mine, Lewis (1996, pp. 150–151) proposes minimum to maximum copper metal values of 205 to 268 tonnes. From the measured dimensions of the 'Great Opencast' and the underground workings, he estimates the 'Run of Mine' (RoM) to be 14,620m³. Using a relative density value of 2.8 for dolomite, this gives the total mass of rock removed by Bronze Age mining as 41,000 tonnes. From this, he calculates the total copper metal production derived from rock extracted in the 'Great Opencast' (142 tonnes) and the underground workings (between 63 to 126 tonnes) using an arbitrary copper metal recovery rate of 0.5% for the former, rising to 1% for the latter. These recovery rate figures are totals for the ratio of ore to 'waste', the grade of copper metal in the ore and the efficiency of concentration and smelting after mining. They are based on his assessment that the percentage of copper-bearing ore in the RoM – the mineralised ore to barren rock ratio – was lower in the 'Great Opencast' than elsewhere in the underground workings.

The measurements on which Lewis based his RoM calculations were correct in 1996. Since then, the area of the 'Great Opencast' has been excavated more fully. Its dimensions may now be slightly larger, although not by any order of magnitude. Williams (e.g., Williams and Le Carlier de Veslud, 2019, pp. 1188–9) appears to have used similar figures in his own production calculations. Like Lewis, he proposes that copper metal recovery rates would have differed depending on which part of the ore deposit was being mined, but the values he uses are dramatically different. He assumes that the overall percentage of mineralised ore in the RoM was around 5%. From chemical analysis of hand samples of malachite-goethite ore surviving in the prehistoric workings, he identified a copper content of 13–38% (Williams, 2019, p. 158).

From these variables, Williams estimates the total minimum copper output of the Bronze Age mine to be 232 tonnes, with a maximum value of 830 tonnes. In direct contrast to Lewis, he suggests that the richest – highest grade – ores were extracted from the 'Great Opencast', together with a large underground chamber (known as Loc.18), with the remaining underground workings yielding only ores of much low copper content. According to his estimate, 90% of the deposit's total Bronze Age copper metal output must have originated from ore extracted from the 'Great Opencast' and from Loc.18: 202 to 756 tonnes, compared to 30 to 74 tonnes from the rest of the prehistoric mine (Williams and Le Carlier de Veslud, 2019, pp. 1188–9).

Williams and Lewis assume that the Bronze Age workers consistently achieved constant mineral and metal recovery rates. As they both acknowledge, however, the actual situation would have been far more variable (Lewis, 1996, p. 151; Williams and Le Carlier de Veslud, 2019, p. 1188). The recovery rate of copper metal during smelting, for example, depends on the complex interplay of numerous parameters, including the atmospheric and temperature conditions in the furnace. These were in turn governed by variables such as furnace size, the number of tuyères and the type of fuel used. As will be discussed in Chapter Three, the efficiency of the simple furnaces found on the Great Orme is still uncertain – and even if known, it may not have been constant between one smelting episode and another, as in a modern, well-regulated industrial process. It has also yet to be established whether these furnaces, which date to the early first millennium BC, are representative of the process used to smelt the bulk of the Great Orme ores throughout the prehistoric life of the mine. Nonetheless, the copper recovery rates of 0.5–1.0% used by Lewis seem somewhat conservative, as recent analytical and taphonomic studies in the Mitterberg have shown that the copper losses involved in Bronze Age metal production could be as low as 13%, indicating higher copper yields than previously assumed (Pernicka, Lutz and Stöllner, 2016, p. 27, table 1).

The copper metal production values for the Great Orme mine by Lewis and Williams are both very high when compared to the quantity of copper and bronze metal objects made of Great Orme ore that have been found in Bronze Age contexts in Britain and elsewhere, a point Lewis makes in relation to his own estimates (Lewis, 1996, p. 151). The maximum output proposed by Williams, for example, could have been used to cast as many as 8,900 bronze palstave-type axes annually (Williams and Le Carlier de Veslud, 2019, p. 1189). Simon Timberlake describes this quantitative difference between the copper and bronze metal estimated to have been in circulation and the amount recovered as the "metalwork enigma" (Timberlake, 2009, pp. 299–302). One probable explanation is that we have so far found only a minute fraction of the total number of deposited copper and bronze objects awaiting discovery and so are underestimating the amount of metal produced at any one time (O'Brien, 2015, p. 272). The recent rare discovery of 90 copper/bronze metal finds at the Must Farm settlement, Cambridgeshire, including tools and weapons, indicates the amount of metalwork that could be present in a single small community (Figure 1.1). This Late Bronze Age settlement comprised four pile dwellings plus an ancillary building and appears to have been occupied for only around one year before it was destroyed by fire (Knight, et al., 2019).

Copper and bronze recycling, a practice to be discussed in Chapter Four, was most probably another important factor, with less 'new' metal being made as existing items were remelted and recast (O'Brien, 2015, p. 271). In addition, an unknown quantity of the ore mined could have been used for other purposes, such as to make pigments (Chapter Two). A further possible explanation is that percentage values for copper-bearing ore in the RoM and the grade of copper in the ore are overly optimistic, resulting in misleadingly high metal production figures for many prehistoric mines (see e.g., O'Brien, 2015, p. 272). With these points in mind, Timberlake (2009, p. 301, table 7.2) revises the total copper metal output for the Great Orme mine to an estimated minimum of less than 30 tonnes, using the latest knowledge of the potentials and limitations of prehistoric mining, processing and smelting techniques rather than ore to barren rock ratios or grade estimates. Although very significantly lower than the previous estimates, this figure still represents more than 50% of the total metal output of 51.55 tonnes that he calculates for all the Bronze Age copper mines in Britain and Ireland. All three researchers (Lewis, Williams and Timberlake) are highly experienced geologists, with extensive field experience in prehistoric and modern mine environments. Lewis and Williams have studied the geological and mineralogical structure of the Great Orme Bronze Age mine in detail, while Timberlake has undertaken excavation, survey and geoarchaeological research at numerous prehistoric mines in Wales. It is therefore challenging to assess independently the assumptions on which each model is based. Research into the five parameters needed for a new output calculation was not part of this present study. Given the metalwork 'gap', however, I suggest that it may be prudent to assume that the total copper output of the Great Orme mine was closest to Lewis's maximum estimate, so around 300 tonnes.

'Life of mine' copper production estimates like those already mentioned are useful for comparing the nominal output of different mines and for developing a general model of their relative scales of production. Even if a convincing output figure for a particular mine can be calculated, however, macro-scale output estimates have limited further interpretative potential. The perspective adopted in this book is that we should instead consider the archaeological evidence at the scale at which people in the past participated in mining as part of their practical, lived experience (Barrett, 1999, p. 498). It is through a concern with the small-scale, local details of the evidence for specific instances of action that we can explore how the practice of the technology of mining produced both the mine and ore and shaped people and their world. This micro level of analysis has the potential to provide insights into the scale, tempo and character of mining that would otherwise be impossible to deduce from a 'life of mine' metal output estimate for a particular source.

When we consider Bronze Age copper production estimates for some British mines on an annual basis, a somewhat different picture of the scale of mining activity emerges. As we have already seen, the total estimated copper output for the Ecton mines is 0.5 tonnes. Bayesian modelling (a method discussed in detail in Chapter Two) indicates that extraction there took place between c.1800–1700 cal BC, and probably over only 20 to 50 years (at 68% probability) (Timberlake, *et al.*, 2014, p. 159). At 50 years' duration, this indicates an annual copper output of 10kg. The total estimated production figures for Mynydd Parys are higher, but Bayesian analysis here points to probable mining activity over 175 to 440 years (at 68% probability) (Jenkins, *et al.*, 2021). Taking the upper values for duration and overall output gives an annual copper production figure of c.9kg, which would point to activity on a similar scale at both Ecton and Mynydd Parys.

Chapter Two considers in detail the development of the Bronze Age Great Orme mine over time. It includes Bayesian modelling of a series of radiocarbon dates from the mine. These indicate that Bronze Age activity at the Great Orme mine probably occurred over a period of 395 to 785 years (at 68% probability). The lower span agrees with Lewis's estimate of 400 years for the likely duration of prehistoric mining at the site, a figure that indicates an annual minimum/maximum copper metal output of 0.51/0.67 tonnes. For the upper duration, annual output is more or less halved, to 0.26/0.34 tonnes of copper metal. If we take Williams's most optimistic total copper production figure (830 tonnes), then the annual minimum and maximum values become just over 2.0 tonnes and 1.0 tonne respectively.

Working out annual metal production estimates like those above assumes that all the ore extracted each year was converted into metal soon after extraction and not stockpiled for later use. As will be discussed in Chapter Three, there is no evidence from the Great Orme mine to show positively that ore was not cached before smelting. Annual estimates also assume a constant output, when there are instead highly likely to have been significant variations in the quantities of ore and hence smelted metal produced over time, on a yearly, decadal or even longer basis. Williams's model of the scale of production takes this variability into account by proposing that the mineral veins in the 'Great Opencast' and Loc.18 were mined out between c.1600 and 1400 cal BC, and that extraction during this 200-year period was responsible for the bulk of the prehistoric mine's output (Williams and Le Carlier de Veslud, 2019, p. 1189). As will be discussed in Chapter Two, however, there is no secure dated evidence from the mine demonstrating that most extraction occurred during this proposed 200-year period. Rather, if the number of dates is taken to indicate mining intensity, as Pernicka, Lutz and Stöllner (2016, p. 26) have suggested for the Mitterberg mines, then there appears to have been more activity at the Great Orme mine in the early 14th century cal BC than in the preceding two centuries. As we will see, a piecemeal pattern of mine-working can be identified at this site, which indicates that ore of different grades from different parts of the source is likely to have been mined at the same time. This situation is best reflected in Lewis's output calculations rather than by Williams's proposed production sequence. A comprehensive scientific dating programme at the mine would, however, help to refine our understanding of when Bronze Age mining-working of this source was at its peak and how the intensity of extraction changed over time. This point will be returned to throughout the book.

Recent research into copper smelting at Colchis, in Western Georgia, provides a useful case study of how focusing on the temporal scale of production can help to build a better picture of the 'as-lived' character of a particular technology. A multi-scale investigation combining landscape survey and archaeometric data reveals how the large total copper output achieved there during the late second and early first millennia BC was the result of numerous small groups of metalworkers acting independently. The researchers describe copper procurement at Colchis as a "slow-motion gold rush" occurring over several centuries (Erb-Satullo, Gilmour and Khakhutaishvili, 2017, p. 124). Considering the Great Orme mine from a similar perspective, the massive amount of physical evidence from the site does not necessarily signify that mining took place on a massive scale throughout the Bronze Age life of the mine. The extensive workings and vast deposits of spoil and finds are the culmination of extraction over at least 400 years, and maybe even nearly twice as long. The impressive physical remains are therefore not a straightforward index of the number of people involved and the amount of ore extracted at any given time. Rather than focusing on the amount of metal produced as an indicator of the scale of a mining event, it may be more useful to consider the amount of RoM material (rock and mineralised ore) that the Bronze Age miners had to excavate to produce, for example, a tonne of copper metal. Using Lewis's (1996, p. 151) measurement of $0.75m^2$ for the average cross-sectional area of an underground passage at the Great Orme mine, it is possible to suggest that just over 11kg of RoM would need to have been excavated every day for an annual output of 1 tonne of copper. To put this into perspective, 11kg of material represents removing just $0.25m^2$ of a given face to a depth of 1cm in a day, a degree of effort in the actual act of material removal that seems feasible given the geological conditions and the mining technology available, both of which will be examined in detail in Chapters Two and Three.

From this, we can suggest that the production of as much as a tonne of copper metal each year would have required the involvement of only a relatively small group of people (Northover, in O'Brien, 2004, p. 538). In this context, the count of Minimum Number of Individuals (MNI) in the animal bone assemblage from the Great Orme mine – only 189 animals – is informative. It is lower than expected given the probable duration of mine-working at the source and so perhaps points to only a small group of people being present on site at any one time (James, 2011, pp. 375–6). Extraction by small groups who needed relatively few tools could also explain why the number of stone mining tools seems small when compared to the duration of mining – although, as will be discussed in Chapter Three, there are other explanations for this, such as the potentially long use-life of stone tools in the prevailing geological environment of much of the mineralised zone. There are no surviving traces of extensive, large-scale settlement, mine work camps, processing or smelting sites of Bronze Age date on or near the Great Orme. While this absence of evidence could be due to research or survival biases, including limited archaeological excavation and disturbance by later mining, it could also indicate that the number of people involved in mining at this source was never large. Similarly, there is no surviving landscape evidence to support a dramatic increase in the scale of production at the mine during the second millennium BC. One possible conclusion therefore is that mining at the Great Orme mine generally involved only small groups of people, rather than a substantial workforce of hundreds of labourers. These groups extracted only small volumes of ore at any one time to take away with them after processing or to smelt at or near the mine. They may even on occasion have stockpiled ore for later smelting.

Despite the Great Orme mine's impressive extent, this model complements that which has emerged from detailed field-based and analytical study of other Bronze Age mines in Wales and England (e.g., Timberlake, 2003a, 2009, 2017; Timberlake and Prag, 2005; Timberlake, et al., 2014). The evidence from these sites also points to ore extraction on a small-scale. Timberlake (2009, p. 275–6) suggests that this activity was most likely undertaken seasonally, by miners who were probably also herders. At the Great Orme mine, however, we will consider in later chapters how there very likely were phases of concentrated activity or heightened output during its Bronze Age lifetime. Rather than describing mining during these periods as 'industrial' in scale or character, following Kienlin's (2013, p. 425) characterisation of copper mining at the Mitterberg during the 16th to 13th centuries BC, I suggest that 'intensive' is a more useful term. These 'intense' phases of working most probably took the form of yearly mining on a consistent, repeated basis, perhaps a more rapid work pace and the occasional contribution of effort by larger groups of people, rather than a permanent marked increase in the number of participants and accompanying infrastructure. The likely tempo of mining and the size of the groups taking part are examined in more detail in Chapters Four and Five, where some of the factors prompting people to mine at this source will also be considered.

The wider archaeological context

To explore how Bronze Age mining on the Great Orme could have been characterised by generally small-scale ways of working, however, we must situate the routines of production at the source within the broader dynamics of contemporary lifeways (e.g., Edmonds, 1990, p. 58; Dobres and Hoffman, 1994; Dobres, 2000, pp. 155–6). This involves asking the following research questions: what were the material conditions of Bronze Age copper ore extraction at the Great Orme mine? What was its practical and social context? How did mine-working there relate to coeval routines of dwelling and subsistence? In particular, how did it articulate with other stages in the local copper production sequence, and with the circulation of copper and bronze, tin and lead within and between communities? The wider social context of activity at the Great Orme mine will be described by drawing on the archaeological evidence for settlement, burial, land use and the circulation of copper and bronze metalwork in Bronze Age society (see the later 'Research Scope' section for the chronological and geographic limits of this study). There are several key trends in these data that are fundamental to understanding the scale, character and tempo of coeval mining at the source. The evidence for these will be discussed in detail in Chapters Four and Five.

In brief, however, at the beginning of the Bronze Age, the patterning of burial and monuments, and of settlement comprising small unenclosed roundhouses, spread across areas of high ground in the hills and mountains, indicates that the social and material conditions of routine life were fairly dispersed. A degree of residential mobility may have been common. Even isolated settlements, however, were part of a wider agricultural landscape. By the end of the period, communities still appear to be relatively small-scale but dwellings are now less scattered, at least in some areas, and there is evidence for the inhabitation of specific places for long periods. Terrace fields appear, indicating new forms of land division and use, and the sustained, systematic cultivation of the same plots of land. The distribution of settlement becomes orientated towards foothills and low-lying areas, such as river valleys and the coast. Buildings are now larger, more nucleated and monumental, with the construction of different types of circular enclosure and the earliest hillforts. The earlier practice of building and using burial and other monuments comes to an end. Some of these changing trends in land use are displayed in Figure 1.6, which shows the distribution of roundhouse settlements, hillforts and field systems on and around the Great Orme. Unenclosed roundhouses can be seen scattered across areas of higher ground, including over the headland itself, with enclosed roundhouse settlement and an early hillfort (with dated Bronze Age activity) clustered in valleys or along the coastal fringe.

The data, however, does not indicate that these changes occurred abruptly nor were they 'watershed' moments. Rather they represent gradual transformations in the scale and patterning of daily life. Throughout the entire Bronze Age, routines of land use and dwelling appear to have revolved predominately around seasonal cycles of activity involving woodland clearance, rearing animals and growing crops. The relevant contextual information for settlement and burial points to a society organised around small-scale 'kin' groups, in which localised lines of authority, and close connections and cooperation within and between affiliated communities, were important. Social hierarchies seem to have been relatively fluid, with a largely horizontal social structure in which ties of actual, socially constituted or mythologised lineage and genealogical ancestry were emphasised.

People in these societies used objects made of copper and bronze alongside those made from a myriad other inorganic and organic materials, such as gold, stone, wood, bone and textiles. They deposited items of copper or bronze metalwork in burials or in fields, quarries and under cliffs or stones, or in rivers, lakes and peat bogs, either singly or as a collection of more than one object (known as a 'hoard'). During its use-life, however, copper/bronze metalwork circulated alongside, for example, gold, amber, flint, jet, animals and textiles in a form of exchange relationship that enhanced the socio-political status of the giver and bonded the receiver to them through ties of debt and obligation. This practice, known as 'gift exchange', was important for creating, maintaining and subverting social relations as, through it, communities were placed within geographically extensive, multi-directional networks of affiliation, obligation and reciprocity. Although the archaeological evidence shows that 'gifts' of copper and bronze, as well as those of other materials such as jet and amber, could move over long distances, each exchange interaction was probably undertaken on a local scale, with most people, most of the time, only journeying over distances of a couple of days' travel.

This brief sketch of the wider archaeological context of Bronze Age mining on the Great Orme complements the recent major reconstruction of contemporary copper and bronze working in Britain, as a typically small-scale, community-organised, ideologically significant activity carried out by people for whom metalworking was only one of numerous skilled tasks (Webley, Adams and Brück, 2020). The narrative of the Great Orme mine as an industrial production complex operating to meet the demands of inter-regional commodity exchange networks sits uneasily against this broader framework, indicating that an alternative reading of the content, character, scale and social significance of mining at this source is both possible and timely.

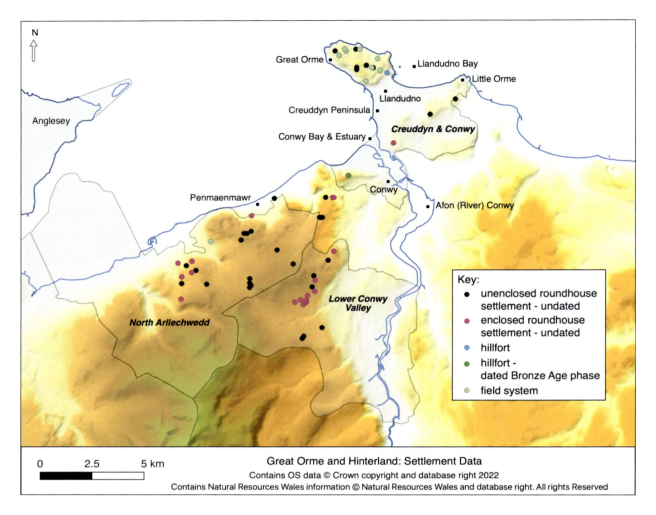

Figure 1.6: Map showing the distribution of settlements, hillforts and field systems on the Great Orme and in its hinterland. Grey-shaded areas (labelled in italics) represent the hinterland areas defined in the Landscapes of Historic Interest in Wales, Part 2.1 (Anon, 1998). Unenclosed roundhouse settlement – both isolated or scattered single roundhouses; enclosed roundhouse settlement – single or double-walled circular or circular concentric enclosures; hillfort types – palisaded double ringwork enclosures, single-walled enclosures or palisaded enclosures.

Bronze Age mining as a social phenomenon

This book aims to think about Bronze Age mining in Britain in a fresh way, using the Great Orme as its example. It takes as its starting point the principle that the gestures involved in even the simplest, most casual and apparently 'natural' acts, such as walking, sleeping and swimming, are not determined simply by the physical facts of matter but rather are socially mediated (Lemonnier, 1992, p. 1). The placing and use of the body in a certain way when, for example, throwing a ball enables the ball to leave the hand and simultaneously expresses and reproduces the thrower's understanding of their world and themselves. It variously encapsulates ideas relating to the nature and structure of social relationships; the identity/ies of both the self and others; and ideologies, cosmologies and worldviews – or cultural ontologies (Dobres, 2000, pp. 96–104). Mauss (1979 [1935]) showed how this concept extends from the simple act of walking or throwing a ball to any technological act. He coined the phrase "total social fact" to describe the enactment of any technology, whether prosaic or spectacular, at whatever material, temporal or spatial scale, and within any historically or culturally situated conditions (Dobres, 2000, pp. 153–5). A recent study of ceramics manufactured in present-day India by expert potters demonstrates this (Harush, et al., 2020). 3-D photogrammetric analysis of seemingly identical water jars produced by two endogamous communities, one Hindu, the other Muslim, revealed consistent differences in vessel shape that were barely visible to the naked eye. The jars embodied the distinct social signature of two different learning networks in ways that the makers themselves were not conscious of. This relationship between making, doing and society is described by Dobres and Hoffman (1994, p. 247) as:

"a 'seamless web' (Hughes, 1979; Latour, 1988 [1986]) that dialectically weaves together social relations, politics, economics, belief systems, ideology, artefact physics, skill, and knowledge."

Viewed from this perspective, the task of copper mining is, like any other technological act, intimately embedded in the ongoing process by which a society – its structures, social mores, identities and worldviews – is created and reproduced. This occurs moment to moment, as a consequence of the recursive dynamic between even the most routine actions on matter and meaning. At any moment of 'doing' mining, a person's actions are a practical and performative interaction with the material world. At the same time, these actions shape and convey a complex range of ideas and meanings, from the personal to the social and political (Edmonds, 1995, chap. 1; Dobres, 2000, p. 131).

Adopting the paradigm of technology as a 'total social fact' extends beyond recognising that prehistoric mining may on occasion have involved ritual or seemingly irrational superstitious actions or beliefs to an awareness that meaning is latent in our every act. The rigid dichotomy between 'ritual' and 'mundane' activities or between 'secular' and 'sacred' areas of practice that we tend to use to classify the world we inhabit today does not in fact exist (Barrett, 1988, p. 31, 1989b, p. 115). These binary oppositions are a creation of recent Western traditions of thought, rather than universal distinctions that would have been understood by and structured the actions of people in the Bronze Age (Brück, 1999b; Harris and Cipolla, 2017, pp. 31–2). Instead, their everyday experience was, like our own, a complex blend of the routine, the ceremonial and the symbolic (Bender, Hamilton and Tilley, 1997, p. 150). This applied as much to mining as it did to other acts of production, such as agriculture.

Over the following chapters, this study unpicks the 'seamless web' entangling technology and society in the context of copper mining on the Great Orme during the second and early first millennia BC. It explores how this activity was about production and the making, reworking, negotiating, contesting, resisting and sustaining of concepts of self and society. It investigates how the tasks making up the production sequence of extraction, processing and, on occasion, smelting on the headland were political as well as technological acts. These tasks provided contexts in which distinctions could be drawn between people. In this way, they were instrumental in the contesting, reaffirming and reworking of individual and group identities and the – sometimes intentional – manipulation and reordering of

social relations. The production sequence also constituted a conceptual framework for understanding the structure of contemporary social relations. It embodied understandings of the character of social life for the groups taking part and provided them with a means of thinking about both biological and social growth and transformation.

This study and its conclusions complement an increasing body of work that O'Brien (2015, pp. 245–9) has recently described as the 'social archaeology' of mining, which aims to understand the role of mines and mining as agents of social reproduction in their specific historic and cultural contexts (e.g., Knapp, Pigott and Herbert, 1998; Stöllner, 2003; O'Brien, 2007; Timberlake, 2009). Although the chosen methodology – the concept of the '*chaîne opératoire*', outlined in Chapter Three – is common in prehistoric mining studies in the UK, Ireland and Continental Europe, albeit often expressed as a focus on 'task structure' or 'work routines' (see e.g., Preuschen and Pittioni, 1954; Weisgerber, 1989a, 1989b; O'Brien, 1994, 2004, 2015; Stöllner, 2003), this study is the first time it has been used to produce a fully contextualised interpretation of Bronze Age mining in Britain from the perspective that this was a fundamentally social activity. While the physical record of mining appears to speak more directly to aspects of technology and the mine environment, rather than to social life, roles and identities, I discuss the ways in which it is possible to access such information despite the incomplete and sometimes challenging nature of the surviving evidence. The ideas presented revisit the established narrative that views the scale and organisation of prehistoric extraction at the Great Orme mine as analogous in many respects to the modern mining industry (e.g., Williams and Le Carlier de Veslud, 2019). They also contribute to current wider discourses on the social role of technology in shaping prehistoric societies, and the character and organisation of Bronze Age life in Britain (e.g., Brück, 2019; Webley, Adams and Brück, 2020).

Accounts of mining for copper, tin and iron worldwide written by historians and ethnographers in both the present and (relatively) recent past are drawn on throughout this book. Such narratives often powerfully convey a sense of the richness and variability of the mining experience (e.g., Herbert, 1984; Nash, 1993; Childs, 1998). They conjure a sense of the complex diversity of the structures of working involved when people physically, materially and conceptually 'do' mining. Without a full awareness of mining as a practical, lived experience in many different cultural and temporal contexts, it is difficult to broaden our focus on mining in prehistory from its technical specifics to the ways in which it was also a 'total social fact'. Any analogy is only ever a possibility, never proof of past action (Webley, Adams and Brück, 2020, p. 6). The perspective to the use of analogy adopted here is to explore the make up of the tasks of mining in contexts other than our own in order to gain a fuller appreciation of the multiple alternative ways there are of being human, thinking and doing (Hind, 2000, p. 32). This can be described as an 'open' approach to the use of ethnohistoric data. Instead of restricting the likely range of explanations to a single option, or providing a direct index of the structure and form of earlier practices, such data 'open up' the range of themes and scales of analysis that can fruitfully be explored through the available archaeological evidence. This can prevent us, as archaeologists, from unthinkingly imposing our own socio-culturally situated expectations onto the past material being studied (Ucko, 1969, p. 262; Rowlands, 1971, p. 210).

Conversely, however, the adoption of an 'open' approach to the use of analogy must be accompanied by recognition of the historical specificity of each case study and the dangers of cross-cultural generalization. Taking this into account requires us to acknowledge the explanatory 'vulnerability' and non-universality of our chosen analogies (Barrett and Fewster, 1998, pp. 848–9; Hind, 2000, p. 33). For example, several of the case studies referred to in later chapters are about iron mining in pre-industrial societies in Uganda and north Cameroon (Herbert, 1984; Childs, 1998). There are apparent similarities in scale between social life in these recent African contexts and the British Bronze Age. In both instances, as we shall see, we appear to be dealing with the making and use of things by small-scale societies. These case studies can therefore raise appreciation of some of the issues and concepts that may have been involved in the practice of a technological activity such as a mining by past communities of similar character (Edmonds, 1995, p. 13).

There are, however, clearly limits to the interpretive potential of these studies as universal explanations. The material, economic and social conditions of post-colonial iron mining in Uganda or Cameroon cannot be considered identical to those of copper mining in the Bronze Age in Britain. The physical properties of iron and its ores are obviously different to those of copper. Their extraction and working involve different processes and techniques (Webley, Adams and Brück, 2020, p. 8). Ethnographic accounts of non- or pre-industrial societies are also no less theory-laden or subjective than those written about recent industrial contexts. Like a sociologist or historical archaeologist, the ethnographer interprets the information from an informant through the filter of their own culturally situated research biases and preconceptions (Bender, 1991). Iles (2018, p. 97) has recently pointed out that a lack of understanding by archaeologists of the effect of these influences on ethnographic case studies has limited the range of interpretations that have been applied to the past. Emphasis on what were considered to be the more 'prestigious' aspects of metalworking by the observer in the field could, for example, be partly responsible for the under-representation of women in ethnographic accounts of this activity – and hence in prehistory (Webley, Adams and Brück, 2020, p. 7).

Any ethnohistorical analogy should always be set alongside a detailed consideration of the specifics of the archaeological data and the socio-historically situated processes through which they were produced. These data may be more limited and hence our accounts of the social role of prehistoric mining less detailed, well-contextualised and vivid than those produced for the more recent period by ethnographers, social historians or historical archaeologists (e.g., Childs, 1998, pp. 134–5, 1999, pp. 38–40). This nevertheless does not invalidate or negate the use of open analogy as a research technique for the study of copper mining in the Bronze Age, whether on the Great Orme or elsewhere. Analogies have the power to open our mind to other, non-Western ways of being and doing, increasing our interpretive possibilities even within the constraints imposed by incomplete archaeological evidence.

Investigating prehistoric mining as a "total social fact" – as advocated by Mauss (1979 [1935]) – is more difficult and controversial than reconstructing its technical sequence. Due to the often ambiguous and partial nature and limited chronological resolution of the surviving evidence, it is considerably easier to express an interest in technical aspects of extraction than it is to address or present any answers to more social issues (Edmonds, 1995, p. 12; O'Brien, 2007, p. 20, 2015, p. 253). This raises the question: why is it important to try to do so? Case studies of modern mining in capitalist contexts, such as that by Nash (1993) in 20th-century AD Bolivia, are particularly useful in this regard, because they demonstrate powerfully how mining is always socially embedded. They reveal the complex ways in which it is driven by economics and is also about the making of concepts of self and society. This reinforces why the contributions of O'Brien, Timberlake and other researchers (e.g., Stöllner, 2014), whose work focuses on investigating prehistoric mining as a technological *and* social phenomenon, are so important. As they rightly recognise, if mining can be seen as a social phenomenon in industrial contexts, we have little choice but to consider it as such in prehistory, despite the vastly different social and material conditions.

This point can be illustrated by briefly considering early/late modern copper mining on the Great Orme during the 18th and 19th centuries AD. From documentary sources (e.g., Smith, 1988; Williams, 1995), we know that, during this period, miners on the headland generally worked in groups of two to eight men, with a leader. The team was known by the name of the leader – 'Peter Jones and partners', for example – who made a bargain with the mine company to work a particular piece of ground:

"...one group might be employed to sink a shaft or drive a level at a certain price per yard, varying according to the difficulty of the work ... Similarly, men raising ore could be paid a certain price per ton ... each partnership got its ore to the foot of the shaft independently, and the ore was kept separate until it was wound to the surface by the company and weighed. The men's wages could fluctuate wildly. On one bargain the work might be easy, and they would make good money; on another it might be so difficult that the money they had to spend on candles and tools might mean that they actually made a loss on it" (Williams, 1995, p. 39).

Occasionally, when a mine was coming to the end of its productive life, each partnership could extract ore wherever they found it. They paid an agreed percentage of the value of the ore raised to the mining company and were allowed to keep the rest, a practice known as the 'tribute system' (Williams, 1995, pp. 39–40; Rule, 1998, pp. 158–9).

The miners followed any veins of ore they found. This typically gave rise to a very complicated series of underground workings. They used iron picks and gunpowder to remove ore and rock from the working face. The miners accessed the part of the mine where they were working by climbing down wooden or chain ladders, or wooden stemples wedged across the shaft or ore vein. Accessing the Old Mine involved climbing down fixed ladders in a footway shaft for 91m, then walking along a level to the main shaft before climbing down that for a further 36m using chain ladders. John Williams, a former miner, recalled in 1911 how two miners had been badly injured falling from a ladder before he started working with his father in the Old Mine in 1849, aged 12. Due to the potential danger, his father refused to allow him to use the ladders to get to work. For three years, he walked along the Penmorfa drainage level every day for about half a mile to reach the workings (Williams, 1995, pp. 33–40).

In shallower workings, the miners removed the ore and waste rock by carrying it to the surface in bags, although some of the latter was also packed into abandoned workings, as 'deads'. Subsequently, as the mines became deeper, mechanical winding methods were used to winch the ore and gangue (barren host rock) to the surface (Williams, 1995, pp. 33–7), where it was cleaned and hand sorted for smelting. By the 1830s, most ore was crushed mechanically, with the waste rock being collected as 'slime' in a big pool located near the workings. John Williams describes how ore was dressed during a period of decline:

"...the copper when brought up the shafts was still kept separate and placed in lots. The men then broke it up with hammers and got it as clean as we could and take [sic] out the stones and wash it in big tubs. Water was scarce and we had to do a lot of work with very little water ... From the yard it was carted (in bags usually) to the ship in Llandudno Bay..."* (cited in Williams, 1995, pp. 42–3).

As there was no landing stage, cargo ships sailed into the bay with the tide and were left on the sands as the water receded. Processed ore was then carted to the vessels for loading before they set sail on the next tide for the smelting works in Swansea, Anglesey or England (Williams, 1995, p. 44).

The mine captains and some of the skilled men tended to be English, specifically Cornish. Most of the miners, however, were local Welshmen, who could start working underground when they were as young as seven. They wore clogs and a felt or leather hat with a wide brim to protect them from dripping water. During their heyday in the mid-19th century AD, the Great Orme mines employed between 300 to 400 men. Most of these combined mining with another activity such as fishing or farming. Some lived in cottages nearby, although John Williams recalls staying with relatives on the Great Orme during the week and then returning with his father to their home at the weekend, c.7km away (Williams, 1995, pp. 39–41). The miners had their own folklore and believed in the existence of the 'Knockers', spirit miners who played practical jokes and who could also signal the danger of collapse by knocking on the mine walls (O'Brien, 2015, p. 254). The Great Orme miners regarded the Knockers as fellow-workmen and good friends, rather than beings to be feared (Hicklin, 1858, p. 73).

It could possibly be inferred from this brief account that the only non-technological or 'social' aspects of mining on the Great Orme during the early/late modern period related to the miners' superstitious beliefs. This enterprise was, however, 'social' in other, less overt ways. As the miners worked together, their shared physical experience created both conscious and tacit opportunities for the relationships between them to be reaffirmed and renegotiated across several scales of interaction. This is illustrated by the way in which the safe running of the enterprise depended upon close cooperation and agreement, both among the miners and between the miners and the management, represented by the mine captain. The organisation of the labour force into teams or 'partnerships' provided plenty of opportunities for social discourse. It played a significant role in the reproduction of social relations such as blood ties, with boys joining their fathers at work in the mine as soon as they were considered old enough. It also appears to have been important for the reworking – and perhaps strengthening – of kinship ties extending across broader spatial scales, as miners coming from some distance away stayed with relatives during the working week. It provided contexts for the interaction of people who worked but did not necessarily live together, as well as those who came from disparate cultural settings (e.g., the Cornish and the Welshmen). The labelling of the teams as, for example, 'Peter Jones and partners' suggests that the identity of individual miners was intimately linked with that of the leader of the group with which they worked. It may likewise have been strongly bound up with the success or failure of that team and a particular area of working or type of labour, such as extracting ore, digging a shaft or driving a level. Each partnership was therefore a social as well as a practical and an economic unit. Differences in skill and experience probably provided further opportunities for distinctions between people to be defined and affirmed through the task of mining. Participation in this activity also appears to have provided a medium for the creation of a broader group identity. This expressed itself bodily, through the characteristic dress of the miners, and conceptually, through a shared set of beliefs about the spirits of the mine. However, the early/late modern miners on the Great Orme were not full-time occupational specialists but variously also farmers and fishermen. In certain times and places, their identity as miners would have been crosscut by others, arising out of their participation in these alternative tasks.

Early/late modern mining on the Great Orme was clearly a commercial enterprise, operating within a capitalist economic system and with the miners themselves subject to specific, historically situated labour and employment conditions. The earlier narrative shows how it was nonetheless also a "total social fact", as defined by Mauss (1979 [1935]), caught up in the reproduction of the miners' identities, beliefs and social structures. Defining mining on the Great Orme in the recent past as a fundamentally social activity urges us to explore the potentials for social reproduction that could have been created by mining there during the second and early first millennia BC, and the ways in which this practice was similarly entangled in concerns about identify, social differentiation and cultural ontologies. The lives and customs of early/late modern miners – from which we can infer aspects of their social world with some confidence – are matters of public record. The challenge that will be taken up in this book is to hypothesise similar aspects of the relationship between technology and society from the physical record of Bronze Age mines.

Research scope

Although this book's focus is the Bronze Age (c.2150–800 BC), the period investigated extends from the later centuries of the third millennium BC (c.2500–2150 BC) to the initial centuries of the first millennium BC (c.800–600 BC). Using the established scheme for organising this period of the past, the time span of interest is from the Chalcolithic, or 'copper-using Neolithic' (Roberts, Uckelmann and Brandherm, 2013,

22–3), to the Earliest Iron Age. The Bronze Age is traditionally divided into three parts: the Early, Middle and Late Bronze Age. As securely dated evidence for Middle Bronze Age activity in north Wales is still comparatively scarce, this study borrows the chronological scheme adopted by Lynch, Aldhouse-Green and Davies (2000) in their comprehensive synthesis of the prehistoric archaeology of Wales. This integrates the evidence from the Early and Middle Bronze Age, a period covering approximately 1,000 years from the 22nd to the 12th centuries BC, into the *Earlier Bronze Age*. The Late Bronze Age, when changes in settlement and land use start to emerge in north Wales, is referred to here as the *Later Bronze Age*. This period lasted for around 350 years, beginning c.1150 BC and ending c.800 BC, followed by the Earliest Iron Age until around 600 BC.

The time frame chosen for this study both pre- and post-dates the dated period of Bronze Age mine-working at the Great Orme mine, as we will see in the following chapter. The expanded chronology takes account of the fact that the social and material conditions of Bronze Age mining on the Great Orme were part of an historical trajectory of action. It allows the emergence of prehistoric activity at the mine in the second millennium BC to be appraised in its socio-culturally specific historic context. It likewise potentially enables variations in the character, scale and significance of ore extraction at this source to be investigated as this practice was coming to an end. This is particularly important because, as will be demonstrated in Chapter Two, we do not yet know exactly when this occurred.

Chapters Two and Three, which look in detail at the structure and development over time of the prehistoric Great Orme mine, and the technology of mining, focus on the evidence from the workings themselves and the surrounding headland. In the remaining three chapters, the lens widens to encompass north Wales, defined here as all of Wales north of northing 300000. Described using current administrative divisions, this area comprises the whole of the island of Anglesey, Conwy and all but the southernmost tip of Gwynedd to the west; and all of Denbighshire, Flintshire and Wrexham and the northernmost quarter of Powys to the east. The Creuddyn Peninsula, tipped by the Great Orme, is situated in Conwy. Physically, the study area is bounded to the north and west by the Irish Sea and to the east by the Cheshire/Shropshire Plain, in England. It merges southwards into the uplands of central Wales (Figure 1.2).

Chapters Four to Six use different spatial scales of analysis throughout, tacking back and forth between the evidence from the entire north Wales region, the Great Orme's hinterland, and the Great Orme and its mine. The hinterland to the Great Orme is defined using the geographical designation of this area in the Landscapes of Historic Interest in Wales, Part 2.1 (Anon, 1998)

(Figure 1.6). It comprises three topographically diverse areas that are archaeologically rich, well-preserved and well-documented: Creuddyn and Conwy, a mainly coastal landscape stretching south and west from the Great and Little Orme to the banks of the Conwy estuary and into the uplands of northern Snowdonia; North Arllechwedd, a narrow strip of coast, uplands and the north fringe of the Carneddau mountain range to the west of the Great Orme; and the Lower Conwy Valley, comprising the low-lying middle reaches of the Afon (River) Conwy and its adjacent uplands (Gwyn and Thompson, 1999).

This study draws on the published excavation reports, monographs and unpublished PhD theses referenced in the bibliography, including Wager (2002), the ideas in which are the basis for this book. Additional information about relevant sites and finds was obtained from the Portable Antiquities Scheme (PAS) online database (www.finds.org.uk), and the Historic Environment Records (HER) for each local authority in the study area. The latter were accessed using the Archwilio web portal (www.archwilio.org.uk). For each site or find, the appropriate report citation (for PAS records) or the HER Primary Reference Number (PRN), and the date the information was accessed online, are given in the text. For HER records, the appropriate regional Archaeological Trust responsible for maintaining the data is also referenced (CPAT – Clwyd-Powys; GAT – Gwynedd). Throughout the text, unless stated otherwise, radiocarbon dates have been calibrated by the computer program OxCal v.4.4.2 (Bronk Ramsey, 2009) using the IntCal20 calibration curve (Reimer, *et al.*, 2020). They are presented as a single range rounded out to the nearest 5 years at 95% and 68% probability (Hamilton and Krus, 2018, p. 9).

The starting point for this study is the archaeological evidence for the chronology of Bronze Age copper mining on the Great Orme, beginning with a more detailed examination of the prehistoric mine's present-day geological and physical setting.

2

The Bronze Age mine: structure and development

In the late 1980s and early 1990s, as part of the programme of underground exploration and survey at the Great Orme mine by Andy Lewis (with others), physical relationships between some parts of the mine were identified (e.g., Lewis, 1990c, 1990d, 1994, 1996). It became possible to demonstrate how the form and layout of the Great Orme's prehistoric workings and their development over time were influenced by their topographic and geological setting. From this study, Lewis concluded that mining for copper ore most probably began on the Great Orme in the last few centuries of the third millennium BC (c.2300–2000 BC), followed by what may have been the main phase of Bronze Age activity occurring half a millennium later, from around 1500 to 1100 BC. He identified a progression over time, starting with mining in shallow workings at or close to the surface, created as the miners followed the brightly coloured secondary copper carbonate mineralisation from the sides and base of the trench workings. This was followed by ore extraction at greater depths and further northwards into the hillside. By the beginning of the Later Bronze Age, he proposed that mining was taking place in underground workings at depths of 30m from the present ground surface, as well as in and around the 'Great Opencast' and its associated tunnels and passages. During this period, there was likely to have been later reworking in parts of the mine from which ore had previously been extracted by earlier mining (Lewis, 1994, p. 35, 1996, pp. 156, 168–70).

In this chapter, the chronology of prehistoric mining on the Great Orme will be explored. What do the form and layout of the workings and the spatial connections between them indicate about how this site is likely to have developed over time? What does modelling the radiocarbon dataset from the mine using Bayesian statistical methods reveal about the timing and span of prehistoric mining on the Great Orme headland? When did it most probably start, how long did it last and when did it end? Can these results be used to enhance our understanding of how the mine developed over time? Was there a steady progression of activity from one part of the source to another? Or was extraction characterised by a less systematic, 'piecemeal' pattern of working across the site? We begin by considering the geological and physical setting of the mine, including how the environs of the present-day mine differed from those encountered by the very first Bronze Age miners. We then move on to examining the physical structure of the workings both underground and at the surface.

The mine's geological and physical setting

Geologically, the Great Orme is an isolated segment of a discontinuous outcrop of predominately hard Lower Carboniferous limestone that fringes the north Wales coastline (Warren, *et al.*, 1984) (Figure 2.1). A cyclical succession of crystalline limestone with minor bands of mudstone and – more rarely – sandstone is exposed underground and at the surface to a total depth of nearly 400m. These strata are gently folded into a shallow north-south trending syncline, the axis of which, when viewed from the south, is

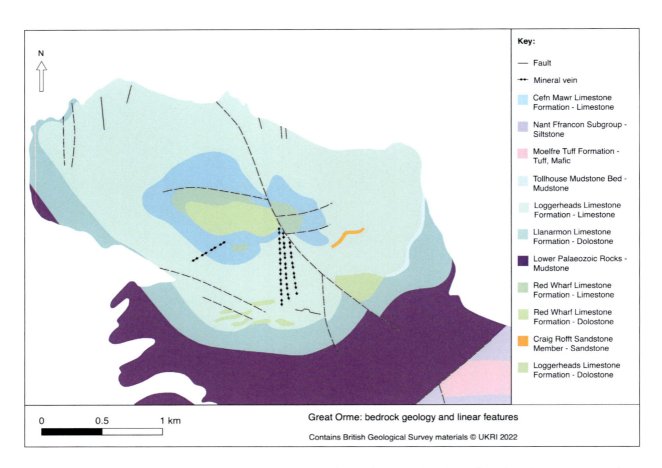

Figure 2.1: The bedrock geology of the Great Orme and associated linear features. The Pyllau Valley, where the mine is situated, is located in the area with the parallel, north-south trending mineral veins in the centre-right of the image (see Figure 2.2 for details).

centred immediately to the east of the headland's summit. When viewed from the east, a similar structure trending east-west can be seen. In places, this has produced dips as great as 25° in the Carboniferous limestone strata, in the cliffs overlooking Llandudno, for example (Jenkins in Dutton, et al., 1994, p. 254; Lewis, 1996, p. 68). There are at least 40 dominant north-south trending faults running through the limestone beds forming a narrow 'fault zone', c.100m wide (Lewis, 1996, fig. 17), concentrated in the eastern part of the headland (at Maes y Facrell) and focused within the Pyllau Valley (Figure 2.2).

The Great Orme's geological structure had a significant impact on the form and size of the ore body. The secondary malachite-goethite ore worked by the Bronze Age miners developed along the major north-south lines of weakness in the limestones of the fault zone – and to a lesser and decreasing extent in the less well-developed, crosscutting east-west trending and oblique joints and fractures. Ore formation occurred in association with dolomitisation of the host limestone rock. This acted as a lithological control on the development of the mineralisation, such that it is restricted to the dolomitised limestones (Jenkins in Dutton, et al., 1994, pp. 254–5; Lewis, 1996, pp. 68–77; Williams, 2014, p. 106). Some of the dolomitised limestones at or near to surface in the mineralised zone are altered or "rotted" (Lewis, 1990d, p. 5) to a loosely aggregated granular rock that is very soft compared to unaltered dolomitic limestone or the surrounding limestone country rock. The rotted dolomite is often so friable that it disintegrates to a sandy material that can easily be removed by gentle scraping with a fingernail. The occurrence of altered dolomite decreases progressively with depth, although some small areas of rotting are known at depths of 130m below the present ground surface (Jenkins in Dutton, et al., 1994, p. 255; Lewis, 1996, p. 78).

Units of shale, mudstone and, less frequently, sandstone up to 50cm thick are interbedded with the dolomitised limestones (Lewis, 1990d, p. 9) (Figure 2.2). They include alternating calcareous mudstone and argillaceous limestone – the 'Pyllau mudstone' – and fine-grained Craig Rofft sandstone (Lewis, 1996, fig. 7). These strata typically restricted the upward migration of mineralising solutions along joints and fractures. They form confining horizons beneath which the mineralisation extends laterally (Jenkins in Dutton, et al., 1994, p. 255; Lewis, 1996, p. 85). Like the dolomitised limestones, mudstone and shale (but

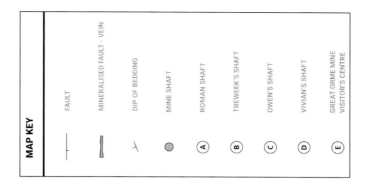
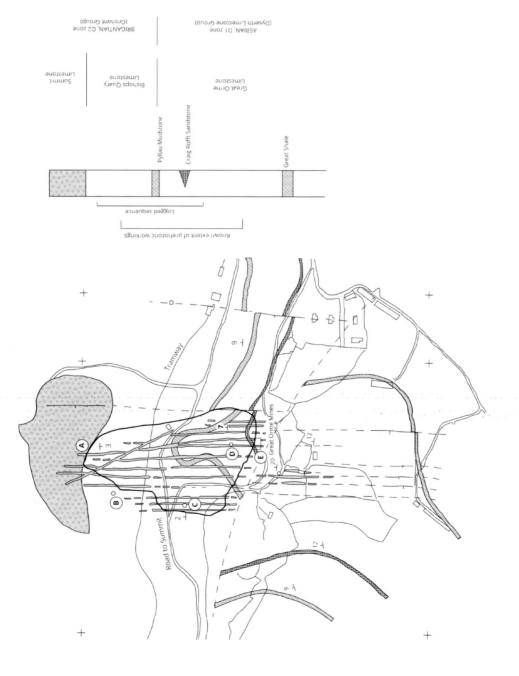

Figure 2.2: Geological plan of the Pyllau Valley, with the known lateral extent of the Bronze Age mine-working (redrawn by M. Andreikos after Lewis, 1996, figs. 17 and 19).

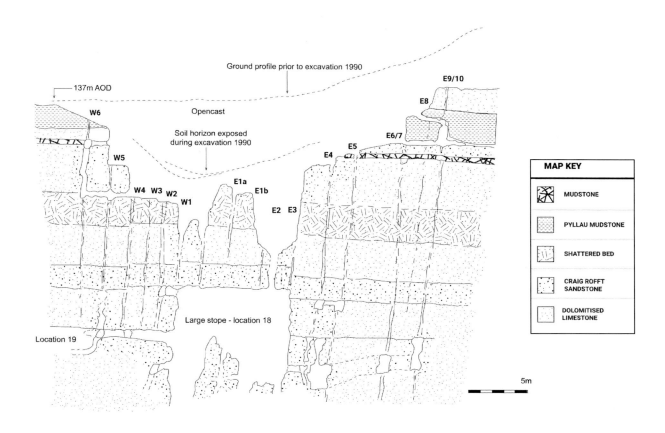

Figure 2.5: North-facing schematic section through the 'Great Opencast' to Loc.18, showing the ore vein numbering system, the mineralised fractures and associated rock strata (redrawn by M. Andreikos after Lewis, 1996, fig. 18).

Figure 2.3 (opposite page, top): The modern landscape setting of the mine, looking north-west along the Pyllau Valley. A – Great Orme mine; B – visitor centre; C – Llety'r Filiast burial mound; D and E – limestone country rock slopes (copyright Gwynedd Archaeological Trust).

Figure 2.4 (opposite page, bottom): View of the Great Orme mine and its landscape setting, facing south-east. A – Tourist Cliff north face, with north-south trending ore veins ('trench workings'); B – 'Great Opencast'; C – cliff on east side of 'Great Opencast', at the location of the Pyllau Mudstone horizon; D – northern slope of mixed prehistoric and recent mining spoil; E – Llety'r Filiast burial mound (photograph author's own).

not sandstone) strata are frequently also altered where they occur in association with the mineral veins. They have often been softened by rehydration to sticky masses of clay. Such altered mudstones and shales can be seen throughout the mineralised zone, although they are less common at the surface than rotted dolomites and are slightly more widespread at depth (Lewis, 1996, p. 80).

The surface traces of prehistoric mining are today visible in the shallow, gentle-sloped Pyllau Valley, located 0.5km south-east of the summit (Figure 2.3). To the south, east and west of this roughly south-east to north-west trending valley, the valley side comprises limestone country rock strata that mainly dip into the area of dolomitised rock, forming a depression (Lewis, 1996, 82). To the east, it now rises in a stepwise series of limestone scarps, overlain by boulder clay and soils. The original form and depth of the valley's northern edge have been extensively modified and obscured by both prehistoric and more recent ore extraction and the ensuing spoil deposits (Lewis, 1996, App. C, 3) (Figure 2.4). Since the late 1980s, more than 100,000 tonnes of spoil have been removed from the Pyllau Valley, centred on the area around the 19th-century AD Vivian's Shaft (Figure 2.5). This clearance was undertaken to excavate the prehistoric workings and landscape the site so that it could be developed as a tourist attraction (Dutton, et al., 1994, p. 266; Lewis, 1996, pp. 43–4). It revealed a north-inclining cliff, exposed to a height of 10.5m, as measured from the present top of the cliff to the floor of the 'tourist entrance'. As the entrances to the modern tourist route through the underground workings are located within this cliff, it is referred to here

Figure 2.6: Tourist Cliff south face, showing north-south trending ore veins ('trench workings') to the east of the 'Central Vein' (the location of Vivian's Shaft). The 'Great Opencast' is immediately behind this cliff, to the north. Entrances into Loc.17, two of which have been extended to create the mine's tourist route, can be seen to the bottom right, below the Craig Rofft sandstone stratum forming the base of the 'Great Opencast' (photograph author's own).

Figure 2.7: Tourist Cliff south face, showing north-south trending ore veins ('trench workings') to the west of the 'Central Vein' (where the metal grille caps Vivian's Shaft). The rest of the western end of the Tourist Cliff is now buried beneath the modern path (photograph author's own).

Figure 2.8: Tourist Cliff north face, showing north-south trending ore veins ('trench workings') and the base of the 'Great Opencast' (centre middle). The rocky slope at the bottom left is part of the northern edge of the 'Great Opencast', which today comprises a steep slope of mixed prehistoric and recent mining spoil (photograph author's own).

Figure 2.9: Llety'r Filiast Earlier Neolithic burial chamber, viewed looking north-west along the Pyllau Valley towards the prehistoric mine (centre), with the Great Orme's summit rising behind (photograph author's own).

for convenience as the 'Tourist Cliff'. There was no trace of the Tourist Cliff at the surface before spoil clearance began (Figures 2.6, 2.7).

The south-facing profile of the Tourist Cliff is marked by a series of stepped scars. The cliff itself is transected by 'trench workings', mainly running north-south. To the north of the Tourist Cliff is a large depression known as the 'Great Opencast'. Entrances into the underground tunnel system have been identified at the base of the 'trench workings' and the 'opencast', although it is unclear whether these all date to the Bronze Age (e.g., Lewis, 1996, p. 168; David, 2000, 2005, p. 146). At the northern end of the 'opencast', the Pyllau Valley's northern slope, which now comprises mixed prehistoric and recent rock waste and shattered bedrock, rises steeply upwards (Dutton, et al., 1994, pp. 258–66) (Figure 2.8).

Sometime in the fourth millennium BC, people constructed the Llety'r Filiast burial chamber on the south side of the Pyllau Valley (Figures 2.4, 2.9). While the tomb has not yet been excavated, it is evidence for the presence of a Neolithic community on the headland. X-Ray Fluorescence (XRF) Analysis detected only background copper levels in soil samples from the chamber's underlying mound, compared to greatly elevated levels in samples taken elsewhere in the valley. This indicates that the monument was constructed

Figure 2.10: The 'North-West Corner', an outcrop of dolomitised limestone above the 'Great Opencast', facing north-west. Surface entrances to underground mine workings at Loc.36 are located in the same area. The wooden structure houses an exhibition about Bronze Age smelting, part of the tourist route around the prehistoric mine (photograph author's own).

before copper mining began, on what was presumably the original ground surface (Lynch, 1995, p. 19; Jenkins, Owen and Lewis, 2001, p. 168). As the monument is located nearly 200m to the south-east of the Tourist Cliff, its position does not, however, securely define the topography of the Pyllau Valley's northern fringe before Bronze Age mining began, nor whether the valley then was roughly the same shape and size as it is now or a more pronounced, deeper feature (Dutton, et al., 1994, p. 266; Lewis, 1996, p. 82). Also relevant are the considerable quantities of deposited material, presumed mining spoil, that field observation shows are still present in the valley between the monument and the Tourist Cliff (Figure 2.3).

Given the nature of the material removed during excavations in the Pyllau Valley from the late 1980s – mining spoil of mixed prehistoric and early/late modern origin – it can be assumed that the Tourist Cliff could be seen as a landscape feature at the surface before any mining began. While exploring underground during the mid-1990s, the edge of this feature was encountered approximately 10m below the then ground surface (David, 1995b). It was probably also formerly overlain by further strata, such as the 'Pyllau mudstone' that can be seen in situ on the eastern and western sides of the 'Great Opencast' (Figures 2.4, 2.5). The distance from the top of the Tourist Cliff to the upper surface of the Craig Rofft sandstone horizon above the tourist entrance could therefore have been rather more than the c.10m exposed today.

The south face of the Tourist Cliff would have defined the north edge of the Pyllau Valley (Dutton, et al., 1994, p. 265). The Tourist Cliff was probably formerly exposed as a rock scarp like those that can now be seen in the limestone country rock to the east (Figure 2.3). The presence of an outcrop of dolomitised limestone in the 'North-West Corner' above the 'Great Opencast' points to the fact that there are likely to have been other such cliffs or rock outcrops on this hillside (Figure 2.10). This is supported by the 2004 excavation at Loc.36 of approximately 150m of previously unknown, characteristically Bronze Age

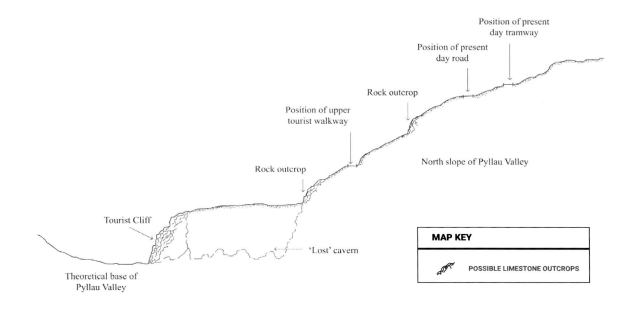

Figure 2.11: Hypothesised east-facing profile of the north slope of the Pyllau Valley in the mineralised area, prior to mining. The Lost Cavern refers to the 'Great Opencast' (drawing by M. Andreikos).

workings on the edge of the 'Great Opencast' immediately to the west of the 'North-West Corner'. At least one point of access into these workings in prehistory appears to have been via an entrance at the foot of a newly discovered vertical cliff of heavily dolomitised limestone, approximately 5m high (David, 2004, p. 144; Jowett, 2017).

To conclude, the Pyllau Valley was probably originally considerably deeper than its current form before its base gradually infilled with successive accumulations of spoil created by copper mining during the Bronze Age and early/late modern periods. The surviving form of the Tourist Cliff likewise indicates that it would have been an imposing feature, stretching roughly east-westwards for at least 50m and with its south face rising to a height of more than 10m. In the mineralised zone, there would have been other similar cliffs or rocky scarps outcropping at the surface further up the hillside (Dutton, et al., 1994, p. 282). The profile of the valley's northern slope is likely to have been very similar to that of the barren limestone country rock that can now be observed further east (Figure 2.11). This idea will be returned to when the location of the surface entrances to the underground workings is considered.

The structure of the prehistoric underground workings

There are interconnecting prehistoric underground workings on all the major north-south and minor east-west and oblique trending mineral veins in the mineralised zone. They extend over an area of at least 240m by 130m, northwards from the Pyllau Valley into part of the headland known as Bryniau Poethion. The most extensive prehistoric mining was concentrated in the area around the 'central vein' running north-south through the Pyllau Valley (identified today by the location of 19th-century AD Vivian's Shaft) (Lewis, 1996, p. 106). It extends to a depth of about 70m below the present ground surface. The deepest workings of likely prehistoric date surveyed so far are located at 106m AOD at Loc.26 (Lewis, 1996, App. C, 14). This is one of 35 locales underground that Andy Lewis identified as being of geological or archaeological interest. He assigned a reference number (Loc.1 to Loc.35) to each location (also used here to refer to areas of working) and described its features and finds in detail. Although some locales (e.g., Locs. 13, 25) displayed only one or two of the attributes considered characteristic of prehistoric mining at this source, it is reasonable to assume that these areas are also Bronze Age in origin, pending the results of further archaeological investigation (Lewis, 1996, App. C).

The structural and lithological controls on the formation of the ore body have resulted in a typically well-defined interface between the rotted and unaltered dolomitised limestone, as well as channels of mineral-bearing rock that narrow and widen both vertically and horizontally along their length (Dutton, et al., 1994, p. 256). As the prehistoric miners almost completely extracted the rotted dolomitic mineral-bearing rock from the faults and joints, passages with a characteristic morphology were

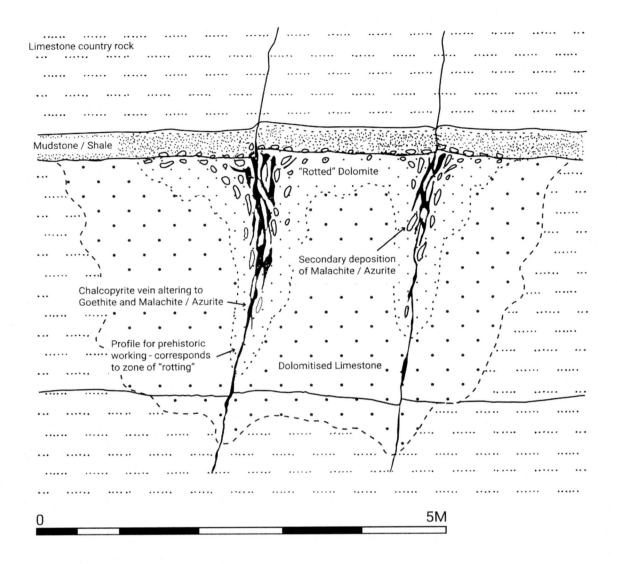

Figure 2.12: Schematic section through mineralised fractures showing the relationship between the occurrence of 'rotted' dolomite and mudstone/shale strata and the morphology of the prehistoric mine workings (redrawn by M. Andreikos after Lewis, 1996, fig. 22).

produced (Figure 2.12). The prehistoric workings typically have hard, smooth, rounded surfaces (Figure 2.13). The most usual passage shape is wide at ceiling level, where the mudstone, shale or – to a lesser extent – sandstone units have formed a confining layer, narrowing to floor level as the vein passed through the host dolomitic rock (Lewis, 1994, fig. 2, 1996, pp. 103–5). The occurrence of multiple veins and lateral ore bodies, together with the superimposition of numerous different phases of activity, has produced a twisting labyrinth of workings of differing heights, widths and lateral and vertical orientations (Lewis, 1996, figs. 20–1) (Figure 2.14). Extraction of ore from the smaller, less pronounced channels, for example, gave rise to what Lewis (1996, p. 105) describes as "tight tortuous tunnels" (Figure 2.15). Some of these are little more than 20cm wide (e.g., at Loc.21). Other workings following the most prominent faults are up to 3m wide (Figure 2.16). There are also some very large chambers, such as at Loc.18 (Lewis, 1996, App. C). Many underground areas cannot be accessed or fully explored as their entrances remain blocked or obscured by backfilled mining spoil and roof collapse debris. In other areas, unstable roof conditions prohibit detailed investigation (Lewis, 1996, App. C). Some passages are too narrow for modern adult excavators to enter (e.g., David, 2004, p. 144).

Like the Pyllau Valley itself, the character and extent of the surviving evidence for prehistoric extraction underground has been significantly affected by later

Figure 2.13: A cobblestone object (possible mining hammer) in situ in the underground mine. The smooth rounded profile of the rock wall is typical of the prehistoric mine workings (photograph courtesy of Andy Lewis).

Figure 2.14: 3-D representation of mine workings around Vivian's Shaft and their associated finds (redrawn by M. Andreikos after Lewis, 1996, fig. 20; the original survey drawing was completed with the assistance of David Jenkins).

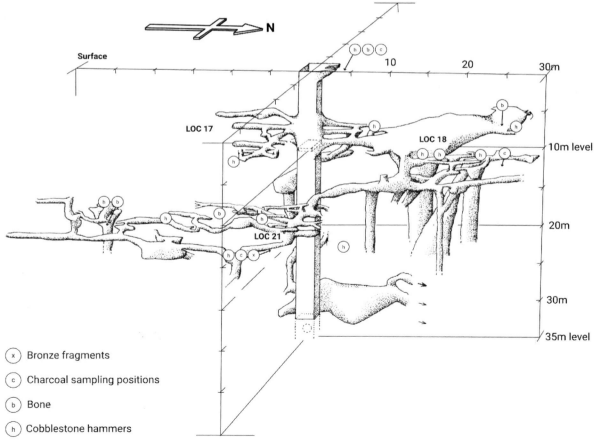

THE BRONZE AGE MINE: STRUCTURE AND DEVELOPMENT

mining (Lewis, 1994, fig. 4). Subsequent activity by miners in the early/late modern periods has often modified and sometimes enlarged the form of the original tunnels, as at Loc.30 (Lewis, 1996, App. C, 15). Elsewhere (e.g., Loc.11), they have driven shafts and crosscuts into prehistoric passages and 'pack' (a miners' term for rock debris that has been backfilled into abandoned mine workings) (Lewis, 1996, App. C, 6–7). Many of the prehistoric workings are in fact today accessed through intersecting tunnels and shafts of early/late modern date (Lewis, 1996, p. 88). Although knowledge of the mine's underground layout and its full extent are still partly conjecture (Lewis, 1996, p. 150), the surviving evidence provides invaluable insights into the chronology, content and scale of mining in this geological environment during the Bronze Age, as we shall see throughout the rest of this book. More workings of definite or probable prehistoric date continue to be discovered. These include, for example, the complex of passages at Loc.36 mentioned above (David, 2004, p. 144; Jowett, 2017, p. 64). In 2003, a

Figure 2.15: The author attempting to squeeze through a narrow opening in the prehistoric underground mine (photograph author's own).

Figure 2.16: A laterally extensive working in the prehistoric underground mine (photograph courtesy of Andy Lewis).

cavern approximately 50m below surface was found and explored (David, 2003, p. 96).

The structure of the prehistoric surface workings

The surviving workings of prehistoric origin visible at the surface today comprise the 'trench workings' and the 'Great Opencast'. The former result from the removal of ore from the major north-south trending, and at least eight less pronounced east-west trending, ore veins crosscutting the rocky outcrop of the Tourist Cliff (Figures 2.6, 2.7). Some of the surface trenches are up to 3m wide (Lewis, 1990c). Like the underground locations of interest, each mined-out vein represented by a surface trench has been numbered for ease of reference (Dutton, *et al.*, 1994, fig. 4). The 'Central Vein' at Vivian's Shaft is datum zero. All the north-south veins to the east and west of this point are then numbered consecutively, e.g., E1, E2, E3, W1, W2, W3, etc. Similar numbering is used for the east-west trending veins, which are labelled S1, N1 etc., depending on their position relative to the central vein (Figure 2.17).

The 'trench workings' have smooth curved surfaces like those observed underground. They have been interpreted as evidence for the earliest initial phase of mining at the source, perhaps created as the miners dug down from the surface to remove the more easily worked ore-bearing mudstones and shales overlying the mineralised dolomites of the Tourist Cliff (Jenkins and Lewis, 1991, p. 158; Dutton, *et al.*, 1994, p. 283). Underground mining may only have started when made necessary by the shape of the ore body (Lewis, 1990d, p. 9). This model does not, however, fully explain the form of the trench workings and the associated sedimentary and geological evidence. Secure evidence for

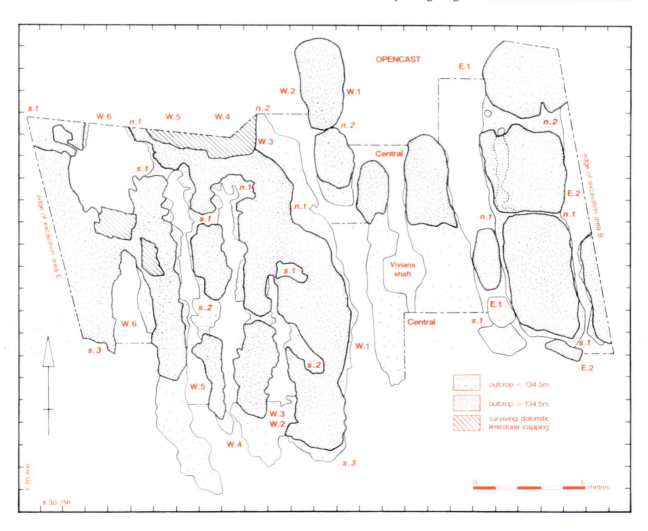

Figure 2.17: Simplified composite plan of the morphology and numbering of the ore veins in the Tourist Cliff around Vivian's Shaft (the 'Central Vein') excavated in 1989. The numbering system is continued to the veins subsequently uncovered beyond the limits shown here (copyright Gwynedd Archaeological Trust, published in Dutton, *et al.*, 1994, fig.4).

Figure 2.18: View of the 'trench-workings', facing south-west. In the left-hand image, mudstone strata capping the West 8 ore vein are in position (highlighted). The right-hand image shows the same location a few years later, after the mudstone capping had been removed. The removal of this overlying layer has revealed a fissure, similar in appearance to the so-called 'trench workings' (photographs author's own).

'true' surface mining at this source in prehistory is in fact difficult to identify. Deposits of boulder clay up to several metres thick are present on the slopes of the Pyllau Valley. The occurrence of similar deposits in the area of the Tourist Cliff and any other crags outcropping in the mineralised zone would have made it difficult to prospect for and extract surface mineralisation (Dutton, *et al.*, 1994, p. 257), other than where it was visible in the exposed faces of the rocky scarps. Numerous processes such as weathering, blasting and the landscaping work carried out since the 1980s could have resulted in the removal of the strata capping the ore veins and so produced the broad trenches visible today. For example, mudstone beds capping the northern end of the W8 ore channel were recently intentionally removed because they were dangerously unstable. Although the mineralisation in this vein had clearly been accessed by underground mining, removal of the capping produced what appeared to be a trench working mined downwards from the surface (Figure 2.18).

This observation shows that the morphology of the other mined-out ore channels transecting the Tourist Cliff does not by itself demonstrate that these features are all prehistoric surface workings. Conversely, thick layers of clays, shales and angular fragments of dolomitic rock have been observed overlying backfilled workings on, for example, the W6 to W4 ore veins crosscutting the Tourist Cliff. These deposits could represent overburden removed by prehistoric miners prospecting for ore from the surface (Dutton, *et al.*, 1994, p. 268). They are not, however, from sealed primary prehistoric contexts and have not yet been dated securely to the Bronze Age. Given the extensive working at this source in the early/late modern periods, it is uncertain whether these deposits were produced during prehistory.

As other researchers have pointed out, the colour of the oxidised copper carbonate minerals outcropping at surface would have been a key indicator of the presence of copper deposits to prehistoric communities (e.g., O'Brien, 2015, p. 196). The black-green and blue of the malachite-goethite and azurite ores infilling rock voids and fissures would have been highly visible against the yellow-orange of the dolomitic limestone of the Pyllau Valley (Lewis, 1990d, p. 9; Williams, 2019, p. 156). This observation, combined with the surviving form and layout of all the workings both at the surface and underground, indicates that the development of the Bronze Age mine most probably did not follow a simple trajectory from the digging of pits and trenches at the surface to tunnelling underground. Rather, as Lewis (1996, p. 168) suggests, the earliest mining on the Great Orme most likely began with extraction of the mineralisation exposed in the south-facing surfaces of the rocky outcrops, such as the Tourist Cliff. This would have produced underground workings running into the hillside, rather than dug down from the surface.

The extent to which the Bronze Age miners subsequently removed ore by digging pits, trenches or shafts to follow the mineralisation downwards from the surface is unclear. Only two workings relating to such activity have yet been identified, both in the Tourist Cliff. The first, Shaft 223, is a narrow, irregularly shaped shaft dug down to a depth of 4m through the soft altered dolomitic rock hosting the S1 ore vein. Marks made using either bone or wooden tools can be seen in places on its rock walls, strongly suggesting that it was formed during the Bronze Age (Dutton, *et al.*, 1994, fig. 11). The second example, also almost certainly Bronze Age in origin, was uncovered at the south-east end of the East 5 Vein (North). This shaft extended underground to a depth of 7m, where it intersected a passage in a previously unexplored area of working. It was unfortunately too narrow along most of its length to excavate fully (David, 2005, p. 146). These shafts could refer to isolated instances of surface ore extraction. It is, however, likely that, as activity at the source progressed, the Bronze Age miners also accessed mineralisation in the Pyllau Valley by mining down from the surface, as well as into the hillside. Such activity could initially have been undertaken to prospect for ore, by exploring the structure of the mineral body and the extent to which the mineralisation visible on the southern face of the Tourist Cliff extended into the hillside. The Bronze Age miners' understanding of the form of this source would have evolved due to experience accumulated by repeated instances of underground working there. As this understanding grew, they may have recognised that mineral deposits could often be reached by digging through the rock and soil of the slope overlying the Tourist Cliff. Either activity could have resulted in the formation of Shaft 223.

The 'Great Opencast' or a 'Lost Cavern'?

The 'trench workings' open out into the southern end of the 'The Great Opencast' (Figures 2.4, 2.8). As its name suggests, this large hollow is interpreted as an extensive opencast surface working. When first uncovered in the early 1990s, this feature was found to be at least 55m long and 23m wide (Dutton, *et al.*, 1994, p. 265; Lewis, 1996, pp. 95, 151). Defined to the east and west by cliffs, the location of its northern edge – and hence its original lateral extent – is unknown because it is still covered by a steep slope of bedrock and mixed prehistoric and recent mining spoil. The vertical extent of the 'opencast' was controlled by the overlying Pyllau mudstone horizon (Lewis, 1996, p. 151) (Figure 2.5). Its lowest point lies more than 25m below the modern road to the Great Orme's summit. The morphology of the rock surfaces at its base and along its edges is characteristic of prehistoric mining on the Great Orme. Lewis (1996, p. 168) suggests that this pit probably began as a series of interconnecting workings. The open hollow was created as supporting rock masses and overburden (both barren and ore bearing) were gradually removed.

There is, however, evidence pointing to this pit instead being a now-collapsed underground cavern rather than a surface working (David, 1995b, 1997; Williams and Le Carlier de Veslud, 2019, p. 1181). The calcite flowstone visible on many of the vertical surfaces of the rock slope on the eastern side of the 'opencast' indicates that these surfaces may originally have formed underground. Some of the prehistoric spoil excavated from within the pit was also cemented by calcite (Dutton, *et al.*, 1994, p. 265; David, 1995b). The rock along the northern side of the opencast is extremely shattered and there are several massive blocks of dolomitised limestone and shale in both the base of the 'opencast' and its north-eastern corner. The strata of these blocks do not conform to those exposed in the nearest rock slope, on the eastern side of the 'opencast'. The southernmost of these blocks is obviously loose and has a large crack in it. The detached blocks lying in the base of the 'opencast' are thought to continue to a depth of c.20–30m below the lowest point of this feature yet excavated. The number, size and condition of these blocks points to the collapse of the roof of an underground chamber, possibly due to blasting in the 19th century AD (David, 1995b; cf. Lewis, 1996, p. 98).

The 'opencast' could in fact be the remains of a cavern discovered and destroyed by blasting in 1849 (Stanley, 1850). The location of this feature was not recorded and it has since been considered lost. Its recorded dimensions – 40 yards (36.56m) long – are comparable to those of the opencast (David, 1995b). Although no drill holes are visible in any of the massive, detached blocks in the base of the 'opencast' (Lewis, 1996, p. 98), a drill hole was discovered at one of the lowest points excavated. This indicates that the base of this hollow must have been largely clear of debris at some time during the early/ late modern periods, when blasting was in use (David, 1995b). This finding supports the hypothesis that roof collapse occurred during a recent phase of mining, before the base of the 'opencast' became filled with huge boulders, rather than during prehistory. This conclusion is not contradicted by the discovery of calcite-cemented prehistoric spoil within the base of the 'opencast'. Spoil is likely to have been present in this feature prior to the roof collapse, due to prehistoric mining in the original underground chamber.

To summarise, the evidence for opencast mining on the Great Orme during the Bronze Age is inconclusive. The 'opencast' is most probably the remains of the 'ancient cavern' recorded by Stanley (1850), which was originally part of the underground workings until its roof was accidentally destroyed by blasting in the mid-19th century AD. The 'opencast' depression is referred to as the 'Lost Cavern' in this study. It is likely to have been similar in form to the large underground chamber lying immediately below it, at Loc.18.

The mine's spatial development

The development of the prehistoric mine had both a spatial and temporal dimension. As mining progressed over centuries, routes of access and movement would have changed repeatedly. A detailed understanding of the physical relationships between different underground areas is, however, a useful starting point from which to assess the mine's stratigraphy, or the sequence in which the Bronze Age miners extracted ore from different sections of the source over time. This study poses the following questions about the mine's spatial layout:

1. What connections originally existed between different locales of working?
2. What were the possible route(s) of access to each underground location?
3. Where were the surface entrances to areas of underground working?

To identify known, likely and possible connections between different areas of mine working, this investigation draws on Andy Lewis's data recorded during his survey of the mine: dip and dip direction; heights above Ordnance Datum; and his two-dimensional plans and section of the prehistoric mine workings (scale 1:500) (Lewis, 1993, 1994, p. 35, 1996, Apps. A, B, C). The conclusions he drew about the physical relationships between different areas of Bronze Age activity at the mine are also taken into consideration. The dip and dip direction of the rock beds in the Maes y

Facrell/Bryniau Poethion area is related to the orientation, location and extent of each area of prehistoric mining and the spot height (AOD) of each location. A key assumption was that the Bronze Age miners usually followed the dip of the mineralised strata into the hillside and did not routinely dig shafts and levels through barren rock to locate and extract ore, for the reasons already outlined.

Based on this evidence, definite, probable or possible routes of access to each location of working are proposed (Figure 2.19) and the likelihood of direct connections to surface from each location are assessed. These interconnections are shown superimposed onto the original mine plans and sections (Table 2.1, Figures 2.20–2.25) and are described in Appendix 1. The results have been compared to verify the likely or possible location of entrances within both the present-day and suggested pre-mining landscape previously proposed by Lewis (1996, App. C). To do this, the reconstructed pre-mining surface topography of the Pyllau Valley and Maes y Facrell/Bryniau Poethion was overlaid onto the present-day surface topography in the immediate vicinity of the mine, as depicted on the 1:500 Ordnance Survey map and in aerial photographs of this part of the headland.

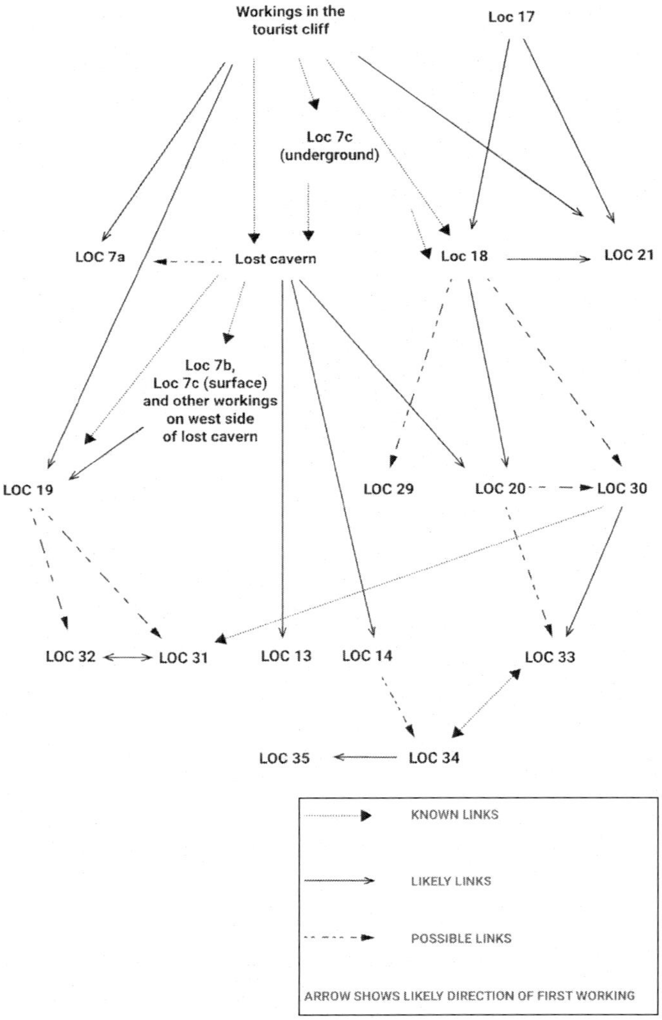

Symbol	Description
	Surveyed workings with evidence for prehistoric mining
	Not surveyed (conjectured)
	Continuing recent workings
	Surveyed workings with evidence for recent mining
	Wall of rock/spoil ('deads')
	Area of backfill or collapse
	Dip and direction of rock strata
	Known, likely or possible area in which the original surface entrances to prehistoric workings were located
	Approximate location of workings to the east or west
	Known links between areas of working
	Likely links between areas of working
	Possible links between areas of working
	Indicator arrow

Table 2.1: Key to the symbols in Figures 2.20 to 2.25.

Figure 2.19: Simplified representation of the known, likely and possible routes of access between prehistoric mine workings associated with the Tourist Cliff (locations with no known connections to these areas of the mine, such as Loc.5, are not shown) (drawing by M. Andreikos).

COMMUNITY, TECHNOLOGY AND TRADITION

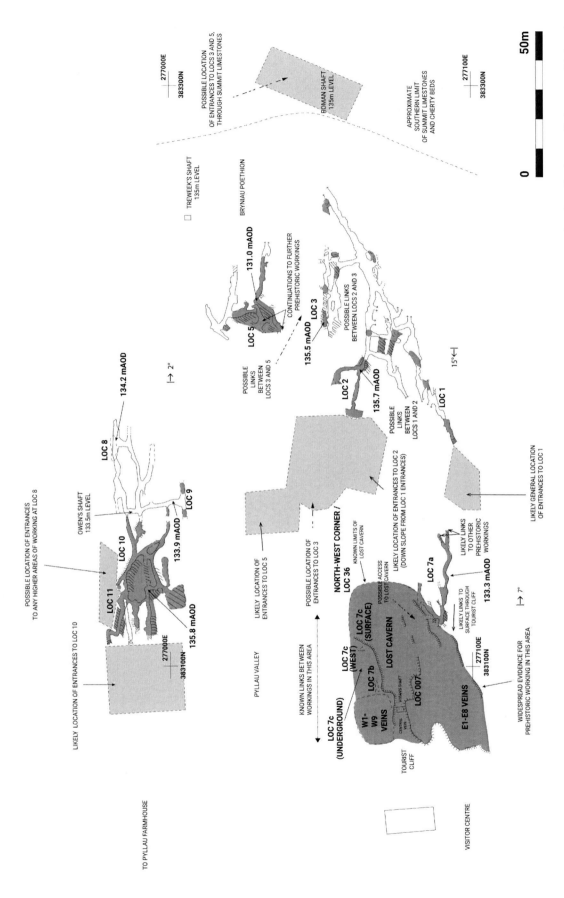

Figure 2.20: Survey plan of mine workings with evidence for prehistoric activity at 133–135m AOD, up to c.12m below the present ground surface. The known, likely or possible locations of the original mine entrances and connections between different areas of working are indicated (redrawn by M. Andreikos after Lewis, 1996, Apps. B, C).

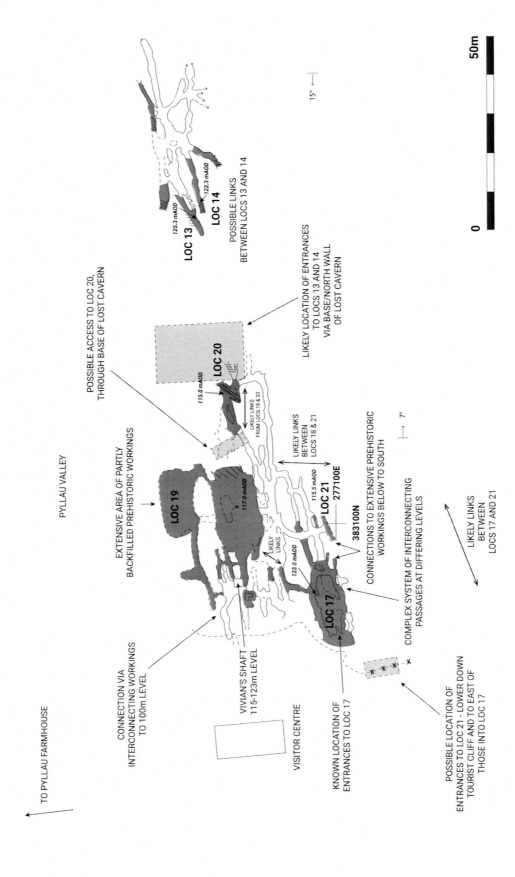

Figure 2.21: Survey plan of mine workings with evidence for prehistoric activity at 115–124m AOD, up to c.30m below the present ground surface. The known, likely or possible locations of the original mine entrances and connections between different areas of working are indicated (redrawn by M. Andreikos after Lewis, 1996, Apps. B, C).

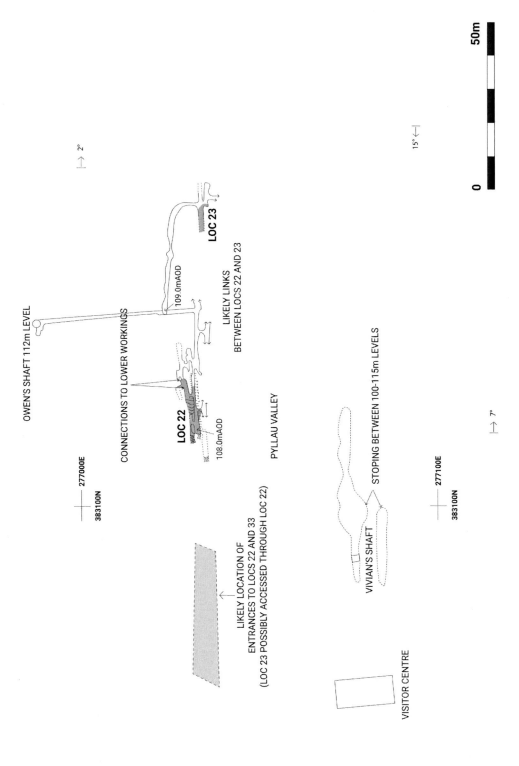

Figure 2.22: Survey plan of mine workings with evidence for prehistoric activity at 106–112m AOD, up to c.40m below the present ground surface. The known, likely or possible locations of the original mine entrances and connections between different areas of working are indicated (redrawn by M. Andreikos after Lewis, 1996, Apps. B, C).

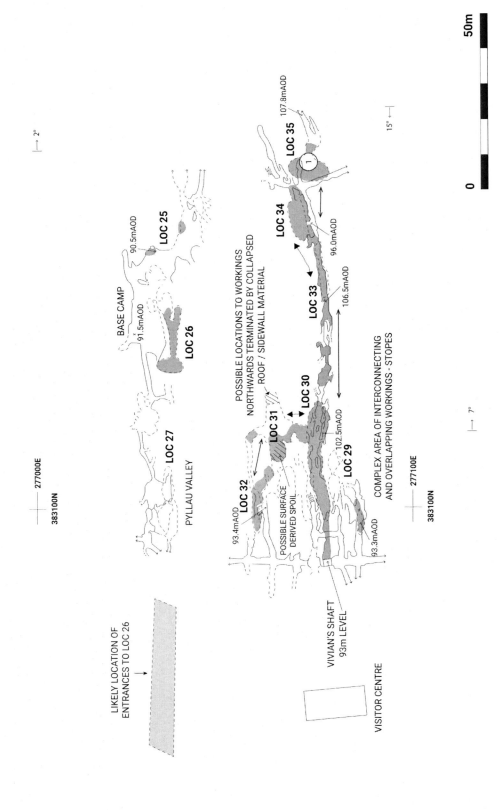

Figure 2.23: Survey plan of mine workings with evidence for prehistoric activity at 90-110m AOD, up to c.55m below the present ground surface. The known, likely or possible locations of the original mine entrances and connections between different areas of working are indicated. 1 – the single radiocarbon measurement from Loc.35 (precise sampled location not shown): BM-3117, 3000±45 BP (for details, see Appendix 3) (redrawn by M. Andreikos after Lewis, 1996, Apps. B, C).

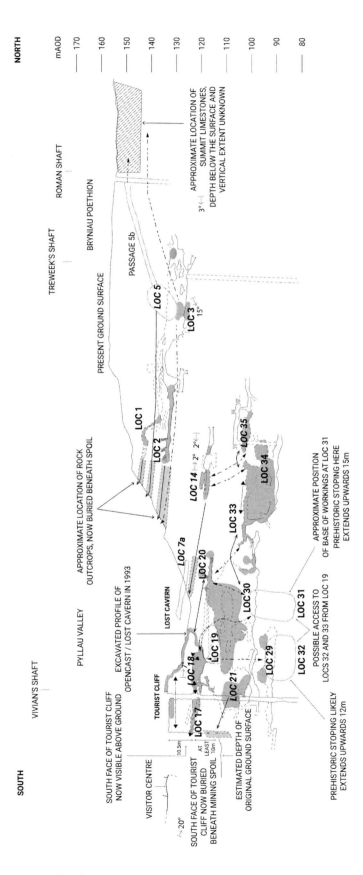

Figure 2.24: Surveyed section through the vein system from Roman Shaft to Vivian's Shaft. The known, likely or possible locations of the original prehistoric mine entrances and connections between different areas of working are indicated. Locations with radiocarbon dates are shown in italics (redrawn by M. Andreikos after Lewis, 1996, Apps. B, C).

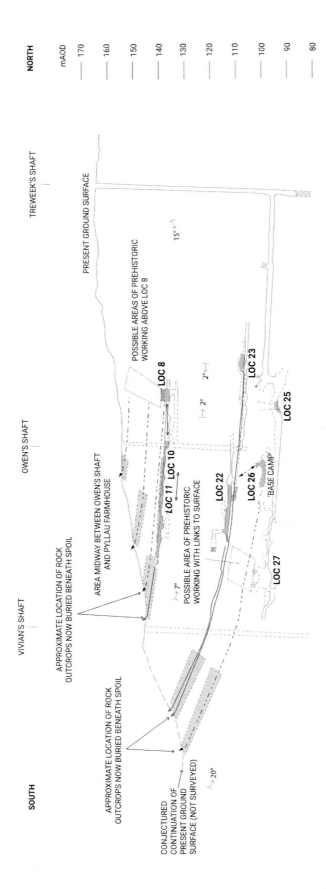

Figure 2.25: Surveyed section through the vein system from Treweek's Shaft to Owen's Shaft. The known, likely or possible locations of the original prehistoric mine entrances and connections between different areas of working are indicated. Locations with radiocarbon dates are shown in italics (redrawn by M. Andreikos after Lewis, 1996, Apps. B, C).

The approximate horizontal distance of each location from the slope of the hillside, as well as the apparent vertical distance of each location from the present ground surface, were then established (Appendix 1). To measure the latter, data on the vertical depth of each location previously recorded by Lewis (1996, App. A) were used. The horizontal distance was measured from each location to the suggested likely point of access at the present ground surface shown on Figures 2.24 and 2.25. It must be borne in mind, however, that neither of these measurements refer to the vertical depth or horizontal distance of each working in from the hillside during the Bronze Age. This is because the topography of the Pyllau Valley has been significantly transformed by mining during both prehistoric and recent periods. Nonetheless, these measurements are useful because they provide a rough guide to the position of each area of working relative to the former ground surface, and a means to compare how difficult each would have been to access.

The two-dimensional 1:500 plans and sections of all the known underground workings, both prehistoric and more recent, give an excellent schematic overview of the known and conjectured location and extent of areas of working (Lewis, 1996, App. B). Two dimensions do not illustrate fully the structural complexity of each location, nor the number and patterning of interconnections that may exist between different locales. Nonetheless, from all the data assessed in this study, five key conclusions emerge about the physical relationships between different areas of working in the Bronze Age mine.

Conclusion One

The proximity and spatial distribution of areas of working indicate that, in some cases, they are probably only the known or surviving sections of what would have been vertically and horizontally more extensive networks of passages. It can be envisaged, for example, that the tunnels and evidence for stoping at Locs. 8 and 9 were originally part of a larger system of working that, with Loc.10, mined the mineralisation on the western fringe of the mineralised area (Figures 2.20, 2.25). ('Stoping' is the removal of ore in a mine by working either upwards ['overhead' or 'overhand' stoping] or downwards ['underhand' stoping]. A 'stope' is the resulting open space or cavity.) The spatial relationship between Locs. 1–3, and the presence of chert at both Locs. 3 and 5, similarly implies that interconnections between all these areas of working once existed and remain to be discovered, even though none have yet been found (Figures 2.20, 2.24). Chert has a limited distribution on the Great Orme. It is confined to the Summit Limestones which, within the mining area, only occur beneath Bryniau Poethion. Here, there are cherty beds in the underground workings at Loc.5. As no chert occurs at Loc.3, its presence in the spoil at that location is intrusive

and points to direct links with Loc.5 (Lewis, 1996, p. 148, App. C, 2). These locations are therefore likely to have been part of an extensive series of workings that could have run into the hillside under Bryniau Poethion for at least 80m and reached vertical extents of at least 15m. The results of this analysis also suggest that the vertical extent of the ore veins between Locs. 26 and 22/23 to the south-west could have been accessed by an as yet undiscovered network of prehistoric workings (Appendix 1, Figure 2.25).

Conclusion Two

A number of locations would not have had 'direct' connections to surface. Instead, they must have been reached through other underground areas of working (Figure 2.19). In each case, this can only have occurred at a certain point in time during the mine's development. This appears to have been the case for Locs. 13, 14, 20 and 29–35 (Figures 2.21, 2.23, 2.24). Locs. 29 to 35, in particular, are only likely to have been accessible via an extensive sequence of stopes and tunnels leading upwards and to the south. In this context, it is important to note that the stoping at, for example, Locs. 29, 30 and 34 was probably carried out upwards. This does not, however, contradict the suggestion that these workings could have been accessed from above. This is because we are not dealing with modern stoping methods, in which extraction is carried out systematically from the base of the ore vein. Instead, as the prehistoric miners were simply following ore-bearing ground, they are likely to have accessed a major ore vein in a particular area at several points along its vertical extent. They may then have chosen to remove its mineralisation by stoping upwards. This appears to have occurred where Loc.33 connects to Loc.34 (Lewis, 1996, App. C, 16).

Describing a working as having a 'direct connection to surface' does not imply the existence of shafts or levels driven straight through barren rock to intercept areas of ore-bearing ground. Rather, this phrase distinguishes between locations that must have been accessed through other known or posited areas of working and those that were not. Even in the latter case, however, we can envisage that each 'direct' connection to surface identified represents the existence of a further complex of as yet unexplored prehistoric workings. This is supported by the evidence from Locs. 26 and 27 (Figures 2.23, 2.25). The likely route to the surface from Loc.26 lies above Loc.27, through an area where Lewis (1996, App. C, 14) suggests further prehistoric mined-out areas remain to be discovered. This situation is likely to be repeated throughout the mine complex. This reinforces the idea that the locations represented on the 1:500 plans are only the known sections of what was a much more extensive network of prehistoric workings. The connections between different areas of working and the routes of access from the surface suggested by the results of this two-dimensional analysis of the mine's layout oversimplify the structure of the workings considerably. A connection between two locations, for example, as recorded in Appendix 1 and on Figure 2.19, may be comprised of multiple tunnels, each with their own interconnections, not the single passageway depicted on Figures 2.24 and 2.25. There could also have been numerous other routes of access to a particular area of working, not simply those suggested here.

Conclusion Three

The (roughly) horizontal distance of a working from its entrance(s) on the hillside, the gradient of that working itself and those through which it must have been reached are as relevant for thinking about the development of the Bronze Age mine as the measurement of vertical distance from the present or original ground surface. Consideration of the distance measurements of each location from the slope of the hillside and the present ground surface leads us to reassess how far a particular area of prehistoric working lies from the surface. Typically, the spot height of each location in metres AOD has been recorded (e.g., Lewis, 1996, App. A). Using this measurement, passages and stopes at Loc.26 are some of the 'deepest' prehistoric workings so far identified on the Great Orme (Appendix 1, Figure 2.25). In absolute terms, this is clearly correct, but it does not consider the fact that, while a number of workings may be positioned at the same absolute height, their relative distance from the present – and hence also the pre-mining – ground surface differs according to the surface topography.

Locs. 3 and 10, for instance, are both situated at around 135m AOD (Appendix 1, Figures 2.24, 2.25). The present ground surface slopes downwards from the north of the Tourist Cliff, above Loc.3, to the area above Loc.10 further west. Loc.3 is consequently c.30m below the present ground surface, while Loc.10 is at a depth of only c.15m. There is likely to have been a comparable difference in depth in relation to the Bronze Age ground surface. In real terms, the vertical distance of an area of working from the present ground surface is a more effective indicator of its depth than height in metres AOD. Passage B at Loc.5 illustrates another issue with a focus on measurement of vertical distance – whether absolute or actual – in the context of thinking about the development of the Bronze Age mine (Figure 2.24). The inclination of this passage appears steep even though it accords with the recorded dip of the rock strata in this area (James, 1990; Lewis, 1993, 1996, p. 67). This represents an inclination of 15° over c.30m. In terms of the vertical height difference from the beginning to the known end of the passage, however, it represents a rise of 0.27m for every metre of horizontal distance. To put this another way, Passage B rises steeply c.2.7m for every 10m mined into the hillside.

It is these data that are important for thinking about the organisation and practical, physical experience of mining at any underground location on the Great Orme, an issue to be returned to in Chapter Three.

Conclusion Four

Taking parameters like horizontal distance into account alters the perception of a location as an accessible area of working to one which would have been much more difficult to access. Loc.20, for example, seems to be a relatively shallow working (Appendix 1, Figure 2.24). It must, however, have been reached through workings that were themselves probably accessed from entrances in the south face of the Tourist Cliff, at least 55m away. Locs. 29–35 must likewise have been accessed through an extensive network of other workings (Figure 2.24). The Bronze Age miners had to follow a particularly arduous route to reach such locations, a factor relevant to reconstructing the character and sequence of working there.

Conclusion Five

A karstic system of natural cave passages may have existed within the mineralised zone and/or the surrounding limestone country rock. This would have facilitated the development of mining underground. The limestone country rock of the Great Orme is riddled with caves and fissures. It has a history of cave exploration and use stretching back into the Upper Palaeolithic (e.g., Davies, 1989; Dinnis and Ebbs, 2013, p. 29; Chamberlain, 2014; Smith, Walker, et al., 2015). The curved, irregular profiles of the prehistoric workings are sometimes suggestive of natural cave passages (Lewis, 1990d, p. 8). As it is difficult visually to distinguish mined-out ore veins from modified natural cave passages within the same host geology, there is debate about whether similar cave systems exist in and/or neighbouring the mineralised zone and whether either or both of these were then utilised by the Bronze Age miners in their search for ore (e.g., Jenkins and Lewis, 1991, p. 157; Budd, Gale, et al., 1992; Jenkins, 1994, p. 255; Lewis, 1996, p. 82; O'Brien, 2015, p. 200).

Potentially the best evidence to support the hypothesis that an active cave system existed in the mineralised zone is the presence of large chambers underground, such as Loc.18 (Figures 2.21, 2.24). Such large masses of mineralisation are in general extremely rare, even where, as is the case on the Great Orme, numerous mineralised faults intersect and occur as sets, rather than individual units. Both Loc.18 and the Lost Cavern may be natural caverns that were modified and enlarged by prehistoric and then subsequently by early/late modern mining. When the Bronze Age miners entered these chambers, either through cave passages leading from the surface and/or through mined-out workings, they would have observed patches of mineralisation on the rock sidewalls and ceilings. The removal of this mineralisation would then have resulted in the myriad tunnels that can today be seen radiating from both locations, as well as the rifts of bedrock apparent in the base of both these features. Alternatively, Williams has proposed that spaces like the Lost Cavern and Loc.18 were in fact produced during the Bronze Age by the extraction of a large concentration of easily mined, ore-bearing ground (Williams, 2019, p. 155; Williams and Le Carlier de Veslud, 2019, p. 1188). Detailed geotechnical investigation at both locations is needed to provide a more definitive answer.

If a karstic system was present, cave passages elsewhere within either the dolomitised limestones in the mineralised zone or the adjacent limestone country rock may have crosscut or intercepted ore veins, which the prehistoric miners could then have followed. In the search for ore, natural cave passages could have been modified and then incorporated into the complex of mined workings, which could explain the existence of the narrowest (<0.20m) tunnels noted in the mine. An underground karst system would potentially have given the Bronze Age miners direct access to areas of mineralisation both at depth and at a considerable distance into the hillside. This has obvious implications for reconstructing the phasing of the mine workings. The stratigraphy apparent from the physical arrangement of the tunnelling system may not actually reflect the sequential development of the workings. Radiocarbon dating could help to confirm the karst hypothesis, if it showed that mineralisation far underground was routinely being worked during an early phase of the mine's development. As we will see in a later section, however, there are insufficient samples from locations at depth and in hard-to-access areas of the mine to verify this securely at present. Until the karst hypothesis is either confirmed or refuted, the data on the spatial layout of the mine are the best starting point for developing a stratigraphic chronology of extraction at this source.

Also relevant to the question of access to different parts of the source are the numerous entrances and routes that formerly connected the underground workings to the surface and which are today hidden beneath the vast accumulations of both prehistoric and later mining spoil modifying the Pyllau Valley (Lewis, 1990d, p. 10). The approximate likely location of entrances at surface to locations underground is represented by the shaded areas on Figures 2.20–2.23. The shaded areas along each suggested access route shown on Figures 2.24 and 2.25 take into account the fact that the profile of the northern slope of the Pyllau Valley during the Bronze Age is not known in detail. Consequently, it was not possible to determine the exact relationship between a particular passage and the former ground surface. The shading therefore represents

the general area along its length at which a particular passage is likely to have breached the surface. This study confirmed Lewis's (1996, App. C) finding that two main locations – in addition to the Tourist Cliff – were formerly the site of entrances (now hidden) into the underground workings (Appendix 1, Figure 2.26):

1. To the west of the Tourist Cliff, between 19th-century AD Owen's Shaft and the present-day Pyllau farmhouse. The underground workings to the north of this location, including those at Locs. 8–11, 22 and 23 (and possibly also Loc.26) would have been accessed from here (Figures 2.20, 2.22, 2.23).

2. To the north of the Tourist Cliff, on what is now a steep, thinly vegetated slope between the hollow of the Lost Cavern and the modern road to the summit. Entrances at this location would have been used to access underground workings to the north, including Locs. 1 and 2 and perhaps also Locs. 3 and 5 (Figure 2.20).

In the reconstruction of the Pyllau Valley prior to mining outlined above it was hypothesised, from geological and morphological evidence, that other rocky outcrops must formerly have been present on the valley's northern slope, in addition to the Tourist Cliff. The location of these other outcrops on this northern slope can now be pinpointed more precisely, from the conjectured surface entrances to the west and north of the Tourist Cliff. As the Bronze Age miners must have begun working mineralisation visible at the surface in the face of a rocky cliff or outcrop, it can be inferred that such scarps – mostly no longer visible – must have been present where these surface entrances were located. They may have occurred as a series of discontinuous outcrops or, alternatively, east-west trending near-vertical cliffs, similar in appearance to the Tourist Cliff and the limestone bluffs that can today be observed further east (Figure 2.3). For example, the presence of chert in prehistoric mining spoil at Locs. 3 and 5, and the dip and dip direction of the rock strata accessed by these workings, indicates that the Summit Limestones at Bryniau Poethion, with their cherty beds, must originally have outcropped at surface as a cliff face to the north of the Lost Cavern, even though they are today hidden beneath overburden and spoil (Lewis, 1996, p. 148, App. C, 3) (Appendix 1, Figures 2.20, 2.26 [no.3]).

Given the complexity of the underground workings, the southern faces of each of these hypothesised cliffs or outcrops must have been dotted with entrances into underground tunnels by the end of the Bronze Age. Single entrances are often likely to have led into an interconnected network of tunnels, produced as the miners followed the ore veins. In other instances,

Figure 2.26: The modern landscape setting of the mine, looking north-west, showing the likely and possible location of former surface entrances to the prehistoric underground workings: 1) likely – Locs. 8–10, 22 and 23, possible – Loc.26; 2) likely – Locs. 1 and 2, possible – Locs. 3 and 5; 3) possible – Locs. 3 and 5 (in outcropping Summit Limestones). Other marked features: A – Bryniau Poethion; B – Roman Shaft; C – Treweek's Shaft; D – Owens Shaft; E – Great Orme Mines visitor centre; F – modern road to summit; G – Pyllau farmhouse; H – Tourist Cliff; I – 'Great Opencast/Lost Cavern' (copyright Gwynedd Archaeological Trust).

there could have been several entrances at the surface into a single area of underground working. The mining of broad, vertically extensive areas of mineralisation may have produced wide channels like those visible in the Tourist Cliff. This is perhaps more likely in the rock face to the north, on the slope above the Tourist Cliff, as this would have been transected by the major ore veins running north-south through the Pyllau area.

The stratigraphic relationships between different parts of the mine

From the definite, probable and possible routes of access and likely surface entrances to each location of working discussed here, it can be seen that some parts of the source were reached only after mining had commenced elsewhere at the site, in another, interconnected, area. This aspect of the mine's spatial arrangement indicates that there was a very broad temporal logic to its development. Three general stratigraphic phases can be proposed, beginning with mining at a cluster of locations centred on the Tourist Cliff and then at increasing distances into the hillside: *Stratigraphic Phases 1 to 3*.

Stratigraphic Phase 1 comprised mining of the ore veins transecting the Tourist Cliff itself (Figures 2.20, 2.24).

Stratigraphic Phase 2 involved removal of the mineralisation at locations immediately to the north, reached through the Tourist Cliff, i.e., the Lost Cavern and Locs. 7a, 7c (Underground), 17, 18, 19 and 21 (Figures 2.20, 2.21, 2.24). All these locations, including the Lost Cavern and Loc.18, are highly likely to have been reached as mineralisation was extracted along the main north-south ore veins transecting the Tourist Cliff. This means that mining on the Tourist Cliff began *before* and was *earlier* than activity in these other areas. It is impossible to determine from the spatial information whether Loc.18 or the Lost Cavern was worked first, or whether they were both worked concurrently. This means that it is unclear whether the locations reached from Loc.18 could have been worked before those reached from the Lost Cavern or vice versa.

Stratigraphic Phase 3 refers to mining of the mineralisation at locations that could only have been reached through existing workings to the north of the Tourist Cliff, i.e., Locs. 13, 14, 20 and 29 to 35 (Figures 2.21, 2.23, 2.24). These workings all have likely or possible connections to Loc.18 and/or the Lost Cavern, but no other known routes to the surface. They can all only have been reached through an extensive series of other workings, which must have been *mined first*. This implies that Locs. 13, 14, 20 and 29 to 35 must all be relatively 'late' areas of working.

The three stratigraphic phases outlined above refer only to workings that were originally accessed from the surface through the Tourist Cliff. It is unclear from the mine's layout how the relative chronology of the workings that must have been reached through the rocky outcrops to the north and west, such as Locations 5, 10–11 and the newly discovered Loc.36, fit into the overall site chronology (Figures 2.20, 2.24, 2.25). Despite this ambiguity, the stratigraphic framework provides a useful macro-level model for the spatial development of the mine over time, alongside which to consider the radiocarbon dating evidence.

The radiocarbon dating evidence

There are twenty-four radiocarbon dates from six locations on or around the Tourist Cliff, as well as from 12 other sites across the source (Figures 2.23–2.25, 2.27, 2.28). These include in the Lost Cavern and Loc.18, the large underground stope immediately to the north of the Tourist Cliff; passages in the rock outcrop at the North-West Corner, to the north of the Lost Cavern; and underground workings in and to the south of Loc.21 and at Locations 5, 11 and 35. Most of these mine samples have been dated to the second millennium BC (at 95% probability) using the OxCal v.4.4.2 calibration programme with the IntCal20 calibration curve (Bronk Ramsey, 2009). The radiocarbon measurements, and calibrated date range for each sample (some of which are previously unpublished) are listed in Appendix 2.

Samples submitted for radiocarbon dating were usually obtained from a deposit of mining spoil considered likely to be of prehistoric origin based on its rock morphology and artefactual inclusions. Details of the sampled archaeological context and the references for each radiocarbon date are presented in Appendix 3. Most of the dated samples comprised wood charcoal and animal bone fragments (Appendix 2). Animal bone is relevant for understanding the mine's prehistoric chronology, as bone tools have been found at mines of known Bronze Age date

elsewhere, including Ecton Mine (England), El Aramo (Spain) and the mines at Kargaly in the Russian Federation, as well as at ore processing sites in the Kartamysh area of the Donbas region, eastern Ukraine (Timberlake, *et al.*, 2014; Zahorodnia, 2014; O'Brien, 2015). At Grimes Graves flint mine in Eastern England, people were using cattle bone picks to extract flint from pits in the Earlier Bronze Age, sometime between the 17th and 15th century BC (Healy, *et al.*, 2018, p. 288) (Figure 1.1). As we shall see in Chapter Three, there is extensive evidence for the similar use of bone as a mining tool to extract ore at the Great Orme mine.

Wood charcoal from this site tends to be unevenly distributed in the mining spoil. It is typically present in minor amounts of very fine fragments less than 1cm across, although there are localised pockets of denser accumulations and larger blocks up to 4cm wide (Lewis, 1990d, pp. 6–7). This material is considered primarily to be a product of fire-setting, a technique for weakening or fracturing a rock face by directing a fire against it (e.g., Lewis, 1990b, p. 55; Timberlake, 1990b, p. 49; O'Brien, 2015, pp. 204–5). As we shall see, a direct functional relationship between this practice and spoil containing charcoal of Bronze Age date can be demonstrated in at least one location at the Great Orme mine. Elsewhere at the site, the association between deposits of charcoal and fire-setting can be inferred.

There are a further six dates from three other sites on the Great Orme headland with evidence for copper metal production activities prior to the early/late modern periods. These include four dates on excavated samples of wood and nutshell charcoal interpreted as the residues of small-scale copper smelting at Pentrwyn, to the east of the mine (Figure 1.5). Three of these samples were dated to the Later Bronze Age, at the beginning of the first millennium BC (SUERC-39896, SUERC-39897, SUERC-44867; 95% probability), while the fourth was dated to the Earlier Bronze Age, in the 19th to 15th centuries BC (BETA-127076; 95% probability) (Smith, Chapman, *et al.*, 2015, table 9). The radiocarbon determination for each of these samples is shown in Appendix 2. There are two additional dates from Pentrwyn: the first, on a sample of nutshell charcoal, lay in the Mesolithic (SUERC-39901; 8775 +/-30 BP; 8160–7615 BC; 95% probability), while the second, a wood charcoal sample from a pit in a higher layer than the Later Bronze Age horizon, yielded a Medieval date (BETA-127077; 840 +/-40 BP; 1050–1275 AD; 95% probability). Neither sample will be discussed further here, although it is worth noting that the excavators suggest that the features associated with the Medieval date could in fact all relate to the Later Bronze Age phase of activity at the site (Smith, Chapman, *et al.*, 2015, p. 68, table 9).

Also relevant to the Great Orme mine are the radiocarbon dates from Ffynnon Rhufeinig (BETA-148793)

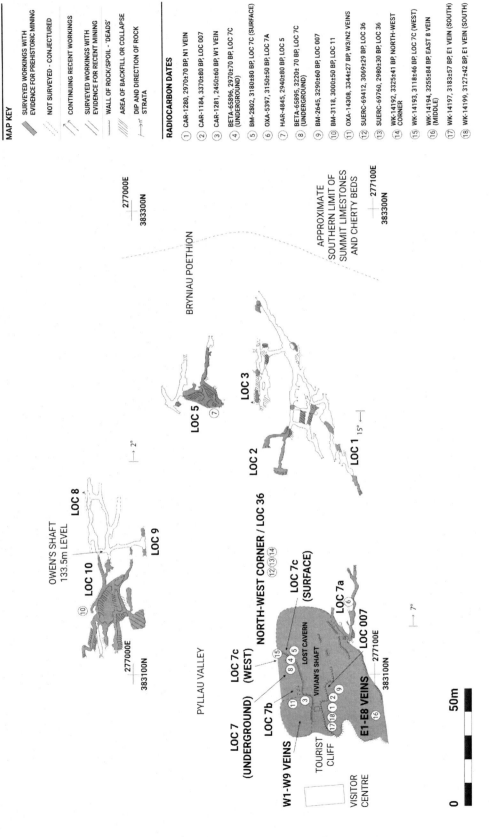

Figure 2.27: Survey plan of mine workings with evidence for prehistoric activity at 133–135m AOD, up to c.12m below the present ground surface, showing sample locations (approximate) for radiocarbon dating, with the associated laboratory number and radiocarbon measurement (for details, see Appendix 2) (redrawn by M. Andreikos after Lewis, 1996, Apps. B, C).

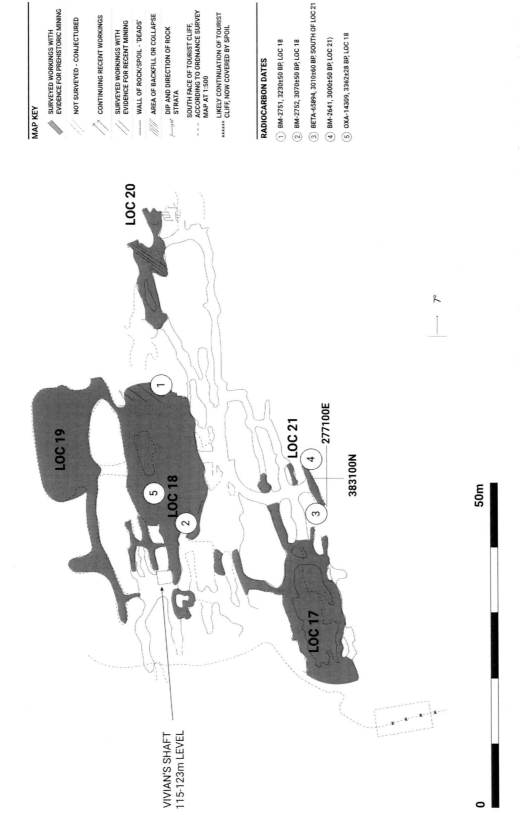

Figure 2.28: Survey plan of mine workings with evidence for prehistoric activity at 115–124m AOD, up to c.30m below the present ground surface, showing sample locations (approximate) for radiocarbon dating, with the associated laboratory number and radiocarbon measurement (for details, see Appendix 2) (redrawn by M. Andreikos after Lewis, 1996, Apps. B, C).

and Ffynnon Galchog (BM-2753), two sites on the headland interpreted as potential locations for prehistoric wet ore processing, a technique which will be discussed in Chapter Three (Figure 1.5). At each location, the dated sample comprised blue-green, copper-stained animal bone fragment(s) excavated from a heap of well-sorted rock, sand and gravel (Lewis, 1990a, 1996, pp. 144–7; Wager and Ottaway, 2019, p. 29). In each case, a sample of the bone finds was selected for radiocarbon dating due to the known functional link between animal bone and prehistoric mining (as tools and for other purposes, as we shall see in Chapter Three). The fact that in each case the bone sampled was stained blue-green by copper carbonates was also considered significant. Copper carbonate staining (with less frequent black discolouration from impregnation by iron and manganese oxides) is a key feature of the bone artefacts from the Great Orme mine, indicating a long period of deposition (Dutton, *et al.*, 1994, p. 275; James, 2011, p. 368). Finding animal bone discoloured in this way within deposits that met the criteria for the field identification of prehistoric mining spoil indicated that the surface features at both Ffynnon Rhufeinig and Ffynnon Galchog comprised material that had originated in the nearby prehistoric mine workings (Lewis, 1996, p. 146; Wager and Ottaway, 2019, p. 29).

The Ffynnon Rhufeinig sample was dated to the first half of the second millennium BC (95% probability). Its significance for the Bronze Age phase of activity at the mine is considered in more detail in a later section. In contrast, the Ffynnon Galchog sample produced a calibrated date in the first millennium AD ranging from 680–980 AD (at 95% probability). It is possible that the date of this sample lies within the narrower range of 705 to 940 AD (68% probability). Given the similarities in the surface features at both Ffynnon Rhufeinig and Ffynnon Galchog, which included flowing water and deposits of graded dolomitic silts, sands and medium gravels, this Early Medieval date from Ffynnon Galchog confounded expectations. It could be the result of a dating error or due to the dating of intrusive material (Lewis, 1996, p. 37). Dating replication – submitting a second sample from this site for radiocarbon dating to a different laboratory (Hamilton and Krus, 2018, p. 13) – could demonstrate more securely that the heaps next to this water source contained material that was the product of Bronze Age activity at the mine. For now, the date from Ffynnon Galchog is omitted from further discussion here as an unexplained data misfit.

Bayesian chronological modelling of the radiocarbon results: Models 1 and 1_v2

Calibrating the notional age and standard uncertainty of a radiocarbon determination in calendar years, using the method described in Chapter One, produces a date range in calendar years for two levels of confidence (95%

and 68% probability) for each individual sample, be it bone or wood. This estimates the date each animal died, or a tree was felled or its branches cut (Bayliss, 2015, p. 689). To estimate the dates of past events in the life of the mine that cannot be directly dated by these samples, such as when Bronze Age mining on the Great Orme started or ended, a simple, site-based Bayesian 'Bounded Phase' statistical model can be constructed (e.g., Timberlake, *et al.*, 2014, pp. 187–90; Hamilton and Krus, 2018, pp. 10–12). This combines calibrated radiocarbon dates with different types of archaeological 'prior' information – what is already known or believed about, for example, the type of site, its stratigraphic sequence and its context – using statistical inference techniques to produce an updated chronology that is more reliable and precise than its individual components (Bayliss, 2015, p. 680; Buck and Meson, 2015, p. 5; Finley, *et al.*, 2020, p. 6).

Such a model (Model 1) for the start and end of prehistoric extraction at the Great Orme mine was created as part of a project to construct a regional chronology for Bronze Age copper mining in Wales and England (Timberlake and Marshall, 2013, 2018, 2019). 17 of the radiocarbon results from the mine, plus one from the Pentrwyn smelting site, were combined with the assumption that prehistoric mining on the Great Orme, once started, continued at a uniform rate and relatively continuously until it stopped (Appendix 4). It was also assumed that all the samples were the outcome of events taking place in a single period or 'phase' of activity. No information about the stratigraphic relationships between samples (i.e., the relative ordering of events) was added to the model. As part of the analysis, 10 of the samples were found to have issues that could potentially affect the reliability of the output, as shown in Appendix 4. Rather than excluding any of these samples from the analysis, they were included as only providing a *Terminus Post Quem* (*TPQ*) date, i.e., the earliest possible date for the activity producing the associated deposit or working. The radiocarbon date calibration and modelling were carried out by the computer program OxCal v.4.1 using the IntCal09 calibration curve (Bronk Ramsey, 2009, P. Marshall, 2020, personal communication, 6 October).

The results from Model 1 indicated that mining at the Great Orme mine probably started sometime between 2000 BC (or earlier) and 1500 BC, with a concentration of dates in the latter part of the second millennium BC. Mining was estimated to have ended sometime between 800 BC to 100 BC, or even later. For the purposes of the current study, the original data from Appendix 4 were rerun in 2021 (Model 1_v2) by Dr Sophia Adams, with the same parameters and using the latest version of the OxCal computer program (OxCal v.4.4.2) (Bronk Ramsey, 2009) and the recently published IntCal20 calibration curve (Reimer, *et al.*, 2020). The

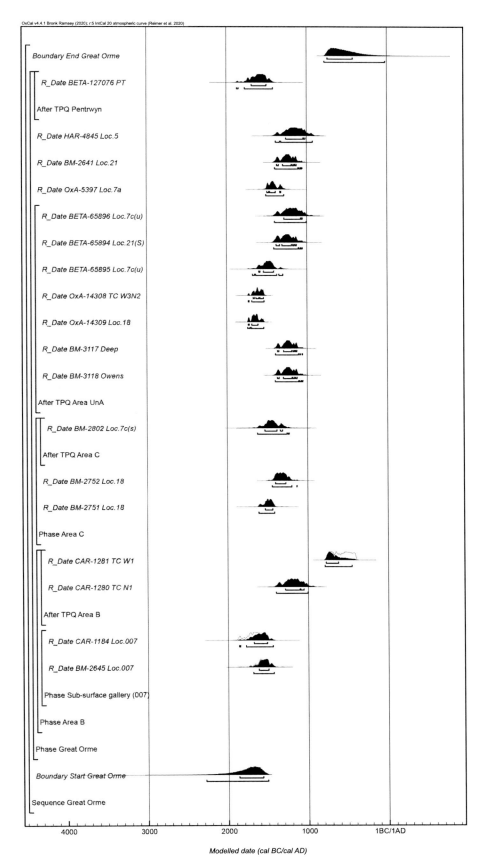

Figure 2.29: Chronological Model 1_v2 for the Great Orme using the Uniform Prior assumption (Bronk Ramsey, 2009). Each distribution represents the relative probability that an event occurred at a particular time. For each of the radiocarbon dates two distributions have been plotted, one in outline, which is the result of simple calibration, and a solid one, which is based on the chronological model used. Distributions other than those relating to particular samples correspond to aspects of the model. For example, the distribution *Start_Great_Orme* is the estimated date when mining at the Great Orme mine started. The large square brackets down the left-hand side along with the OxCal keywords define the model exactly (illustration by S. Adams).

updated probability distribution estimates in Model 1_v2 (shown here in italics) were found to agree closely with the original output (*start_Great Orme, 2280–1510 BC, 95% probability, probably 1870–1570 BC, 68% probability; end_Great Orme, 780–25 BC, probability 95%, probably 740–430 BC, 68% probability*) (Figure 2.29).

The output for Model 1_v2 shows the modelled dates for each sample, based on the spread of calibrated dates in a single, continuous phase of activity. Each calculated probability distribution or 'posterior density estimate' is a more accurate indication of a sample's date than its simple calibrated date range (Bayliss, 2015, p. 683; Hamilton and Krus, 2018, p. 9). As with all Bayesian statistical models, Model 1_v2 gives an interpretive estimate of the currency of mining activity on the Great Orme, not absolute dates (Timberlake and Marshall, 2013, p.63).

Refining the Bayesian chronological model: Model 2

More samples have been submitted for radiocarbon dating since Model 1 was first produced. Dates are now available for the mine's North-West Corner, Loc.36 and various additional locations on and around the Tourist Cliff, as well as from the Pentrwyn smelting site (Appendix 2). A revised version of the model combining the original dataset with these new data points using the same 'Uniform Prior' assumption – i.e., that prehistoric mining at the Great Orme mine represented a single continuous phase of activity that did not vary significantly in intensity over time – would be likely to give a narrower probability distribution and so a tighter estimated date range for the currency of mining on the Great Orme. The probable start and end of mining at a site-based level (Timberlake, *et al.*, 2014, pp. 187–90), or the sequence of activity over time from one mine to another on a regional scale (e.g., Timberlake and Marshall, 2013, 2018, 2019), are the broadest chronological questions we can ask of the data. At this macro level of analysis, whether a sample was residual or intrusive – i.e., redeposited from a different context or even another location at the mine – does not affect the overall site-level chronology, if we are confident that all the samples are a product of Bronze Age mining there. To address more complex questions about past events at the mine, such as which parts of the source were mined before or after others, we also need to evaluate the strength of the link between the dated sample of bone or wood charcoal and the formation of the specific deposit or working where it was found. For a robust chronological model, only samples for which a secure prehistoric provenance can be inferred or – even better – demonstrated should be included in the Bayesian analysis (e.g., Taylor and Bar-Yosef, 2014, p. 43; Bayliss, 2015, pp. 689–90; Hamilton and Krus, 2018, p. 5).

The archaeological information about the recovery context of every dated sample from the mine and the associated production sites was re-evaluated for the purposes of the present study. Each sample was scored on a scale of 0 to 2, where 0 indicated no secure functional association between the sample and its context, 1 an inferred association, and 2 where this relationship could be demonstrated archaeologically. This evaluation was undertaken without referring to each sample's simple calibrated date range. Undertaking this evaluation is particularly challenging at a large, complex multi-period site such as the Great Orme mine. During the intense phase of mining in the 18th and 19th centuries AD, miners and prospectors often moved prehistoric spoil considerable distances within the mine from its original place of deposition. They deliberately cleared areas of Bronze Age working, subsequently redepositing the spoil at the surface or in neighbouring tunnels underground, sometimes behind carefully constructed dry-stone revetments (e.g., at Loc.10) (Figure 2.27).

Andy Lewis (1996, tables 5–8) has, however, developed detailed criteria for the comparative field identification of prehistoric and early/late modern phases of working at the Great Orme mine. For samples where traces of 18th/19th-century AD mining, such as broken speleotherms and morphologically recent spoil or artefacts, had been recorded nearby using these criteria, particular care was taken to consider the probable impact of this subsequent activity on the provenance of the dated material. It is also likely that the prehistoric miners moved and redeposited spoil and tools during the Bronze Age. They may also have reprocessed spoil of earlier date in later workings (Dutton, *et al.*, 1994, pp. 256–7; Lewis, 1996, p. 156). Again, care was taken to assess the recorded archaeological information for each sample to evaluate whether any evidence for this could be identified. The outcomes are presented in Appendix 5.

Four of the mine samples were found to have no or an uncertain functional association with their recovery context (indicated by an 'archaeological association' score of 0 in Appendix 5). Two samples were of human, not animal, bone (OxA-14308, OxA-14309). Although, as we will see in a later chapter, it is likely that both fragments were the remains of a single individual, they were found disarticulated and in unstratified spoil deposits at two different mine locations (OxA-14308 from the junction of the W3/N2 ore vein on the Tourist Cliff and OxA-14309 from Loc.18) (Figures 2.27, 2.28) (Lewis, 1996, p. 126; James, 2011, table 1.2), not from a single clearly defined context such as a burial pit. Were these dates to be modelled despite their contextual uncertainties, it would also be necessary when constructing the model to consider the Great Orme's coastal location and the possibility of a dietary age off-set caused by the regular consumption of marine animals (Hamilton and Krus, 2018, p. 13) – particularly given that, as we shall see later, there is evidence that this individual was local

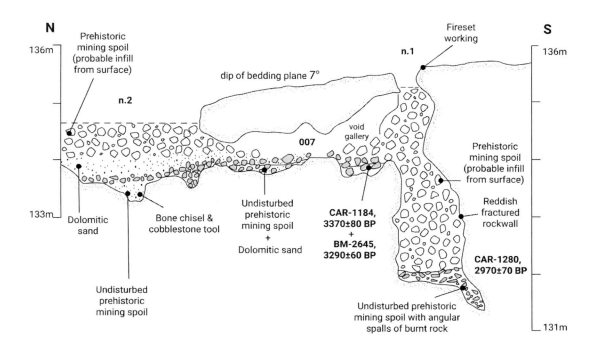

Figure 2.30: Semi-diagrammatic section through Loc.007, on the East 1 ore vein, and the fire-set mine working on the North 1 ore vein (redrawn by M. Andreikos after Dutton, *et al.*, 1994, fig. 8a; copyright Gwynedd Archaeological Trust).

to the area. Dating (potentially) the same individual more than once in the same chronological model is also likely to bias the output towards a shorter estimated duration (Bayliss, 2015, p. 688).

In an underground passage at Loc.36 (Figure 2.27), a compacted deposit of mining waste (005) was found sealed by an overlying layer (004) that was rich in charcoal, pieces of heavily fire-reddened rock and fragments of bronze metal. Using Lewis's (1996, table 6) criteria for the identification of fire-setting at the mine, there was, however, no indication in the rock walls of the passage for the use of this technique at that location. The excavators surmised that at least some of the spoil in context 004 derived from an earlier and as yet unidentified working further into the hillside. This was subsequently redeposited over the residues from an in-situ phase of extraction – represented by layer 005 – that had taken place in the intervening period (Jowett, 2017, fig. 8). The dated wood charcoal sample from 004 (SUERC-69412) was therefore most probably redeposited. While it is more likely that the dated wood charcoal retrieved from layer 005 (SUERC-69760) was a product of extraction at its find location, the recorded occurrence of a few bronze fragments in this lower layer raises the possibility that the sampled charcoal was, like the bronze fragments, also intrusive. We cannot therefore be confident that either sample dates the formation of the working in which it was found. This uncertainty contrasts with the very secure functional association that can be demonstrated for wood charcoal sample CAR-1280, which was obtained from a deposit of fire-cracked rock found immediately beneath a fire-reddened rock wall on the N1 ore vein in the Tourist Cliff (Figures 2.27, 2.30). There is no recorded evidence for early/late modern disturbance at this location and the fire-setting residue was sealed by a layer of dolomitic sand interpreted as evidence of hiatus (Dutton, *et al.*, 1994, p. 261, table 1). In this case, we can be confident that the sampled wood charcoal dates the fire-setting event (although as we shall see, the sample's composition raises other uncertainties that need to be taken into consideration when using it to build a Bayesian chronology).

Leaving the mine, the next contextual relationship to consider is that between the dated sample of animal bone fragments from Ffynnon Rhufeinig (BETA-148793) and the heap of well-sorted clay, dolomitic sand and gravel from which they were excavated. Interpreting the significance of this single date is complicated by the documented evidence for extensive early/late modern activity at this location (Wager and Ottaway, 2019, p. 26), which makes the functional association between the sample and the date of formation of the specific deposit where it was found difficult to demonstrate securely. At the site-based level, however, a relationship between the dated bone and prehistoric ore processing can be inferred by tying together different strands of contextual evidence.

As discussed by Wager and Ottaway (2019), these include the proximity to the mine of numerous water sources on the headland; the relative material properties, particularly density, of the mined copper minerals and host rock; the nature of the features and deposits at this site, which are characteristic of prehistoric wet ore processing; and finally the comparative evidence from Bronze Age sites elsewhere for this practice.

Turning finally to the nearby Pentrwyn smelting site, a secure archaeological relationship can be either demonstrated or inferred for the three wood and nutshell charcoal samples associated with the evidence for metallurgical activity at this location (SUERC-39896, SUERC-39897, SUERC-44867), despite this small site's complex, multi-phase formation history (Smith, Chapman, *et al.*, 2015, p. 54, table 9). Burrowing by rabbits or other small mammals had, however, disturbed the feature with the final Bronze Age date from Pentrwyn (sample BETA-127076). Given the likelihood that sample BETA-127076 was residual, no functional association between it and the pit infill from which it was recovered could be inferred and so it was assigned an archaeological association score of 0. This sample also comprised unidentified wood charcoal from the pit infill that had been bulked together (Smith, Chapman, *et al.*, 2015, table 9). This lowers the probability that the result is representative of the date of the feature in which it was found and hence that it links to the prehistoric phase of smelting at Pentrwyn. Other potential sources of a radiocarbon age offset include dating long-lived species, such as oak (*Quercus* sp.), and 'old' wood from mature trees, rather than twigs or small branches (e.g., Bayliss, 2015, p. 687; Hamilton and Krus, 2018, p. 13).

Leaving aside all the samples with an archaeological association score of 0 (which, as we have already seen, indicates that they are unsuitable for chronological modelling), the remaining 20 mine and four non-mine samples were scored according to whether they comprised only single-entity and short-lived material. Those fulfilling these criteria were assigned a score of '1' in each case. By combining these results with the archaeological association scores, an overall value for each sample was obtained (Appendix 5). This allowed the samples to be loosely ranked in terms of their date 'quality' (following Bayliss, 2015) and hence their suitability for chronological modelling. Using this sample hierarchy, a total of 17 samples (14 from the mine and three from the Pentrwyn smelting site) were selected for inclusion in a new Bayesian 'bounded phase' chronological model (Model 2). Model 2 had the same 'Uniform Prior' assumption as in Models 1 and 1_v2. Unlike these previous models, however, in which all the selected samples from both the mine and the Pentrwyn smelting site were considered to belong to a single phase of activity ('*Phase Great Orme*'), Model 2 assumed that the dated samples represent events in a single, continuous phase of activity at each site ('*Phase Mine*' and '*Phase Pentrwyn Smelting Site*'), but that the relative sequence of the dates of the samples and the sites was unknown.

As we saw earlier, the mine's physical structure indicates that there were three very broad stratigraphic phases of activity at this site over time, beginning with the mining of the ore veins transecting the Tourist Cliff (Stratigraphic Phase 1); then the removal of the mineralisation in locations immediately to the north of this feature (Stratigraphic Phase 2); followed by extraction of ore from locations further into the hillside (Stratigraphic Phase 3). There were, however, unfortunately too few 'quality' radiocarbon dated samples associated with Stratigraphic Phases 2 and 3 for this relative ordering to be included in the '*Mine Phase*' in Model 2, as at least five samples corresponding to each stratigraphic phase are needed for robust modelling (Buck and Meson, 2015, p. 8).

A further issue considered during assessment of the data for inclusion in Model 2 was the presence of potential age-offset samples. Six samples of wood charcoal from the mine had an inferred functional association with their recovery context but were of unidentified species and form (indicated by a 'total sample quality' score of '1id' in Appendix 5). Following Hamilton and Krus (2018, p. 13), these samples were included in the '*Mine Phase*' in Model 2 as providing a TPQ for the date of formation of their recovery context. A further two mine samples with a 'quality' score of 2 were also assessed as suitable for inclusion in Model 2, both as TPQ dates for the deposits in which they were found. In the case of BETA-65894, a sample of oak wood charcoal from a restricted underground working to the south of Loc.21 (Figure 2.28), its inclusion as a TPQ was to minimise the effect on the model of potentially incorporating an 'old-wood-offset' date that was older than its associated activity. Wood charcoal sample CAR-1280 (Figure 2.27) similarly included oak and so potentially had an old wood offset. This, however, was not a single-entity sample as hazel wood charcoal was also dated alongside that from oak. Given this sample's very secure contextual association, described earlier, it was considered probable that this oak/hazel wood charcoal mixture was the residue from a single event and that, rather than being redeposited material of different date, it included more than one species due to the Bronze Age miners' intentional choice of wood for use in fire-setting, a topic to be returned to in Chapter Three.

At the Pentrwyn smelting site, sample SUERC-44867 was recovered from the fill (112) in the base of a small pit (feature 111), which as we shall see in Chapter Three, is interpreted as a simple smelting furnace. Feature 111 had been dug into an occupation layer (107), from which samples SUERC-39896 and SUERC-39897 were recovered during excavation (Figure 2.31). The results for samples SUERC-39896 and SUERC-39897 therefore represent an earliest possible date for the digging of the pit and the subsequent smelt. Consequently, the dates for SUERC-39896 and SUERC-39897 were included in the 'Pentrwyn Smelting Site Phase' in Model 2 as TPQs for the formation of the charcoal-rich smelting residue 112 infilling the smelting pit.

Model 2 (Figure 2.32) has a satisfactory 'dynamic index of agreement' (Amodel=103.9), indicating good overall agreement with the assumption that the dated material is from a 'uniform' distribution. There is also good agreement (A>60) for each of the individual results. Model 2 estimates that the start of activity at the mine itself occurred from *2085 to 1455 BC (95% probability; Boundary Start Mine)*, most probably from *1770 to 1520 BC (68% probability)*. It is estimated to have ended from *1235 to 640 BC (95% probability; Boundary End Mine)*, most probably from *1165 to 935 BC (68% probability)*. Although, like Model 1_v2, this latest model estimates that mining probably started during the first half of the second millennium BC, in the Earlier Bronze Age, the range between the probable start and end of activity at the mine in Model 2 is considerably more tightly constrained than that in Model 1_v2, with an estimated ending in the Later Bronze Age, rather than the Earliest Iron Age. None of the mine samples appear to represent a different phase of mine-working. In the next few sections, we will briefly assess these results alongside other sources of archaeological evidence for when mining on the Great Orme began and ended, before looking in more detail at possible patterns of activity at the mine over time in the intervening period.

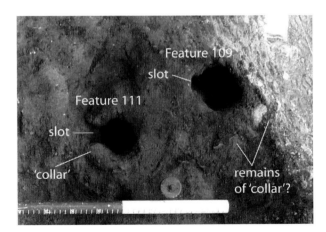

Figure 2.31: Excavated features at Pentrwyn interpreted as two small smelting pits, after removal of layer 107. A sample for radiocarbon dating was recovered from the base of the fill in feature 111 (SUERC-44867, 2727±33 BP). Two further samples were obtained from layer 107 (SUERC-39896, 2730±30 BP; SUERC-39897, 2780±30 BP) (copyright Gwynedd Archaeological Trust, published in Smith, Chapman, *et al.*, 2015a, fig.9).

The beginning of mining

The estimate that mining probably began sometime during the 18th–16th centuries BC corresponds well with the likely start date of around 1700 BC proposed by Timberlake and Marshall (2013, 2018, 2019) for extraction at this source. It is, however, somewhat later than Lewis's suggestion for the earliest mining occurring sometime between the 24th and 21st century BC (Lewis, 1996, p. 168), perhaps towards the end of the Chalcolithic. Recent work by Williams has detected copper metal produced with ore from the Great Orme mine in a small number of metal objects with an estimated date of deposition from c.2150 BC onwards (Williams and Le Carlier de Veslud, 2019, fig. 6). This indicates that, by the end of the third millennium BC, people were certainly mining ore on the Great Orme to make copper, despite there being as yet no dated evidence from the mine for activity in this early period. Local contextual evidence also hints that the dateable mining information may not be telling the whole story about the beginnings of mineral extraction on the Great Orme. Human activity in this area has a long history stretching back to the Upper Palaeolithic (e.g., Smith, Walker, *et al.*, 2015). As already discussed, the surface copper mineralisation was most probably an obvious feature of the Pyllau Valley, particularly as its northern slopes are likely to have been more sparsely vegetated than those further east, due to toxicity caused by the elevated copper content of the soils overlying the copper-bearing rock (Dutton, *et al.*, 1994, p. 255). Charcoal samples from outcropping mineral vein exposures at several sites elsewhere with similarly strongly coloured minerals and associated rock have returned occasional radiocarbon dates in the fourth and third millennia BC. This includes two locations in mid-Wales: the Bronze Age copper mine of Copa Hill, as well as from a scree-covered mound at the Nantyricket Mine. At the Alderley Edge mines in north-west central England, campsites occupied in the Mesolithic period (c.8500–4000 BC) have been identified from scatters of worked stone tools (Timberlake and Prag, 2005, pp. 8–9, chap. 3; Timberlake, 2009, pp. 239–40, 2015a; Timberlake and Marshall, 2018, p. 428). The activity represented by all these findings significantly predates the earliest known instances of copper ore mining for smelting in the British Isles. It hints that well before people began digging out copper minerals to make metal, they were attracted to these sites, perhaps to extract minerals on a very small-scale to make pigments (e.g., Timberlake, 2015a), or even medicines or jewellery – although there is no conclusive evidence for such uses from prehistoric sites in Britain and Ireland.

In the absence of direct, dated information, we should at least be open to the possibility that the prehistory of copper mineral extraction on the Great Orme predates the Bronze Age, given that the mineralisation

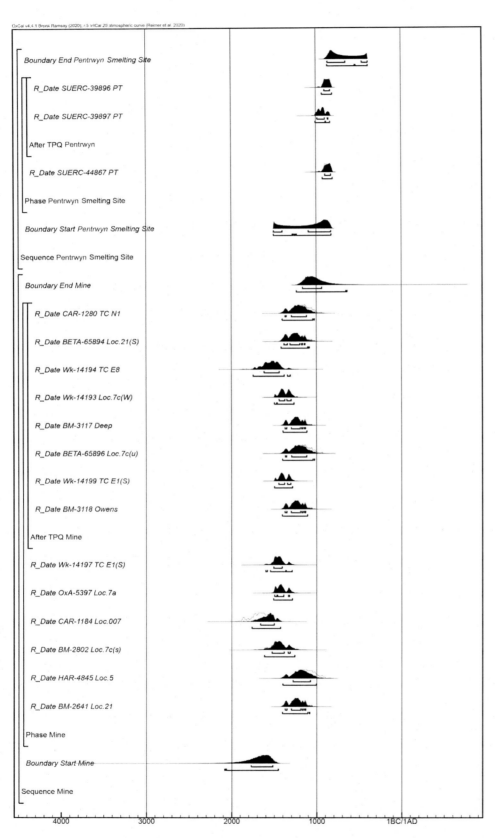

Figure 2.32: Chronological Model 2 for the Great Orme using the Uniform Prior assumption (Bronk Ramsey, 2009). Each distribution represents the relative probability that an event occurred at a particular time. For each of the radiocarbon dates two distributions have been plotted, one in outline, which is the result of simple calibration, and a solid one, which is based on the chronological model used. Distributions other than those relating to particular samples correspond to aspects of the model. For example, the distribution *Start_Mine* is the estimated date when mining at the Great Orme mine started. The large square brackets down the left-hand side along with the OxCal keywords define the model exactly (illustration by S. Adams).

outcropping at the surface would have been visually striking and easy to remove. The Neolithic community who built and used the burial chamber of Llety'r Filiast next to the source during the fourth millennium BC (Figure 2.9) must have been aware of the exposed seams of highly coloured minerals in the southern faces of the valley's rocky outcrops. Could they even have explored areas underground by accessing natural caves in a karst system extending into the mineralised zone? There is certainly evidence for activity during the Neolithic and Mesolithic periods in some of the caves and rock shelters located in the limestones of the headland outside this zone (e.g., Smith, Walker, *et al.*, 2015). Even if Neolithic groups were never actively involved in mineral extraction on the Great Orme, the proximity of the Llety'r Filiast burial chamber to the copper deposit suggests that the siting of this monument was not coincidental. The choice of location may at least partly have been due to an awareness of the presence of the unusual rocks and minerals nearby, a cognisance that Timberlake (2009, pp. 239–40) refers to as a "mental geological map" of the landscape by prehistoric communities. This deep and subtle knowledge of topographic and geological variations by local Neolithic communities can perhaps be considered the foundation for the beginnings of copper mining on the Great Orme in the Bronze Age. This is not to say that the people who built and used the Llety'r Filiast tomb during the fourth millennium BC understood that the minerals they encountered in the Pyllau Valley could be smelted into copper metal. Rather, knowledge of the existence of these substances set up the necessary conditions for this link to be made once the idea and practice of copper smelting started to gain currency at the end of the third millennium BC. Mining on the Great Orme during the Bronze Age most probably therefore arose out of the gradual reworking of historically situated and detailed understandings of the headland's landscape and its materiality: its form, vegetation, rocks and minerals.

The end of mining

The estimate that the Bronze Age phase of mining at the Great Orme mine most probably ceased around 900 BC agrees with the end date proposed by Williams and Le Carlier de Veslud (2019, p. 1180). In this context, the probability density estimates from the Pentrwyn smelting site are intriguing. Model 2 estimates that the start of metallurgical activity there occurred from *1505 to 825 BC (95% probability; Boundary Start Pentrwyn Smelting Site)*, most probably from *1505 to 830 BC (68% probability)*. The end of this activity is, however, estimated to most probably have taken place from *865 to 390 BC, at both 95% and 68% probability; Boundary End Pentrwyn Smelting Site)*. Model 2 used only three dated samples from this

site, which as we saw above, is an insufficient number for robust modelling. At least five quality samples directly associated with the metallurgical activity are needed to establish its chronology more firmly. For the moment, it remains an open question as to whether the smelting evidence from Pentrwyn is representative of the methods used during the period of customary Bronze Age mine-working on the Great Orme – when people regularly and repeatedly mined there – or whether it relates to a different phase of activity, as other researchers have previously suggested (Timberlake and Marshall, 2013, p. 65). This topic will be returned to in the next chapter. What we can say for now is that the three modelled dates from this site cluster between 985 to 830 BC (68% probability). Although radiocarbon modelling cannot prove exact contemporaneity (Timberlake, *et al.*, 2014, p. 190), when considered alongside the information about the sequence of events uncovered during excavation (described earlier), these results most probably represent the same, short-lived smelting event.

The single sample (CAR-1281) of Earliest/Early Iron Age date (*790–455 BC, 95% probability;* most probably *775–625 BC, 68% probability*; modelled probability estimates, Model 1_v2) from a working on the W1 ore vein in the Tourist Cliff was excluded from Model 2, because it comprised a bulk sample of highly disseminated wood charcoal (Figure 2.27). Timberlake and Marshall (2013, p. 64) point out that the probability that this sample is part of the Bronze Age phase of dated mining activity is 0%. Submitting for radiocarbon dating the cattle rib gouge recovered from a small pocket-working associated with the working floor (Dutton, *et al.*, 1994, p. 258) may help to address the still open issue of how this location fits into the mine's chronology and whether any post-Bronze Age extraction took place there.

The main or 'customary' mining phase

Although there were possibly earlier or later instances of mining on the Great Orme, the dated evidence from the mine supports the idea developed from consideration of the broader archaeological context that the customary phase of mine-working at this source occurred during the Bronze Age, from probably the late 18th to the mid-10th century BC. Its probable duration was *395–785 years (at 68% probability)*. This does not mean that this activity always took place continuously, without any breaks or interruptions. As all the radiocarbon results from the mine have errors of at least +/-25 years BP (and in most cases considerably longer) (Appendix 2), a pattern of activity in which people mined at the site seasonally on an annual or even decadal basis for short periods can be modelled statistically as 'full and continuous use' (Hamilton and Krus, 2018, p. 5, D. Hamilton 2021, personal communication, 21 January). The archaeological evidence for this tempo of working at the mine will be examined in the following chapters.

As we saw above, Model 2 assumes no activity at the source before the start of the dated phase of mining and that, once ore extraction began, it occurred at maximum intensity without any breaks or interruptions until it stopped, to be followed again by no activity (Fitzpatrick, Hamilton and Haselgrove, 2017, p. 369). Between these start and end points, for the 'Uniform Prior' assumption to be valid, the intensity of mining should not vary significantly over time (Hamilton and Krus, 2018, p. 5, D. Hamilton 2021, personal communication, 21 January). The chronological model produced in this case fits an archaeological scenario in which the Great Orme source was 'discovered' by prospectors searching for copper and that extraction began only after this discovery event, even though, as Timberlake and Marshall (2013, p. 65) point out, there could have been a time lag between this event and when mining started. It likewise assumes that the end of use of the source was defined by a distinct moment of 'closure' or 'abandonment'. As we shall see in Chapter Four, however, the archaeological evidence for the scale, intensity and patterning of settlement and other forms of land use; the mobility of people, ideas and things; and the making of copper and bronze in Bronze Age north Wales does not support the idea of such abrupt changes, either at the beginning nor at the end of this period. Rather, transformations in these aspects of life seem to have taken place only gradually from the late third to the early first millennia BC. Against this backdrop of incremental change, it is unlikely that the beginnings of mining on the Great Orme were characterised by the start of mining operations as a one-off event. Instead, going to this site to mine copper ore to smelt into metal most probably only became customary gradually over a period of time. Likewise, the 'end' of mining on the headland was most likely a process rather than an event, for reasons to be discussed in Chapter Five.

It has been suggested that the period of peak productivity at the mine during the Bronze Age, during which the largest quantity of copper metal was produced, occurred for approximately 200 years from c.1600 BC. This main phase of production was then followed by a marked decrease in output in the following centuries until Bronze Age activity at the mine ended c.900 BC (Timberlake and Marshall, 2013, 2018, 2019; Williams, 2014, 2017; Williams and Le Carlier de Veslud, 2019). This model of varying intensification is based on the finding that the chemical and lead isotope signature of Great Orme ore correlates strongly to the chemical composition of bronze metal objects assigned typologically to the Earlier Bronze Age 'Acton Park metalwork phase' (c.1600/1500–1400 BC) (Needham, 1996). 96% of 22 'Acton Park' objects sampled displayed this match. It was identified in far fewer bronze

items attributed to the periods immediately before and after: in only 36% or 14 of the objects sampled from the preceding metalwork phase ('Arreton') and in 37% of artefacts (17 items) from the following 'Taunton' phase. This value drops to only 10%, or 10 artefacts, by the 'Ewart Park' phase (c.1020–800 BC) (Williams and Le Carlier de Veslud, 2019, pp. 1185–8, fig. 6).

These results do seem to show marked usage of Great Orme ore to make Acton Park-type objects. Detailed consideration of the relationship between copper production and consumption for Bronze Age communities is beyond the scope of this book, but it is worth noting, however, that there is debate around when Acton Park-type metalwork was being made and used. The modelled probability distributions in Model 2 also have no obvious clustering during the 16th-15th centuries BC. Rather, six of the fourteen modelled dates have later start dates, in the early 14th century BC, compared to two in the 16th century BC and three in the 15th century BC (all at 68% probability). As we will see throughout this book, there is little evidence for distinct changes in people's lifeways on and around the Great Orme, and further afield across north Wales, corresponding specifically to the Acton Park period. Nor can the development of specific parts of the mine yet be securely tied to particular chronological phases of activity at the source, as discussed in the following section.

While the archaeological evidence certainly agrees with a chronological model characterised by a change in mining intensity over the course of the Bronze Age, with a sustained period of customary working preceded and followed by periods when the mine was less routinely visited, I argue that these changes are unlikely to have been sudden 'watershed' moments. Willis and colleagues (2016, p. 349), in their study of Neolithic cremated human remains from Stonehenge, suggest that use of the Trapezium Prior date distribution model is preferable to the Uniform Prior assumption in cases where the archaeological information does not point to abrupt change. The former is designed to show how fast the introduction of a dated event was. With the small number of quality radiocarbon samples currently available from the mine, the chronology produced by the Trapezium assumption would be very similar to that seen in Model 2. Using this model in the future, however, as more quality samples become available, will help to refine the start and end of activity (D. Hamilton 2021, personal communication, 18 August). It would also allow the tempo of activity between the first and last uses of the mine, plus the duration of the period when activity was at its peak, to be estimated (Fitzpatrick, Hamilton and Haselgrove, 2017, p. 369). For the present, however, there is insufficient information to identify more securely the customary mining period, or when people habitually visited the Great Orme during the Bronze Age to mine copper ore for smelting into metal.

Reassessing the development of the mine over time

Due to the data limitations already discussed, Model 2 was not structured according to the stratigraphic phases deduced from the mine's layout. Although few definitive conclusions can be reached using such a small dataset, the output does, however, provide some interesting hints of how the mine developed over time, whilst acknowledging the inconsistencies due to the sampling background to these dates. Beginning with the earliest modelled dates, these are from Loc.007, a confined tubular channel (now exposed at the surface), at the junction of the East 1/ North 1 ore veins (Figure 2.30), and a small working on the East 8 vein (CAR-1184, probably *1660–1500 BC, 68% probability*; Wk-14194, *1615–1440 BC, 68% probability*) (Figure 2.27). Both locations are on the Tourist Cliff, suggesting that the ore veins transecting this outcrop were a focus for initial Bronze Age activity at the source. Loc.007 and the East 8 vein may not have been the very first areas to be mined, however, despite their early date. Located in the broad ore veins above the 'Craig Rofft' sandstone horizon (Figure 2.5), they worked exposures of mineralisation that were, like that which once infilled the entrances to Loc.17, higher up the cliff face and so potentially more difficult to reach than the ore-bearing rock lower down (Figure 2.24). It is possible that the very earliest Bronze Age miners first extracted the mineralisation outcropping at the exposed base of the Tourist Cliff – which they could reach simply by standing at its foot – although there is no dated evidence for this. The miners would then have followed the soft ore-bearing ground as it ran into the hillside. This activity would have created entrances close to the exposed base of the Tourist Cliff that eventually led into the workings at Loc.21.

Eventually, numerous entrances into underground passages would have been formed all over the Tourist Cliff's southern face. This may have occurred soon after Bronze Age mining began, but none of the available dates refer directly to this initial phase of extraction. As the mine's physical structure is so complex, with many locations comprising a network of workings, not a single passage or stope, the time depth of working at each of the dated mine locations is far from clear. Only the working on the East 1 (South) vein on the Tourist Cliff has two 'quality' samples (Figure 2.27); for all the other dated mine locations, only one sample is available (Appendix 5). A single dated sample refers only to the activity that produced the specific deposit in which it was found. Its probability estimate could represent an individual mining event (as was probably the case for sample CAR-1280, above) or just one instance of activity in an extended episode of unknown duration. It does not necessarily indicate the age of all the other workings, nor all the episodes of activity, at the same location, nor when extraction leading to that location first began. For example, there is dated mining activity at Loc.21, probably starting around *1375 BC (BM-2641, 68% probability)* (Figure 2.28). This single radiocarbon sample, from a restricted underground working at a vertical depth of c.20m from the present ground surface, is however unlikely to indicate when extraction first began there, nor when the passages leading to it through the Tourist Cliff were formed.

One of the earliest modelled dates from the mine is from an overhang towards the base of the north-west side of the Lost Cavern, at Loc.7c (Surface) (BM-2802, *1525–1315 BC, 68% probability*) (Figure 2.27). Using stratigraphic information, extraction could only have begun there once a route had been opened through the Tourist Cliff mineralisation and into the Lost Cavern itself (Stratigraphic Phase 3). It would have taken time, using available tools and techniques, to reach this part of the source, which may lend support to the hypothesis that the Tourist Cliff was a focus for activity from the very beginning of the customary Bronze Age mining phase. Alternatively, it is possible that the miners found a more direct route to Loc.7c (Surface), either by digging pits, trenches and shafts downwards from the surface as they began prospecting across the hillside for new ore veins, or through as yet unidentified prehistoric workings or even natural cave passages.

The dated evidence hints that the mineralisation of the Tourist Cliff and Lost Cavern continued to be worked repeatedly over the Bronze Age life of the mine. Fire-setting at the junction of the North 1/East 1 ore veins, for example, took place most probably between *1370 and 1115 BC (CAR-1280, 68% probability)*, some three centuries after the earliest dated instance of extraction from the same veins close by (CAR-1184). There is also dated mining activity from Loc.7c (Underground), a small underground gallery at the western edge of the Lost Cavern, within the same period (BETA-65896, *1370–1150 BC, 68% probability*) (Figure 2.27). Model 2 produced other sets of identical results, although the significance of these overlaps is ambiguous. The probability estimate for a wood charcoal sample from the southern end of the East 1 ore vein (Wk-14199, *1445–1305 BC, 68% probability*) is the same as that for wood charcoal from a narrow passage near the base of the deep West 8 ore vein in the cliff overlooking the south-west corner of the Lost Cavern (Wk-14193). The modelled date from Loc.21 is the same as those from a small, backfilled passage at Loc.11 accessed today from the 18m deep level off Owens Shaft (Figure 2.28) and from the confined and unstable working at Loc.35 (BM-3118, BM-3117, *1375–1130 BC, 68% probability*) (Figure 2.23). While these results could point to contemporary activity

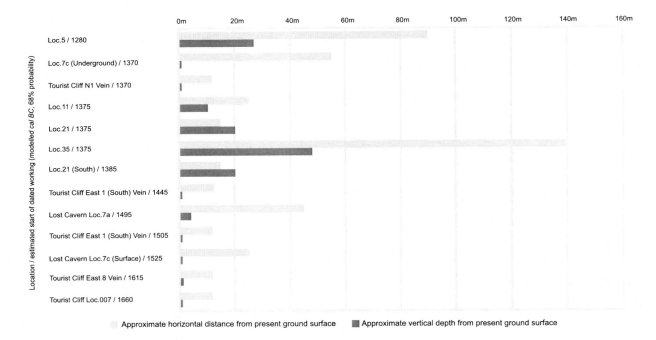

Figure 2.33: Modelled start estimates for different mine-working locations at the Great Orme mine, in chronological order from latest (top) to earliest, with their approximate depth and horizontal distance from the present ground surface (drawing by M. Andreikos).

Figure 2.34: Modelled start estimates for different mine-working locations at the Great Orme mine, in chronological order from latest (top) to earliest, with their Stratigraphic Phase (drawing by M. Andreikos).

at the different locations, there is no direct archaeological evidence to show that they were part of the same mining episodes. The fire-setting that produced charcoal sample CAR-1280, for example, could have taken place at any time during a 250-year period (at 68% probability).

As extraction progressed, each series of tunnels and stopes opened up as ore was extracted from the Tourist Cliff and Lost Cavern typically led into a further network of passages at other locations. Over time, this created the Bronze Age mine's complex, sinuous structure (Figure 2.14). There is very limited dating information available for the development of the mine beyond the mineralisation associated with the Tourist Cliff and Lost Cavern. For example, the chronological relationship between Loc.18 and the Lost Cavern – for which there are no clear stratigraphic markers – is still undefined, as there are as yet no 'quality' radiocarbon samples from the former. Considering the modelled data in terms of the general relative vertical depths of the sample find locations from the present ground surface, a trend for deeper workings to be later is evident (Figure 2.33). This is unsurprising, given that extraction at depths of c.50m, at Loc.35 for example, could only have occurred sometime after mining began at the surface (unless the miners found direct routes of access through existing cave passages) (Figure 2.23). In other areas of deeper working, such as the complexes of passages at Locs. 23 and 26 c.55m below the present ground surface, the chronology of activity forming these workings is still unclear (Figures 2.22, 2.23, 2.25). Only two of the modelled dates are from locations at more than 20m below the present ground surface: BM-3117 (*1375–1130 BC, 68% probability*) from Loc.35, an area of narrow and restricted workings produced by following minor ore veins through rotted dolomite (Lewis 1996, App. C, 16–17) (Figure 2.23); and HAR-4845 (*1280–1075 BC, 68% probability*) from Loc.5, a complex system of prehistoric workings at a depth of c.27m below the present ground surface (Figures 2.24, 2.27). The dated working in the Tourist Cliff and Lost Cavern sometime during the 14th–12th centuries BC (CAR-1280, BETA-65896 Loc.7c (Underground)) also hints that the trend for deeper working to be later may not be absolute (Lewis, 1996, p. 155) (Figure 2.27).

When the modelled data are considered in terms of the approximate relative horizontal distances of the sample find locations from the slope of the hillside (Figure 2.33), as might be expected the very earliest probability estimates are from workings at least distance (less than 12m), while the workings at the greatest distance (more than c.25m) are also the latest. There is a very general trend for later mining to take place further into the hillside, although as with the vertical depth data, this trend is not absolute. As we saw above, a working such as Loc.7c (Surface), which has an early date in the history of the mine's development despite being at a considerable horizontal distance from the present ground surface, could have been accessed by other, as yet unidentified, means.

The early chronology of working associated with the Tourist Cliff and the Lost Cavern can be modelled with more confidence than other areas of the mine because there are more data points from this part of the site, as discussed earlier. Of the 14 mine samples in Model 2, five were taken from workings in ore veins transecting the Tourist Cliff itself and four were from areas around the Lost Cavern. A further three dates were from parts of the mine that the prehistoric miners most probably reached via existing workings associated with the Tourist Cliff, as there were no other direct connections to surface from these workings. This clustering of dates creates the impression that the

mineralisation associated with this outcrop was the focus of Bronze Age activity and that it was worked earlier, and for longer, than the ore accessed from other outcrops to the north and west in the mineralised zone. The apparent concentration of working here could in part be because the Tourist Cliff and Lost Cavern have been researched far more extensively than other parts of the mine, as indicated by Geoff David's summary excavation reports published annually in Archaeology in Wales (David, 1992, 1993, 1994, 1995a, 1996, 1997, 1998, 1999, 2000, 2001, 2002, 2003, 2004, 2005). What is still missing from our understanding of the mine's internal development is how extraction proceeded at the other nearby rocky outcrops in the mineralised zone. There are only two modelled dates from these areas. One is from Loc.11, a small underground working with a surface entrance that was probably located in a hollow between Owen's Shaft and the present-day Pyllau farmhouse, to the south-west of the Tourist Cliff (BM-3118, *1375–1130 BC, 68% probability*). The other is from Loc. 5, a working which was most probably reached from entrances in an outcrop to the north of the Tourist Cliff, below the modern road to the summit (Figure 2.27). The probability estimate from the latter shows that, by the last few centuries of the second millennium BC, the miners had accessed the northern end of the mineralised zone (HAR-4845, probably *1280–1075 BC, 68% probability*). In neither case, however, do the modelled dates indicate when each entrance location was first worked in the Bronze Age nor the subsequent chronology of extraction. While it currently appears that the areas of rich mineralisation in and around the Tourist Cliff and the Lost Cavern were the exclusive focus for mining during the 17th to 15th centuries BC, it is unlikely that this is the full picture of the Bronze Age mine's development.

Finally, the stratigraphic model proposed earlier in this chapter assumes that, after beginning on the Tourist Cliff, Bronze Age mining at this source proceeded sequentially over time to other parts of the mineralised zone, with new areas of the source being reached and 'opened up' in turn. In fact, as the structure of the mineral body is so convoluted, it is highly unlikely that mining progressed at the same rate at every location. For reasons that will become clear in the following chapter, when we look in more detail at how mining was organised, we also cannot simply assume that the Bronze Age miners followed what now appear to be the most 'obvious' or 'direct' routes from one location to another. Comparing the chronology of the modelled dates, from earliest to latest, against the stratigraphic phasing (Figure 2.34), also shows that the sequence of the mine's development was far more complex than indicated by this simple stratigraphic sequence. While the miners gradually accessed new areas of mineralisation, they also continued to extract ore from locations that had already been worked. The rich, easily accessible ores on the Tourist Cliff were extracted throughout the Bronze Age life of the mine, with the later miners returning sometimes to rework ore veins that had been a focus for earlier activity. The ores of the Lost Cavern likewise were repeatedly reworked. This suggests that the copper mineralisation that the miners were seeking was not completely mined-out or 'exhausted' in these areas before extraction began elsewhere. This may not have been the only or even the main driver for the development of new workings in other parts of the source.

The sequence of mining over time: conclusions

The small size of the radiocarbon dataset associated with the Great Orme mine (a total of 29 samples, including five from the nearby sites of Pentrwyn and Ffynnon Rhufeinig) contrasts with the 23 radiocarbon measurements obtained from samples of charcoal at the far smaller prehistoric copper mine of Copa Hill, Cwmystwyth, in mid-Wales (Timberlake, 2017, table 1), or the 75 radiocarbon age estimates available for the second-millennium BC mines on the main lode of the Mitterberg mining region in the Eastern Alps, Austria (Pernicka, Lutz and Stöllner, 2016, pp. 23–5). The number of 'quality' samples is even smaller, with only eight samples being ranked highly (with an overall score of 3) in the sample hierarchy used in this study (Appendix 5). As Hamilton and Krus (2018, p. 8) point out, however, the small size of a data set is not necessarily a problem for Bayesian modelling, if the quality of the data is high.

We can therefore be confident that Model 2 provides the most accurate estimate of the currency of Bronze Age mining at the Great Orme mine, at least for working of the mineralisation associated with the Tourist Cliff and Lost Cavern areas. It provides an accurate estimate of age for those parts of the mine with a secure archaeological association between the radiocarbon samples and their find context, which, as we have seen, does not include its lower reaches. The relative number of samples compared to the size of the mine and their uneven distribution means that there are still insufficient data to address many key questions about the mine's chronology, such as when extraction associated with the other mineral exposures in the Pyllau Valley began and ended; the sequence of working associated with these other areas of the mine; any changes in the intensity of mining from one part of the source to another and over time; and how the smelting identified at Pentrwyn fits into the chronology of the Bronze Age mine. The data are currently too limited to model robustly the chronology of mining in relation to both vertical depth and horizontal distance from the present ground surface. More data are also needed to demonstrate whether, as has recently been proposed (Williams and Le Carlier de Veslud, 2019, p. 1188), all earlier mining was focused solely on the areas of rich mineralisation, with later phases involving extraction from the narrower, low-grade veins. The available dated mining evidence suggests that this model, like the hypothesis of unidirectional progress from shallower to deeper working, oversimplifies the complexity of the mine's development.

Model 2 uses a legacy dataset: 'old' radiocarbon dates previously obtained by other researchers (Hamilton and Krus, 2018, p. 7). Many of these radiocarbon dates have large standard deviations and are from samples which were not originally specifically selected with Bayesian chronological analysis in mind. Nonetheless, this study shows how, by using only 'quality' dates (with a demonstrated or inferred direct functional relationship between each sample and the working in which it was found), it is possible to be reasonably confident that the output of the modelling refers to the patterning of extraction over the site through time. Within the limitations due to a small dataset, it is unlikely that the modelled results are the outcome of practices – such as spoil reprocessing or redeposition – which may have led to residual material becoming incorporated in workings of later date. Excluding bulk samples, such as BETA-127076 from the Pentrwyn smelting site, which was made up of charcoal from two different contexts, improved the reliability of the chronological model by ensuring that it only included samples without any likelihood of an offset between the radiocarbon date and the date of the formation of the deposit from which they were recovered. This study also shows that creating a sample hierarchy is a very useful tool for establishing the 'best' and 'worst' samples for Bayesian analysis. Assessing the value of mid-ranking samples, however, requires more detailed and archaeologically informed relevant prior knowledge, such that some samples may be considered for analysis while others with the same score will be rejected as having low or no use for addressing more complex and nuanced chronological questions.

To conclude, Lewis's (1996) model for the mine's development, outlined at the beginning of this chapter, still seems to offer the best overall fit with the chronological and other archaeological information from the mine. The activity sequence he proposed can be refined somewhat, such that the customary mining phase most probably started in the early 18th century BC and continued until around the mid-10th century BC (68% probability). It is also suggested here that trench working did not play a major role in the mine's development, while the 'Great Opencast' is most likely to be a cavern, mined underground in prehistory, which subsequently collapsed or was enlarged by mining in the early/late modern periods. At a micro scale, broadly contemporaneous extraction in many different areas and later reworking to remove ore in locations that had been a focus of earlier activity (as proposed by Lewis, 1996) seems to have been a feature of the customary phase of Bronze Age mining at this source. There is no definite evidence from the modelled chronology for the systematic and progressive exhaustion of one part of the source after another. This pattern of activity can best be described as 'piecemeal'. Throughout the rest of this book, we will examine some of the possible reasons why this may have been the case, and why this pattern of working is significant for understanding mining on the Great Orme as a social technology.

3

The technology of mining: its sequence and content

In this chapter, we will investigate the physical form and layout of the prehistoric workings and the form and distribution of the associated artefacts to identify each step in the mining process at the Great Orme mine, from initial preparations to the dispersal of the end products. As mentioned in the opening chapter, the focus of this book is copper ore mining and processing: the first stage in the making of objects of copper or bronze. Each activity in this initial stage can be broken down into further process steps (e.g., O'Brien, 1994, p. 179, fig. 82). Beginning with mining, in prehistory this potentially involved a range of tasks, from collecting raw materials – like fuel – and preparing or repairing tools and other equipment, such as mining hammers, wooden ladders or launders, to implementing various methods to extract the ore-bearing rock (O'Brien, 2015, pp. 199–220). The second step, ore processing, sometimes called 'beneficiation', is the process by which the desired copper ore minerals were separated from the barren host rock, or gangue, to produce a concentrate suitable for smelting (Wager and Ottaway, 2019, p. 22). This could involve sequences of coarse and fine crushing of the mine extract, combined with dry separation techniques and those using water. On the Great Orme, there is also evidence for smelting, the second stage of metal production. Like mining, smelting required compiling the necessary raw materials, particularly fuel and the copper ore to be smelted; preparing tools and other equipment, such as crucibles and the furnace itself; followed by carrying out the actual smelt to make raw copper metal (O'Brien, 2015, pp. 221–33).

With this basic technological sequence in mind, in this chapter we will pose the following questions: what specific tasks did Bronze Age mining and processing at the Great Orme mine involve? Were there variations in the type and distribution of activity across the site? How did mining and processing articulate with smelting, practically, spatially and temporally? How did the content and structure of the entire technological sequence of metal production on the Great Orme change over time? And in what form did the 'end product' of this sequence leave the headland? We will begin by examining in depth three key components of the methodology used here to investigate these issues – the '*chaîne opératoire*', the notion of technological choice and Ingold's concept of the 'taskscape' – alongside some of the empirical challenges with applying these methods at this specific site.

Introducing the *chaîne opératoire*, technological choice and the taskscape

The *chaîne opératoire* is a task-based, empirical research method (e.g., Leroi-Gourhan, 1971 [1943]; Lemonnier, 1992). It describes the step-by-step physical movements, gestures and contacts with materials involved in all our practical engagements with material culture, from its initial procurement and preparation to final disposal. It includes processes of modification, alteration, shaping, use, repair and recycling. The *chaîne opératoire* potentially gives detailed and quantifiable information on the sequence of technical actions involved in any interaction between people and their physical world, in both the present and recent or distant past, and at any scale – from knapping an individual piece of flint to a complex practical sequence, enacted over time and space, such as building a jumbo jet (Lemonnier, 1992, pp. 25–6; Dobres, 2000, pp. 167–8). In this study, to address the questions about the Great Orme mine posed above, each step is, where possible, broken down into its sequential material and gestural components. The aim is to identify explicitly the structure or logic involved at each stage of the mining sequence, in terms of its specific constituents (e.g., tools, materials, bodily movements) and outcomes (e.g., items of equipment, mine extract) (Lemonnier, 1992, pp. 5–6). This includes defining the patterning of each task across space and through time.

To reconstruct any *chaîne opératoire* in such detail requires fine-grained, relevant empirical data on all aspects of each stage in the technical sequence (Dobres, 2000, p. 168). At the Great Orme mine, the evidence for Bronze Age working is characterised by poor chronological resolution, as we saw in the preceding chapter. The number, type and distribution of the finds could also have been affected by preservation and recovery biases. While the neutral to slightly alkaline ground water conditions in the workings have resulted in the excellent survival of animal bone, they are unfortunately not conducive to the preservation of timber, nor other organic items such as leather or hide. Very little prehistoric wood has survived. Timber impressions – tentatively interpreted as the remains of unburnt fuel – have been identified in, for example, the base of a thick deposit of calcite flowstone overlying prehistoric mining spoil at Loc.7a (Lewis, 1996, pp. 134–5). Radiocarbon dating of charcoal embedded in the base of the speleothem gives a modelled probability estimate of *1495–1320 cal BC* (Model 2: OxA-5397; 3150±50 BP; 68% probability). Other than bone, wood charcoal is the organic material most commonly found throughout the mine (Lewis, 1990d, pp. 6–7).

Natural processes, such as roof collapse and occasional flooding, as well as later mining are also very likely to have affected the evidence for prehistoric action underground and particularly at the surface, close to the mine. In 1993, for example, 60% of the prehistoric workings accessed from Loc.21 were blocked by quantities of spoil deposited by flooding in a severe storm (Wood and Campbell, 1995; Lewis, 1996, App. C, 11–12; James, 2011, fig. 9.2). This does not mean, however, that no useful information about Bronze Age practices can now be retrieved. A recent study concluded that the impact of taphonomic factors – weathering, trampling, acid erosion and gnawing by carnivores – on the composition of the mine's bone assemblage was not significant. The species and anatomical parts recovered can therefore be considered representative of the faunal remains deposited in the mine workings by the activities of the Bronze Age miners (James, 2011, p. 365).

Most of the prehistoric underground workings at a distance from the surface, although not systematically excavated nor recorded in detail in situ, have been explored and/or cleared of finds. Underground locations where more detailed recording of archaeological features and finds has been completed include Locs. 5, 7a, 21 and parts of Loc.18, as well as the newly discovered Loc.36 (e.g., James, 1990; Lewis, 1996, App. C; Jowett, 2017). Many of the bone and stone finds were, however, recovered during bulk surface machine clearance of redeposited mining spoil and so their provenance is uncertain (Lewis, 1996, p. 121). Consequently, more than 75% of the nearly 400 cobblestone mining tools discovered before 1995 lack a detailed provenance (Gale, 1995, p. 136). Further in situ recording of finds with the details of their recovery context is still needed (James, 2011, p. 435).

Taken together, all these issues mean that it is not possible to produce a granular *chaîne opératoire* for every specific stage in the sequence of mining, ore processing and metal production. The *chaîne opératoire* methodology, however, provides useful information whether employed at the micro or macro scale. Even generalising sketches of the components and outcomes of working enable conclusions to be reached about the content of the mining sequence and – to a limited extent – how this differed through time and in different parts of the source. We will see that some features of Bronze Age mining at this source cannot be explained simply in terms of the physical constraints imposed by material properties, mechanics or thermodynamics. This is because a practitioner also applies a complex series of decisions and mental filters to the practice of any technological task (Edmonds, 1990, pp. 57–8, 1994, p. 474; Lemonnier, 1992, p. 6, chap. 3). This can be described as the concept of 'technological choice'. As in the unfolding of other technologies, it played a significant role in determining the process of prehistoric mining (O'Brien, 2015, p. 195). The *chaîne opératoire* of mining at the Great Orme mine involved the making of 'technological choices', which gave rise to a particular sequence of activity or the intentional selection of specific resources. The issues involved in the making of such choices can be wide-ranging and complex, or subtle and routine, but are often caught up in ideas about the self, skill and social relations (as the articles in an edited volume by Lemonnier indicate) (Lemonnier, 1993). A practitioner's choice of tools or technique can be as much about expressing their own identity, for example, as about successfully accomplishing a task (Ingold, 1993a, p. 124). Conversely, the choice of which tools to use for which activity, and how they are then deployed, are also part of the way in which an individual's identity is moulded, often unconsciously, through the practice of their everyday activities (Edmonds, 1994, p. 474). The *chaîne opératoire* approach is ideally suited to the study of technological choice, as it enables us to consider explicitly the ideas, choices and decisions bound up in each stage of the mining sequence. As we will see throughout this study, the Great Orme miners would have made choices about the 'best way' to 'do' mining based as much on their socially constituted understandings of how things should be done as on the actual physical properties of matter. Viewing the evidence from this perspective can help to make sense of why, for example, certain materials were used to make mining tools while others, though physically suitable and available, were not chosen.

Finally, this study borrows Ingold's (1993b) terminology of the 'taskscape' to describe the content of Bronze Age mining at the Great Orme mine. Ingold (1993b, p. 158) defines a 'task' as:

> *"any practical operation, carried out by a skilled agent in an environment, as part of his or her normal business of life".*

In pre-industrial or traditional societies, tasks include activities such as tilling a field, milking an animal or carding wool. A set of related tasks form a 'field of activity', like farming, cooking or weaving. Together these fields constitute the 'taskscape', which Ingold (1993b, p. 158) describes as: "the total ensemble of tasks, in their mutual interlocking" that make up the pattern of activity in a given community. In relation to prehistoric mining, as discussed at the beginning of the chapter, tasks included toolmaking and repair, and ore extraction and processing. The *chaîne opératoire* methodology can be used to identify and describe how each of these specific tasks unfolded in practice. All these tasks taking place at the ore source together comprise the 'mining field of activity' (or the initial stage of the copper metal production sequence). This in turn is part of the taskscape of all the other fields of activity – such as animal husbandry and dwelling – that make up the routine of life for the people mining at the source.

Ingold (1995) discusses how, in traditional societies, people typically draw no distinction between 'work' and leisure (non-work), but rather between different fields of activity. People in such communities do not 'work' (in the sense this term is used in the industrial world) but are instead perpetually engaged in socially embedded activities – which are generally understood by the participants to be as much about the negotiation of social relations as they are about the making and doing of things (Ingold, 1995, p. 6). Like recognising that the strict dichotomy between 'ritual' and 'production' activities or between 'secular' and 'sacred' areas of practice does not in fact exist, Ingold's observation has profound implications for how we conceptualise the content of life in non- and pre-industrial societies, including those in the prehistoric past. It supports the paradigm of technology as a "total social fact" (Mauss, 1979 [1935]), as discussed in Chapter One. In summary, the 'taskscape' construct provides a useful vocabulary to describe what is happening at the Great Orme mine during the Bronze Age. As we will go on to see in detail in Chapters Four to Six, when applied in combination with a *chaîne opératoire* research methodology, it also provides a theoretical basis for investigating the ways in which prehistoric mining at this source was a 'total social fact'. Let us now turn our attention to unpicking the details of the task sequence making up the mining field of activity at the Great Orme mine, beginning with the making and repair of bone tools.

Making and repairing bone tools

Technological choices

The onset of a mining event would have involved considerable planning and practical preparation. Much labour must have revolved around the making and repair of tools and other items of mining equipment, as well as the acquisition of the needed raw materials. We know the Bronze Age miners used implements made of bone. More than 30,000 bone fragments have so far been recovered from the workings, a few examples of which, as discussed in Chapter Two, have been radiocarbon dated to the second millennium BC (James, 2011, table 8.1). As we shall see in a later section, while these remains are food waste, a significant number of the bone finds from the mine are also definite and possible bone tools (Figures 3.1, 3.2).

From this assemblage of bone objects, it is possible to sketch the *chaînes opératoires* of bone mining toolmaking and repair at this site during the customary Bronze Age phase of activity there, beginning with the technological choices involved in making a tool from this material. As we shall see, there is limited direct evidence for the activity areas where bone tools were made. Such details must be inferred from the objects themselves. In the assemblage studied by Sian James, which comprised 56% (16,793) of the total bone finds from the mine, 5% were identified as recognisable tools, with a further 4% classed as possible tools (James, 2011, p. 169). Tools can be identified from use-wear traces, including scratches, fractures and rounded or shaped ends; a distinctive polished appearance caused by handling; and evidence of intentional breakage, e.g., vertical fracturing along the length of a limb-bone shaft, leaving one articular end or joint. The main tool types present are scrapers, chisels, wedge-like implements and awls (Dutton, *et al.*, 1994, pp. 275–6; James, 2011, p. 168).

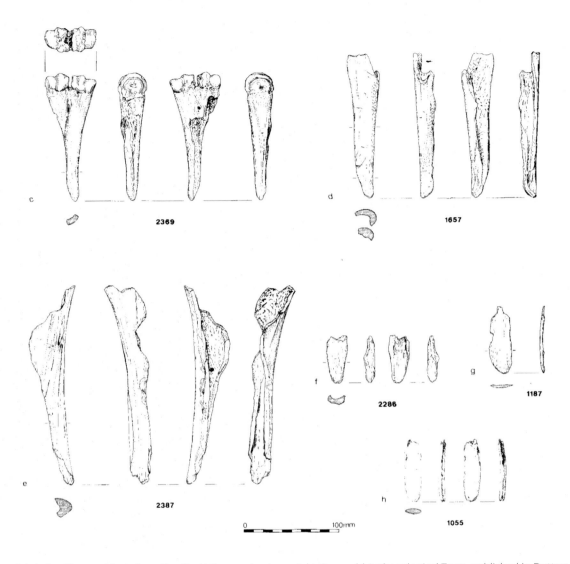

Figure 3.1: Animal bone objects from the Great Orme mine (copyright Gwynedd Archaeological Trust, published in Dutton, et al., 1994, fig.12).

The technical sequence of bone tool manufacture involved a selection step, in which the appropriate raw material – in terms of both the animal and anatomical element – was chosen. The faunal assemblage from the Great Orme mine is dominated by the remains of cow and cow-sized animals (53%) with, in order of decreasing frequency, sheep/goat (and similar-sized animals), pig, red and roe deer, plus very minor amounts of horse (possibly not Bronze Age), hare and dog (James, 2011, fig. 5.1). As at the Bronze Age copper and flint mines of Ecton and Grimes Graves (Timberlake, et al., 2014, p. 198; Healy, et al., 2018, p. 288), however, certain species and anatomical elements were deliberately selected for use as tools (Figures 1.1, 3.3).

There was a preference for ribs and limb bones (primarily tibiae and radii) from cattle or another similar large ungulate. Scapulae were less common, as were implements from pig and sheep/goat (James, 2011, p. 171). James's (2011, pp. 348–50) study of the tool assemblage revealed some interesting details about the choices being made by the Bronze Age miners. There was preferential element selection at species level, with tibiae from all animals (cattle, sheep/goat, pigs) represented, while radii were primarily from cattle and sheep/goat, ulnae from cattle or pigs, and scapulae from sheep/goat or pigs. The preference for cattle long bones over ribs was found to be statistically significant, but the representation of ribs and humeri was almost exactly as predicted if the sample of these elements was random.

There are several possible utilitarian reasons why the Great Orme's Bronze Age miners chose bone as a material

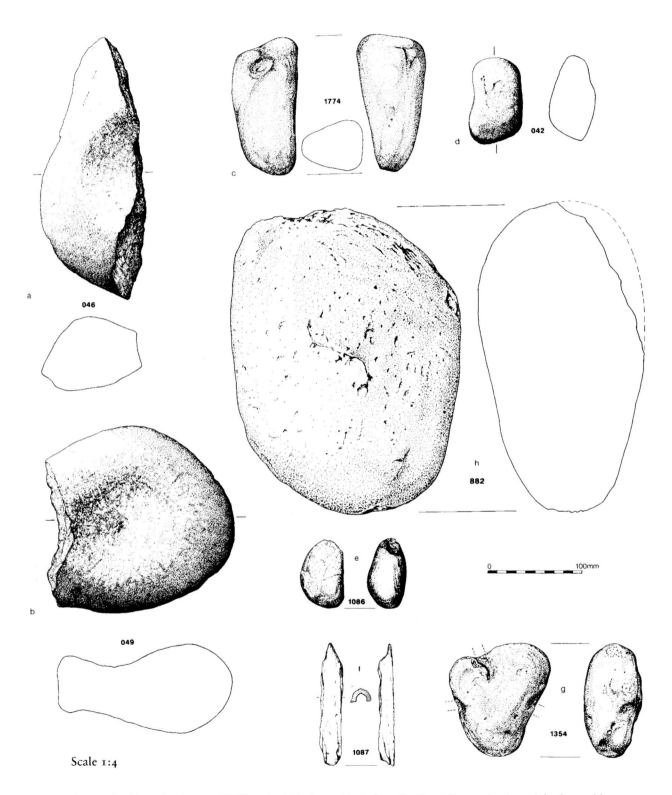

Figure 3.2: An animal bone implement (1087) and cobblestone objects from the Great Orme mine (copyright Gwynedd Archaeological Trust, published in Dutton, *et al.*, 1994, fig.14).

THE TECHNOLOGY OF MINING: ITS SEQUENCE AND CONTENT

for making mining tools. Tough and resilient, replication experiments show that it would have been highly effective at removing the clay-like rehydrated mudstones and friable rotted dolomitic limestones and sandstones hosting the copper mineralisation (e.g., James, 2011, pp. 180–3). It would likewise have been suited to extract ore in the similar geology at Ecton Mine, as well as flint from the soft chalk at Grimes Graves (Timberlake, *et al.*, 2014; Healy, *et al.*, 2018), perhaps explaining its selection as a mining tool at these locations. Bone was also readily available and easily worked. Finds of bone pins (Lynch, 2000, p. 112) from cremation burials in north Wales, for example, indicate that Bronze Age communities were familiar with the physical properties of bone and would have known how to work it effectively to achieve a desired result.

The potential variability of animal bone would also have allowed for the selection of a particular species and skeletal element that were most suited – in terms of size, shape, weight, strength or robustness, ease of handling and degree of modification required – to a specific task (e.g., Dutton, 1990, p. 13; James, 2011, p. 171). The degree of care and deliberation that appears to have been involved in the fabrication of bone tools reinforces the idea that they were made with a particular purpose in mind. The lower part of the fore and hind limbs could have been preferentially selected for tool use because they can be comfortably gripped at the articular end, by their joint. Other elements, such as skull fragments, mandibles and vertebrae may have been either too small, insufficiently robust or have the wrong shape to be useful tools, with or without modification. Functionality could explain why there are so few implements made from these body parts (Hunt, 1993). The lower limbs of sheep/goat may have been the only parts of these animals robust enough for use as mining tools, hence the statistically significant preference for sheep/goat tibiae and radii. The purposeful selection of sheep/goat and pig scapulae, instead of larger implements from cattle, perhaps

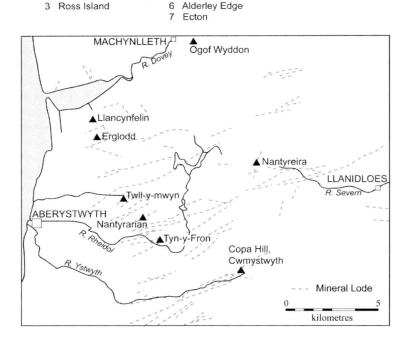

Figure 3.3: Map showing (top) finds of cobblestone mining tools and Bronze Age mines in Britain and Ireland and (bottom) prehistoric mines and mineral veins of central Wales. Derrycahoon mine is one of the cluster of sites shown around Mt Gabriel mine in south-west Ireland (drawing by B. Craddock, Early Mines Research Group).

relates to the intended context of use. Were the smaller scapulae from these animals more effective as scoops in the cramped workings underground (James, 2011, p. 349)? Continued systematic replication experiments with tools from pigs and sheep/goat, not only cattle, combined with quantitative microscopic use-wear analysis, should eventually demonstrate whether sheep/goat scapulae were particularly suited to use as scoops in the mine. It will also add to our understanding of the physical suitability of particular species or skeletal elements for other specific tools or functions at this site.

The characteristics of the bone tool assemblage, however, clearly show that implement selection was not determined solely by material or functional constraints. Rather, the participants made active technological choices about which species and body part to use, which are difficult to understand from a strictly utilitarian standpoint. Replication experiments, for example, have already shown that ribs are as effective as tibiae for use as picks or scrapers in the softer rock areas within the mine. The miners did not, however, preferentially and intentionally select ribs for use as tools but used them only if they happened to be to hand (James, 2011, pp. 348–9). The reasons for this bias are unclear. Recent replication experiments found that processed or 'old' bone (which had been buried for six months) is easier to use as a tool than a freshly butchered one (James, 2011, pp. 183–4). Could the use of ribs as mining tools relate to a different, later instance of working, perhaps by a group who did not come prepared with limb-bone tools but instead used whatever materials they found already present at the mine? The small number of scapula and antler tools is similarly surprising, given the physical suitability of these elements for use as mining tools. Cattle shoulder-blade scoops have been found at Ross Island copper mine, Ireland, and are well documented in European Neolithic flint mines (James, 2011, p. 185; O'Brien, 2015, p. 127) (Figure 3.3). Only 0.4% of the possible bone tools from the Great Orme mine were of antler (James, 2011, p. 383). In contrast, there is extensive evidence for the use of antler for ore extraction at other prehistoric copper mines in Britain and Europe, including Rudna Glava (Serbia), Ai Bunar (Bulgaria), El Aramo (Spain), Copa Hill (mid-Wales) and at Ecton Mine (England) (O'Brien, 2015). Replication experiments on the Great Orme and elsewhere have demonstrated that it is a highly serviceable and versatile mining tool (e.g., Lewis, 1990b). Despite this – and the assumed availability of red and roe deer in the wider landscape – the Bronze Age workforce on the Great Orme did not choose antler as a material for making mining tools.

Consideration of the miners' preference for cattle bones for tool manufacture, rather than those of other species, highlights some of the non-utilitarian factors that may have influenced their choice of raw materials. This choice could perhaps have been grounded in understandings of the living animal. Bones are, after all, the surviving elements of once-living animals. The perceived character of these animals may have defined the types of objects that could be made from their bones. These understandings could also have imbued the items into which they were made with specific qualities and associations. Edmonds (1999b, p. 488) suggests that the task of herding could have generated connections with and claims to particular areas of land for Bronze Age communities. The age profile and restricted size range of the cattle assemblage from the mine is indicative of a single-sex group of animals reared for meat, breeding and/or traction, rather than dairy (James, 2011, pp. 345, 403). These animals, which would have been valuable as both a nutritional and material resource, may also have represented wealth. More generally, the constitution of herds could have offered a rich source of metaphor for the identities, social relations and lifecycles of a community and its members (Edmonds, 1999a, pp. 27–8). They may have been potent symbols of the order of things. As such, their remains could have been considered particularly appropriate for use in the mine, a place that perhaps seemed 'other' and at a great actual and conceptual distance from the land of the living. Cattle bones may have been chosen because they provided both the physical and the conceptual strength needed to dig ore from the ground.

Butchery

Selection of the appropriate raw material was followed by butchery to disarticulate and deflesh the preferred body part. The bone tool assemblage provides compelling evidence that the *chaîne opératoire* of tool manufacture was well organised and characterised by considerable planning and forethought. The task of butchery seems to have been undertaken with the eventual use of the animals' body parts as mining tools in mind. The scarcity of metapodials, femora, phalanges, teeth and articulated animal skeletons at the mine indicates that the primary slaughter and butchery of animals selected for future use as tools were carried out elsewhere (James, 2011, pp. 177, 191). In this scenario, cuts of meat comprising the elements preferred as tools, such as the tibiae and radii, would have been selected for transportation to the source, either as joints of manageable size or – possibly – as defleshed bones (Dutton, *et al.*, 1994, p. 276; James, 2011, p. 349). James (2011, p. 387) points out that metapodials are one of the most common body parts used as tools at other prehistoric sites, due to their durability and robustness. The miners' decision not to bring whole carcasses to the mine could have been partly due to wanting to retain the metapodials for crafting into, for example, needles for use in other fields of activity like leatherwork, taking place elsewhere in the taskscape.

Bone tools could have been brought to the source ready prepared, with selection, butchery and modification happening elsewhere. The total bone assemblage is, however, highly fragmented, with a significant proportion of the finds broken to less than 3/4 of their original size (James, 2011, p. 193). James (2011, p. 369) argues that this is not a result of post-depositional processes, as the faunal assemblage displays a high degree of fragmentation across the entire site, including in confined areas little affected by trampling. Traces of gnawing by carnivores were identified in only 0.1% of the assemblage (James, 2011, p. 404). As we shall see in a later section, the degree of breakage is also unlikely to have been caused during mining. Replication experiments show that only small fragments of bone, similar in appearance to those from the archaeological assemblage, fracture off the main shaft when it is hit by a stone axe (James, 2011, pp. 174, 369). At least some of the fragmentary bone recovered at the site could therefore be indicative of in-situ tool manufacture, perhaps after further butchery.

A bone tool – even one made from an animal and body part preferred for tool manufacture – was at different stages in its 'lifecycle' an implement and food. It is probable that the faunal evidence does not just refer to the steps in the *chaîne opératoire* of tool production and use, but to a range of practices around the preparation and consumption of food, an issue we will return to. All the bones (apart from any shed antler) must have started off as food 'waste' even if they were not the most meat-bearing elements of an animal – and even if, as James (2011, pp. 350–8) argues, the total bone assemblage primarily reflects either actual mining tools or those intended for future use as tools, all brought to the source because they were material constituents of the mining field of activity. We do not know whether the butchered joints were consumed cooked or raw before the residual bones were modified into tools and then used for mining; the bone assemblage displays little evidence of burning and cooking practices in any case may not have caused the bone remains to char or burn (James, 2011, p. 347). The disarticulation and defleshing of the skeletal parts (steps which perhaps even involved the partial weathering or boiling of the bones) may have been enmeshed into a sequence of activities concerned with the processing of the carcass for other products valued by Bronze Age communities, such as the marrow, fat, hide, meat and sinews. Marrow, for example, is a nutritious food and can also be used as grease or fuel. It can be scooped out once the bone has been fractured as part of the process of tool manufacture. It is in fact necessary to remove the marrow to break the bone completely. This could be a further explanation for the high degree of fragmentation of the bone assemblage (James, 2011, p. 174).

Bone working

Many bones were used as tools without further modification, but there is evidence that some bones were worked or 'knapped' to change their shape to make a suitable tool (James, 2011, pp. 359, 381, fig. 9.9). The technical sequence of modification of an animal bone into a mining tool of the required shape and sharpness seems to have involved a series of cutting and splitting steps, perhaps using stone axe-type tools (although more experimental work is needed to determine whether any of the cobblestone tool-types identified at the mine could have been used for this purpose). Some skeletal elements, such as the ulna, tend to have been extensively modified. There is also some evidence to indicate that the Bronze Age miners deliberately exploited differences in the form of each skeletal element to produce tools with a particular shape. The tibia and radius, for example, often appear to have been fractured to produce tools utilising the 'corners' in the cross-section of the original shaft. These parts of the bone are relatively thick and robust. Fracturing also appears to have been designed to produce a point at one end (e.g., Dutton, 1990, p. 12; Dutton, *et al.*, 1994, p. 276). There is no evidence, such as traces of birch-bark glue, to show that the bone tools used at the mine were hafted. Replication experiments show that hand-held bone tools are effective mining implements, indicating that hafting would not have been necessary (Dutton, 1990, p. 12).

Tool manufacture is a dynamic process, in which the form of the finished item emerges out of what Ingold (1997, p. 112) describes as a "sensuous" engagement between the maker and the material being worked. It is not simply the solidification of a pre-imagined mental design of what the completed object should look like. The Bronze Age practitioner would have started with an idea of the eventual purpose of the tool and their own perception of the form that would achieve that task effectively. The latter would not just have been grounded in their own understandings of the physical properties of animal bone (which, as we have seen, may have been related to perceptions of the living animal). It would also have been conditioned by their socially constructed notions of what made a 'good' tool. These notions could have related to a host of factors, including whether they were right- or left-handed, their body shape and size, the type of grip they preferred and the degree of skill and dexterity they possessed. The evidence that the initial stages of bone tool selection took place away from the mine reinforces the idea that there was considerable forward planning in bone mining tool manufacture. It also hints that the toolmaker was intimately involved in some way in the *chaînes opératoires* of ore extraction and selection and possibly also the slaughter of the living animals to be made into tools. As working progressed, the mental design they were following could have shifted, perhaps as the bone failed to split the way it was intended. They may have examined splinters produced during the process, discarding some and putting others to one side, perhaps as potential mining tools, or for further modification for another use. Some of the bone fragments from the mine, possible by-products of the tool manufacturing process, also have use-wear traces suggesting that they too were used as tools (James, 2011, p. 381). A similar dynamic would also have been played out when broken or blunt implements where modified or repaired during mining.

Even if intended from the outset for use in the mine, each tool was the outcome of a complex task sequence. The fabrication of a bone tool – or any other items of mining equipment – would have been a fluid process that gave rise to a slightly different end-product each time, even if the form aimed for was the same. The mining field of activity was, through the *chaîne opératoire* of tool manufacture, entangled in other intersecting fields of activity within the wider taskscape, all centred around animals and including animal husbandry, raw material procurement and food preparation. Some of these fields of action may have been only indirectly and on occasion caught up in the *chaîne opératoire* of a given mining event. Some – such as animal herding and slaughter – seem to have usually unfolded away from the mine. It is not possible to determine conclusively from the evidence whether all the steps in other tasks, such as butchery, always took place elsewhere. While the representation of the skeletal elements from the mine points to primary animal slaughter and butchery taking place off-site, we may be missing information about what was discarded at the surface. There is no Bronze Age settlement associated with the mine, nor other sites on the Great Orme and in its hinterland where we might find direct evidence for contemporary local practices such as animal rearing and butchery. It is, however, likely that some of these activities were undertaken at the source, as joints of meat or defleshed bones were further reduced by food preparation, processing to extract useful substances like marrow, and working to make serviceable tools. The distribution of the faunal evidence does not speak to where exactly across the site these tasks took place. They are perhaps more likely to have been carried out outside the workings rather than underground, to take advantage of the better light conditions at the surface, but specific, distinct tool working or other activity areas cannot be identified.

Making and repairing cobblestone tools

Technological choices

Stone mining tools are a common feature of prehistoric copper mines across Europe, including at all the known Bronze Age mines in Britain and Ireland (O'Brien, 2015) (Figure 3.3). There are more than 2,500 cobblestone finds at the Great Orme mine (Jowett, 2017, p. 63). Following the terminology suggested by Timberlake and Craddock (2013, p. 39), the generic term 'cobblestone mining tools' is used here to refer to all these finds, as this recognises the uncertainties over their function and typology. Approximately 15% (391) of the total number of stone finds so far recovered from the Great Orme mine have been the subject of detailed published study (Gale, 1995, p. 135). The majority (90–95%) are pebble-, cobble- and boulder-sized stones (Figures 3.2, 3.4). They vary greatly in size and shape, from examples 5cm long weighing 0.25kg to those 40cm long and 29kg in weight. Most are interpreted as hammers (mauls) or pounders for extracting ore-bearing rock (Dutton, *et al.*, 1994, p. 269, figs. 14, 15; Lewis, 1996, p. 118). The remainder of the assemblage are objects interpreted as ore "crushing surfaces" (Dutton, *et al.*, 1994, p. 269), such as mortars, anvils or dressing stones, although their precise typology and function are open to definition (see the various classification systems for the Great Orme stone tools proposed by, for example, Gale, 1995; Lewis, 1988 and Dutton, *et al.*, 1994). As with bone tools, there is limited direct evidence for the activity areas where cobblestone mining tools were collected, modified and hafted.

The rock types used as cobblestone mining tools at the Great Orme mine have been identified from the surface examination of each tool. They are predominately of igneous origin and include basalt, dolerite, gabbro, rhyolite, granite and microdiorite. With the exception of the latter, which originates locally from the intrusion forming the Penmaenmawr headland approximately 10km away on the other side of Conwy Bay (Lewis, 1996, p. 119) (Figure 1.6), the rock types represented have a wide distribution, around the Creuddyn Peninsula and along the north Wales coastline, as well as in glacial till deposits further inland (e.g., Gale, 1995, pp. 148–9). There are today several nearby pebble beaches – immediately to the east of the Great Orme, at Llandudno Bay, and around the southern fringes of the headland – comprising cobbles of the rock types found at the mine. The Bronze Age miners may have picked up cobblestones from these beaches, as well as from riverine and till deposits on or around the Creuddyn Peninsula, on the way to the source (Gale, 1995, pp. 151–2) and as needed once they had started mining.

19th-century AD accounts (cited in Gale, 1995, pp. 151–2) also highlight the existence of alternative locations, now vanished, where the various rock types discovered at the mine could once be found. The Bronze Age miners may therefore have known about and used other cobble sources to those that today appear most obvious and accessible. The apparently deliberate addition of sparse and uncommon igneous rock types to the clay mixtures from which Earlier Bronze Age communities on

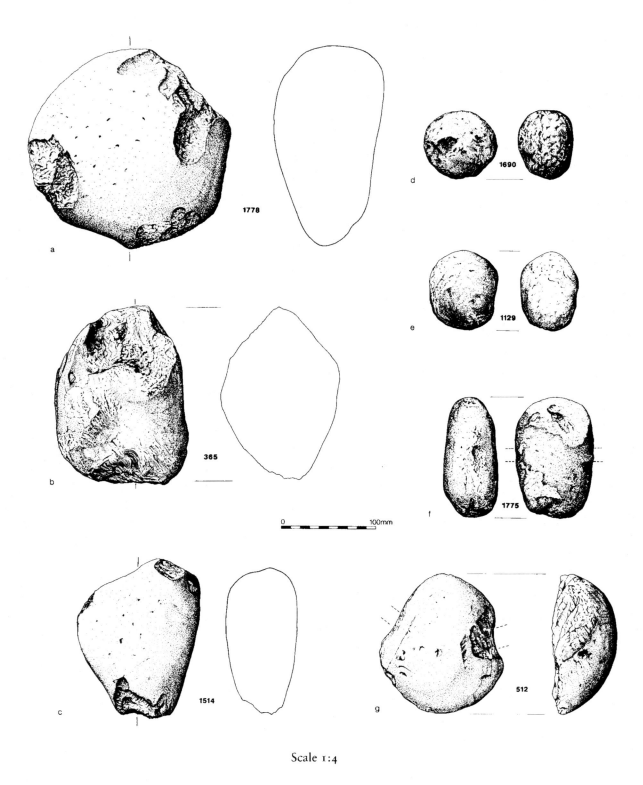

Figure 3.4: Cobblestone objects from the Great Orme mine (copyright Gwynedd Archaeological Trust, published in Dutton, *et al.*, 1994, fig.15).

the island of Anglesey made funerary pots, for example, points to the in-depth knowledge such groups could hold about local stone sources (Williams and Jenkins, 1999, p. 219). Traditions of stone procurement, not just physical availability, are also likely to have influenced the decision about which cobble source to use. These traditions could have been enmeshed into the *chaînes opératoires* of other tasks, such as the gathering of edible shellfish and other coastal foodstuffs. It is possible that certain beaches, rivers or till deposits, even those at a distance from the Great Orme, may have been considered more 'appropriate' and hence preferred as a source of suitable cobblestones than those located closer to the mine. The stone finds from the mine could therefore have originated from numerous sites and not necessarily those nearest to the headland.

The cobblestone mining tool assemblage shares certain common characteristics and conforms to a narrower lithological and morphological range than would be expected if it represented a random selection of all the rocks available at any beach or deposit (Gale, 1995, pp. 250–1). The Bronze Age miners preferentially selected hard igneous rocks, such as basalt, dolerite and microdiorite, for use as mining hammers. The important selection criteria appear to have included physical properties like shape, hardness, surface form, lack of mechanical flaws and weight, as well as ease of handling (Gale, 1995, p. 254; Timberlake and Craddock, 2013). Particular sizes of cobblestone, for example, could have been selected to perform specific functions in particular areas of the mine. Smaller stones could have been picked as percussive implements to use with tools of bone, wood or metal (Dutton, *et al.*, 1994, p. 269), or for deployment in a working of restricted size that required a tool to be used in one hand. (Other possible interpretations for cobblestones of different size and weight are discussed in Chapter Five.) At the beach or other pebble source, however, the Bronze Age miners may have selected a range of potentially useful cobbles of different size, weight and hardness for transportation to the mine, only making a final choice about which to use from this cache as extraction progressed.

A small number of the studied assemblage of cobblestone mining tools are softer sedimentary rocks such as sandstone and siltstone (Gale, 1995, table 7.1; Lewis, 1996, p. 119). When assessed solely from a functional perspective, these items cannot be considered particularly effective mining or processing implements, even within a rotted host rock environment. Although they may have had another, as yet unidentified, function, their presence at the mine hints that its cobblestone mining tool technology was not entirely utilitarian. In the same way that the metaphysical properties of cattle bones may have played a role in their selection as mining tools, did culturally situated understandings of the material

qualities of different rocks likewise influence which were picked out and then brought to the site? We will touch on this theme again in Chapter Five.

The number of cobblestone mining tools recovered so far (c.2,500) is small compared to the duration of Bronze Age mining at the source. This could largely be due to the utility of bone tools for extracting minerals from the soft, rotten, dolomitised rock. There is also possible evidence for 'stockpiling' of these tools at the site, in which case the miners could on occasion have made use of the cobblestones already present at the mine. Perhaps only a small number of cobbles, if any, had to be collected and brought to the source for a given mining event? The task of collecting and bringing cobblestones to the mine need not, therefore, have been particularly laborious (cf. Gale, 1995, p. 151). It is worth noting in this context that the Great Orme is not as isolated, nor as difficult to access as is sometimes envisaged (e.g., Budd, Pollard, *et al.*, 1992, p. 682), despite the steepness of its cliffs and the likelihood that it was almost an island with only a narrow strip of marshland separating it from the mainland (Smith, *et al.*, 2012, p. 3). There are numerous less arduous – albeit slightly less direct – routes to the summit than the steep modern road. Direct access by sea was also possible, into the sheltered natural quay close to Pigeons' Cave on the north-western headland (Figure 1.5).

Cobblestone tool manufacture and repair

Once selected and brought to the mine, most of the cobblestones appear to have been used as mining hammers or pounders without any knapping or working to modify their shape for use and/or hafting. Only c.5% of the total assemblage has light pecking on either side of the medial axis, perhaps to secure a haft (Lewis, 1996, p. 118). In the case of similar side-pecked tools from other Bronze Age mines, such as those at Derrycahoon and Mount Gabriel in south-west Ireland, while it is usually very clear where the abrasions are related to haft modification (e.g., O'Brien, 1994, p. 127, 2022, p. 68), it can sometimes be difficult to distinguish minor pecking for hafting from side wear in a used hand-held tool (O'Brien, 1994, p. 127). Similar slight pecking to that observed on the Great Orme mine examples has been observed on cobblestones used as pounders in ore crushing experiments (Doonan, 1994, p. 87), highlighting the challenges that can be involved in accurately reconstructing the *chaîne opératoire* of manufacture and use of these apparently rudimentary artefacts. Only a handful of cobbles have definite grooves that are highly likely to be connected to hafting (Gale, 1995, p. 160; David, 1999, p. 79). Unmodified cobbles were probably used in the hand, though replication experiments have shown that cobbles can be hafted with no, or very little, prior modification, particularly if they have natural notches or abrasions, or are an irregular shape (Dutton, *et al.*, 1994, p. 272; Timberlake and Craddock, 2013, p. 50). Brenda Craddock (1990), for example, has shown that hitting a hafted trapezoidal cobblestone against a rock face causes it to become more firmly wedged into the handle.

We cannot yet identify locales within the Great Orme mine where the task sequence of cobblestone hafting took place. No hafts have been recovered from the workings, although twisted withy handles have been found at the Mount Gabriel and Copa Hill Bronze Age copper mines, where waterlogged conditions favour the preservation of organic material (O'Brien, 1994, p. 129; Timberlake, 2003b, fig. 91). As we shall see later in this chapter, a haft or sling would have been needed to wield the larger cobblestones from the Great Orme against the rock face, if these tools were in fact used for mining rather than ore crushing. Perhaps the absence of hafts at this site is therefore due to taphonomic bias and hafting was, at least on occasion, part of the *chaîne opératoire* of cobblestone toolmaking – although it may not have been a major component of the practical preparations for mining at this source. This contrasts with the Bronze Age copper workings at Ross Island (south-west Ireland) and Alderley Edge (England), where most of the cobblestone mining tools are notched or grooved for hafting (Timberlake and Craddock, 2013, p. 39) (Figure 3.3).

Replication experiments show that hazel withies used with rawhide ties form the most effective hafts. Crafting a secure and effective handle needs considerable skill and experience – a degree of technical know-how that has taken some experimental archaeologists "many, many hours of repairing hafts" to achieve (Craddock, 1990; O'Brien, 1994, p. 129; Timberlake and Craddock, 2013, pp. 50–1). Many of the Great Orme cobblestones display varying amounts of use-wear and damage, but typically have fewer signs of wear and breakage than examples from other early mines (Gale, 1995, p. 172; Lewis, 1996, p. 120). This suggests that they did not break easily when used as mining hammers. Hafts may also have lasted longer before needing repair. This contrasts to the situation in other prehistoric copper mines, where ongoing repair of broken hammer hafts was probably an important part of the mining process (O'Brien, 2015, p. 236). At Copa Hill (mid-Wales), for example, it is estimated that repairs would have been carried out as frequently as every 30 minutes, by someone on hand dedicated to that particular task (Timberlake and Craddock, 2013, p. 51).

Making and repairing tools from other materials

The Bronze Age miners most probably also regularly used tools and equipment made from a range of other organic and inorganic materials. The array of artefacts – picks, wedges, shovels, drainage launders, baskets, buckets, planks, withy ropes and handles, stakes and staves, possibly from ladders, staging and stempling – of wood and leather from primary prehistoric contexts at other Bronze Age mines in Britain and Europe hints at the diverse types of objects that are likely to have figured in the mining field of activity on the Great Orme and which no longer survive (e.g., Timberlake, 2003b; O'Brien, 2013; Thomas, 2015). Although less well known than the exceptionally well-preserved finds of mining equipment from the Early Iron Age salt mines at Hallstatt, in Austria (Kern, *et al.*, 2016), these earlier assemblages of organic mining materials are as varied and informative. They show how wood (including charcoal) would have been needed for mine supports; as fuel for fire-setting and cooking fires; to provide a source of illumination underground; and to make and repair items of mining equipment, such as shovels, buckets and ladders. Animal skins and leather may have been needed for clothes, containers and perhaps for shelter. Plant fibres could have been woven into baskets, ropes and used for hafting stone tools.

Alongside mining tools of bone and stone, the miners may have used sharpened wooden sticks and wedges as mining implements, like those known from Mount Gabriel Bronze Age copper mine, in south-west Ireland (O'Brien, 1994, pp. 147–50, fig. 70). These would have been as suitable as bone for extracting ore from the Great Orme's rotted dolomitic limestone (Lewis, 1996, p. 134). Bone tool marks are typically groove-like, between 1–5cm long and up to 2cm wide, and usually have a concave profile with rounded ends (Lewis, 1996, p. 128, table 6). Replication experiments at the mine have shown that macroscopically identical indentations can be produced by wooden picks (D. Chapman, 2002, personal communication). Some of the 'bone' tool marks found underground could therefore have been made by wooden implements (Lewis, 1996, p. 135). 'Bone' tool marks have been noted on the rock wall of Shaft 223 in the Tourist Cliff (Dutton, *et al.*, 1994, p. 268), for example, although no bone is recorded from this working. If this really is evidence of absence, it points strongly to the use of wooden tools to dig this shaft.

In terms of inorganic materials, finds of bronze metal implements from the mine hint that their use was part of the mining field of activity on the Great Orme – at least on occasion and in some areas of working – although, as we shall see in the next section, interpreting what they were used for is still far from straightforward. During the 19th century AD, at least three objects of copper or bronze were found at the Great Orme mine, in what were reported to be 'old men's workings' (Lewis, 1996, p. 130). They include a copper implement with a curved blade, possibly part of a billhook or knife, and two bronze 'picks' or 'wedges': one c.7.5cm long, the other c.2.5cm long (Lewis, 1990d, p. 7, 1996, p. 40). Excavations since 2016 in dated Bronze Age mining spoil in an underground passage at Loc.36 have uncovered more than 1,000 fragments of bronze (Figure 2.20). Recovered by dry and then wet sieving excavated rock waste (Jowett, 2017, p. 67), these fragments are typically very small, less than 1cm in length. A further c.150 metal fragments, mostly bronze, have been excavated from other prehistoric workings underground, mainly at Loc.21, c.20m below the present ground surface, and in a passage at Loc.17 (e.g., David, 1992; Lewis, 1996, App. C, 12; Jowett, 2017) (Figure 2.24). Enhanced excavation techniques (detailed by Jowett, 2017) are leading to the ongoing recovery of more bronze fragments from other areas of working (N. Jowett 2018, personal communication, June). Chemical and lead isotope analysis of the bronze fragments indicates that they are likely to have been made using ore from the mine (Williams and Le Carlier de Veslud, 2019, p. 1182). We will examine the first steps in the production sequence transforming this ore into metal later in this chapter.

No Bronze Age pottery has been found in the prehistoric workings. While it is known from a few Bronze Age mines in Europe (e.g., Rudna Glava in Serbia, Ai Bunar in Bulgaria, Kelchalm/Bachalm in north-central Austria and Ross Island in south-west Ireland), it does not appear to have been produced on site at these locations (O'Brien, 2015, p. 240). The discovery of sherds of Beaker and Peterborough ware-type vessels on the Great Orme and in its immediate hinterland indicates that use of ceramics was part of the coeval taskscape in this region during the Chalcolithic and Earlier Bronze Age, at least on occasion (e.g., Roberts, 1909; Davies and Stone, 1977; Smith, Chapman, *et al.*, 2015; Smith, Walker, *et al.*, 2015; Rees and Nash, 2017). Preservation and recovery biases could therefore be responsible for the absence of pottery of this date at the mine (Dutton, *et al.*, 1994, p. 279). In the Later Bronze Age, pottery finds in north Wales are relatively rare and communities in north Wales were aceramic by the Iron Age (Lynch, 2000, p. 121; Waddington, 2013, p. 18). We can expect that the use of ceramics at the mine over time would have reflected this apparent transformation in practice in the wider taskscape.

Without any physical finds, the *chaînes opératoires* of the procurement, making and repair of items of wood, plant material, leather, hides or pottery on the Great Orme must be inferred. Even metal tools such as picks, chisels, axes, billhooks or knives would have needed wooden hafts or handles to be wielded effectively. The making and repair of hafts for metal tools may also therefore have been part of the activities at this site. Such tasks – like the similar ones for bone and cobblestone tools – are likely to have exemplified care, planning and a high degree of technical know-how. Evidence from a mining context for how skilled Bronze Age communities could be at working with wood, for example, comes from numerous sites, including the Mount Gabriel and Copa Hill copper mines, as well as the prehistoric copper and salt mines in the

Eastern Alps (O'Brien, 1994; Timberlake, 2003b; Thomas, 2015). These practical preparations for mining were most probably grounded in contemporary fields of activity for local woodland use, animal rearing, food preparation and consumption, as they unfolded across spatial and temporal scales greater than the mine and the actual act of extraction.

Mining with tools

When the necessary tools and equipment had been readied, agreements perhaps had to be reached about when to begin mine-working and who would participate. Other fields of activity in the taskscape may have had to be rescheduled, necessitating negotiation with others. The *chaînes opératoires* of ore extraction and processing could then begin. The Bronze Age miners typically extracted ore from the rotted dolomitic limestones, sandstones and mudstones in the mineralised zone by chiselling, gouging, scraping, prising and digging using bone (and most probably also wooden) tools. As they followed the soft ore-bearing ground into the hillside, passages were created, some of which retain on their rock walls marks and striations from the mining tools used (e.g., Lewis, 1994, p. 33, plate 3). In areas of mudstone or rotted dolomitic limestone, the pointed ends of bone tools have been found to match indentations in the walls of the working in which they were found (Dutton, *et al.*, 1994, p. 275). Scratched and polished bones tools are sometimes even found 'in position', stuck into the sidewall or ceiling of a passage running through softer rock (Dutton, 1988).

As we saw in a previous section, the miners often made and selected bone tools of a specific form, robustness and feel in the hand with a clearly defined function in mind, for use in particular ways. Replication experiments demonstrate, for example, that limb bones, with their tapered ends and joints that can be easily held in the hand, would have been ideal for gouging, chiselling and scraping. Visual examination of use-wear marks indicates that they were in fact used in this way (Dutton, *et al.*, 1994, p. 276; Lewis, 1996, p. 126) Tibia tools are particularly suited to digging and scraping, while it has been suggested that rib tools may have been chosen when a slicing action was needed (James, 2011, pp. 178, 183). Mining experiments, however, indicate that a specific tool can perform multiple functions. Tibia tools, for example, can be used both as digging implements and as hammers. James (2011, p. 184) suggests that this potential for multifunctionality may explain why the tibia is one of the most commonly chosen elements for use as a tool at the mine – and this functional flexibility could also have contributed to the Bronze Age miners' choice of bone as a material for making mining tools and the apparent longevity of this tool type for mining on the Great Orme. It is difficult to break a bone mining tool in this soft rock environment and so a single bone tool could have been used repeatedly (James, 2011, p. 382) and sometimes for more than one purpose.

Bone tools were probably often used on their own but they may also have been used in conjunction with small cobblestones. Held in one hand, these would have been ideal for use as hammers to beat the point of the bone (or wooden) implement further into the rock face (Dutton, et al., 1994, p. 269). The miners may first have wrapped the cobblestones in hide or some other soft material, to prevent damage to the bone implement. This could explain the minimal wear marks on some of the smaller cobblestones found at the mine. Perhaps the miners, on occasion, chose to use cobblestone tools to extract ore without bone or wooden implements when they encountered copper mineralisation within soft, rotted dolomitic limestone, even though these other tool types would have been equally effective mining implements in that rock environment? This practice could explain why some of the cobblestone mining tools – 8% of the 391 stone finds from the mine examined by Gale (1995, p. 150) – appear completely unused, with no macroscopic wear traces at all. Repeated percussion against softer rocks may not have broken or damaged the cobblestone tools. Alternatively, such finds could represent stockpiled tools that were, for whatever reason, never used. Marks produced by battering of the rock face using stone cobbles are, however, typically seen in the harder, unaltered dolomitic rock at its junction with more rotted material. Such indentations occur less frequently than bone tool marks and are shallow, almost circular, hollows up to 6cm in diameter. They often contain the powdered remains of dolomitic rock crushed by the percussion of the stone maul against the rock face (Lewis, 1996, table 6).

Cobblestones interpreted as mining hammers are mostly small and are damaged at either end or along their edges from use (Gale, 1995, p. 168). From this use-wear evidence and the geological structure of the mineral deposit, we can make several observations about how the miners used these hammers to extract ore. At Locs. 007 and 21, for example, where the ore veins are restricted in size (Lewis, 1996, plate 11), the miners must have pounded the hard dolomitic rock using smaller stones gripped in one hand, as they worked in places in a crouched or even recumbent position (Figures 2.13, 2.30). In more open underground locations, such as Loc.18 or the Lost Cavern (Lewis, 1996, plate 13), it would have been possible to stand upright and use larger implements. These could have been swung by a handle underarm against the rock face although no cobblestone hammers modified for hafting have been conclusively identified underground (Gale, 1995, p. 155). Replication experiments show that this technique uses the weight and momentum of a hammer during both its initial swing and subsequent recoil to loosen rock from the face (Timberlake and Craddock, 2013,

p. 51). If the haft failed, a cobblestone could subsequently have been held in either one or both hands, depending on its size and weight. The latter method may even have been preferred, given that it affords greater blow precision (Pickin and Timberlake, 1988, p. 165) (and given that, as we have already seen, evidence for cobblestone tool hafting at this site is limited). The few very large cobblestones found, weighing up to 29kg, can only have been wielded against the rock face when in some form of sling or cradle (Lewis, 1996, pp. 118–9). Replication experiments undertaken on the Great Orme suggest that only one or two people would have been required for this, depending on the support method used. Using a sling passed over the shoulder, for example, a large cobblestone could have been wielded against the rock face by a single practitioner. Alternatively, use of a timber support would probably have demanded two people, one to regulate and fix the height of the cradle holding the cobblestone and the other to guide the hammer in both hands to direct blows against the face (Lewis, 1990b).

Using cobblestone hammers to extract mineralisation in hard, unaltered dolomitic rock would have been a laborious and generally ineffective process. Continuous use of hand-held cobblestones is physically exhausting (O'Brien, 1994, p. 134). The cobbles would also have deteriorated rapidly due to spalling caused by repeated percussion against the hard rock (Lewis, 1990b, p. 55; Dutton, et al., 1994, p. 272). While some cobblestone implements may have had a clearly defined function, use-wear evidence indicates that the Bronze Age miners reused used, worn or broken cobblestone hammers for different tasks. Most of these tools seem to have been multifunctional and were probably used, reused and recycled spontaneously in a variety of fluid and opportunistic ways. A broken mining hammer could have been repurposed as a pestle or anvil for ore crushing, for example, or a previously discarded hammer spall as a chisel for prising out ore (Timberlake and Craddock, 2013, p. 52). As the cobblestones with some apparent modification for hafting display the same use-wear patterns as those with no modification, hammers that were previously hafted may subsequently have been used in the hand (Gale, 1995, p. 173), or vice versa.

Given the relatively small number of cobblestone tools from the mine and the significant proportion without any evidence of use, it is possible that the Bronze Age miners only occasionally used these implements for mining (Lewis, 1996, pp. 120–1). When areas of harder rock were encountered, bronze picks or chisels would have comprised far more effective mining implements than cobblestone ones. It is unclear, however, how prevalent use of bronze tools for ore extraction actually was, as no complete or near-complete examples have been found in this or other Bronze Age copper mines in Britain. The ongoing discovery of bronze fragments in the underground workings hints that it may have been more common than previously thought. These fragments are fractured and heavily corroded, with irregular profiles. The surface form of two of the larger fragments (c.1.5–3cm) is similar to the tip of either a bronze pick or chisel (Jowett, 2017, p. 69, figs. 9, 10). The Bronze Age miners could have used such implements either on their own or with fire-setting when they encountered mineralisation in hard, unaltered dolomitic rock, as we shall see.

In Loc.36, for example, numerous bronze fragments were excavated from a spoil layer (004) containing heavily fire-reddened pieces of rock and abundant charcoal (Figure 2.20). As the passage in which this deposit was found had been dug out of softer rock, the excavators surmise that this was an earlier working that the prehistoric miners used to access mineralisation further into the hillside, where they then encountered harder rock requiring extraction by fire-setting in combination with bronze tools (Jowett, 2017, p. 64, fig. 8). They subsequently moved the resultant rock spoil back towards the entrance into the original access passage and deposited it there (presumably after mining at both locations had ended or as other direct routes into the later working were opened). As discussed in Chapter Two, however, the complex stratigraphic sequence at this location, with older dated deposits overlying ones of later date, suggests that this is not the whole picture of activity in this part of the ore deposit. Striations produced by a tool of uniform shape have been found in the rock sidewalls at Loc.36. They are interpreted as supporting evidence that the bronze fragments at this location are from the use of bronze mining tools. The indentations are, however, very similar in form to marks previously thought to have been created by bone (or wooden) mining tools (Jowett, 2017, p. 68, fig. 15), which are in turn very different to the sharp, angular profile of striations caused by metal (bronze or iron) picks or chisels (Lewis, 1996, p. 103). Interpretation is further complicated by the fact that the Bronze Age miners would have encountered relatively soft rock as they worked. Use of bronze picks or chisels instead of those of bone or wood was not functionally necessary at this location. Nonetheless, the contextual association between the wall marks and the bronze fragments is certainly suggestive and will hopefully be demonstrated conclusively by the planned research into the tool marks at the mine (Jowett, 2017, p. 68).

The archaeological team at the mine is also continuing research into the metal fragments themselves (Jowett, 2019, pp. 60–1). It is uncertain whether they could have arisen from prolonged use of a bronze metal mining tool against the rock walls of the Great Orme mine, as bronze may deform rather than fracture when struck cold, depending on its composition. Fragments like those from Loc.36 and elsewhere in the mine have been produced experimentally by hammering heated bronze (Lewis, 1996, p. 132; Knight, 2021). What role such a process may have had in the mining *chaîne opératoire* is, however, unclear. Discoveries of bronze metal mining tools in other prehistoric copper mines in Britain and Europe are relatively rare. Axes, rather than picks, are the typical metal find type from these sites. Such tools may have been primarily embedded in a field of activity concerned with woodland husbandry, and so used to cut timber and work wood needed for a range of mining purposes, such as making or repairing equipment or chopping fuel for fire-setting (O'Brien, 2015, pp. 210–12). At the Mount Gabriel mines (south-west Ireland), for example, although no actual axes were discovered, metal tooling marks have been identified on mine timbers, roundwood fuel and items of wooden equipment (O'Brien, 2015, p. 212). Some of the best examples of tool marks come from the Bronze Age mines in mid-Wales, particularly those from flat metal axes used in making the wooden launders found at the Copa Hill mine (Timberlake, 2003b, p. 81–2). Could some of the bronze fragments from the Great Orme mine be the by-products of similar *chaînes opératoires*? In this context, it is interesting that one of the copper objects from the mine is a possible billhook or knife, an item more likely to be associated with agriculture or dwelling than mining.

In summary, each miner may have had a personal 'toolkit', comprising bone, wood and possibly bronze tools of various forms, as well as small cobblestone mining tools. Some tools could have been selected for specific purposes while others were intended to be multifunctional (and selected with this utility in mind). Some may have acquired significance to those who used them – and who had perhaps also had a hand in their fabrication or modification. As we have seen, each person would have had their own individual understanding of the characteristics of a 'good' tool. Tools also represent the means of accomplishing a task. The Bronze Age miners may therefore have prized an implement that sat well in the hand, that had helped them extract mineralisation from a particularly bountiful vein, or that was made from a material, like cattle bone, with powerful metaphysical associations. Such items may even have been thought of as 'showing the way' to the ore.

Mining by fire-setting

Occasional finds of reddened, spalled rock-wall surfaces, in association with heat-discoloured, fire-cracked, charcoal-containing spoil indicate that Bronze Age miners on the Great Orme sometimes used fire-setting as an ore extraction technique when they encountered hard unaltered dolomitic rock (e.g., Lewis, 1996, table 6; Jowett, 2017, p. 64). Fire-setting weakens and micro fractures hard rock. At many other Bronze Age mines, the rock was then removed relatively easily by pounding with stone hammers (O'Brien, 2015, pp. 204–8; Timberlake, 2015b).

At the Great Orme mine, however, there is no evidence that cobblestone mining tools were specifically used with fire-setting, even in those parts of the mine where fire was used. Replication experiments have shown that the mineralisation at this site can usually be prised from the fire-cracked face using fingers or pointed bone, wood or bronze implements (O'Brien, 2015, p. 204). Cobblestone tools used for this purpose quickly become abraded after only a few hours of use and displayed a degree of wear not found in the archaeological material (Lewis, 1990b, p. 55, 1996, p. 120).

Convincing evidence for fire-setting comes from a narrow working (less than 0.2m wide at its mid-point) on the N1 ore vein in the Tourist Cliff (Dutton, *et al.*, 1994, p. 261, fig. 8a) (Figure 2.30). Here, a 0.35m thick deposit comprising angular spalls of burnt rock mixed with mixed oak/hazel charcoal was found immediately below a heat-discoloured, fractured, concave rock wall. A sample of this charcoal has a modelled probability estimate of *1370 to 1115 cal BC* (Model 2: CAR-1280; 2970±70 BP; 68% probability). No stone finds were recorded, but a bone implement was discovered wedged into the highest point of the roof of the working. From replication experiments carried out on the Great Orme (Lewis, 1990b), it is possible to sketch the *chaîne opératoire* of fire-setting at this location. The ore vein appears to have been worked inwards from either end. Where it is widest, at its eastern and western end, a small fire must have been built under or against the rock surface. This could have consisted of wood and/or charcoal (on a base of timber to ignite it). Use of charcoal would have produced an intense, localised heat (Lewis, 1990b, p. 56). This would have been particularly appropriate for extraction within the confined working on the N1 ore vein. Although burning charcoal produces much less smoke than wood, considerable heat and poisonous fumes would still have been produced (Craddock, 1992, pp. 149–50). Assuming that the working on N1 was a closed tunnel rather than the open trench it is now (for the reasons outlined in Chapter Two), it is therefore likely that the miners – and everybody working in the adjacent area – retreated the short distance to the surface through the working on the E1 or E2 ore vein once the fire was lit, whatever type of fuel they used. If the timber and/or charcoal had been carefully stacked during lighting, it may not have been necessary to add further fuel and the fire could have been left to burn out (O'Brien, 1994, p. 172). In such a small working, this could have taken as little as 1.5 to 2 hours (Lewis, 1990b, p. 56).

The working area could have been doused with water to cool down the rock face and extinguish any glowing embers (e.g., Timberlake, 1990a, p. 53). Alternatively, and more likely, it may have been left, perhaps overnight, until cool, providing a pause in the rhythm of extraction at that location (Crew, 1990, p. 57; O'Brien, 2015, p. 204).

The remains of the fire could then have been cleared away from the face, to allow more room for working. Cobblestone hammers were probably used to compact and pulverise the weakened dolomitic rock. Although no stone finds were discovered in the N1 ore vein itself, they are recorded in workings nearby, such as the Central Vein (David, 2000, p. 74). Given that the N1 working is little more than 1m wide at its widest point, the cobblestones must have been wielded in either one or both hands, rather than in a handle. The fractured rock then appears to have been prised away using bone implements like the point found in the working, as well as by hand. It is unclear whether there was one or repeated episodes of fire-setting at this location. The fire-set area extends to a height of nearly 2m (Dutton, *et al.*, 1994, fig. 8a). If this was produced in a single event, it would have been possible to have mined while standing upright and crouching. Two people, at most, could have taken turns to work at each end of the passage, where the ore vein was widest. More may have been involved, to carry fuel to the location before fire-setting began, to help clear away the spoil extracted from the rock face and to remove it to the surface as mining progressed.

To summarise, physical evidence for fire-setting in the prehistoric workings is generally scarce. Even considering the probability that evidence of this practice has been destroyed or obscured by later mining, Bronze Age miners on the Great Orme appear to have made the technological choice to use fire-setting as an extraction method less regularly than seems to have been the case at other Bronze Age mines in Britain and Ireland (e.g., Lewis, 1990d, p. 7, 1996, p. 155). One reason for this is the geology of the copper deposit: while most other mines in Britain and Ireland involved hard rock mining, there was no practical need for fire-setting in the many areas of soft, dolomitised rock that the Great Orme miners encountered. Another constraint could have been the logistical difficulties involved in setting fires in constricted workings located at considerable distances underground, with no direct connections to surface. The probability that, as we have seen, there was fire-setting at a point further into the hillside from Loc.36, a working that is itself c.10m from the present-day hillslope (Jowett, 2017, p. 64), is intriguing. Was fire-setting in the unaltered dolomitic rock at, for example, Loc.34, c.130m into the hillside, even a practical possibility (Figure 2.24)?

Processing ore

The quantity of spoil found in the prehistoric workings indicates that, after extracting the copper mineralisation using tools and/or fire-setting, the Bronze Age miners began the *chaîne opératoire* of ore processing. Underground, the miners are only likely to have achieved a preliminary degree of separation, hand picking any identifiable nodules of ore and discarding the largest fragments of apparently barren rock. In the poor light in many sections of the mine (an issue we will return to), they may only have been able to do this by touch, distinguishing the copper-bearing section from the unwanted host rock due to its differing texture, weight and hardness. There was no mineralogical basis by which to identify the ore from its taste or smell. The spatial layout of the mine would also have constrained the methods and extent of processing undertaken underground. The mine workings are often too confined for the miners to have attempted much separation as the ore was being mined. While working crouched or lying in the narrow passages at Locs. 21 and 007 (Figure 2.30), for example, they may only have been able to break nodules of rock with their hands or using a small hand-held stone hammer or pestle so that it could then be sorted. They may have used a specific tool for this or, more probably, a cobblestone serving the dual or secondary function as a pestle/mining hammer (e.g., Lewis, 1988, p. 46; Dutton, 1990, p. 13). In more expansive – and possibly better illuminated – underground areas, such as the low but laterally extensive workings at Loc.10 (e.g., Figure 2.16) and the large chambers at Loc.18 and the Lost Cavern (Figure 2.24), the miners would have been able to pound the ore more effectively by wielding cobblestones gripped double- or single-handedly.

This process probably did not typically reduce the mined ore to a fine debris that would have been difficult to handle and remove to the surface. The Bronze Age miners are likely to have used scapula scoops and/or wooden shovels to shift the partially crushed mine extract into some form of wooden, basketwork or leather container for hauling laboriously to the surface by hand (O'Brien, 1994, p. 172). The possible remains of shovels, woven baskets and wooden troughs – perhaps for rock haulage – have been found at other European Bronze Age copper mines (O'Brien, 2015, p. 217). The miners would have had to select containers suited to movement over frequent changes in gradient and through confined spaces, on hands and knees or by squirming along on the stomach. When excavating the narrow working at Loc.21, Geoff David (1995b) found that the most effective method of rock removal was to fill a closed steel box with spoil at the working face and then to pull it by a rope attached to its handle along a plank as he crawled to the passage entrance. The closed design of the box prevented spoil being spilled while manhandling it up the steep slope to the opening. The Bronze Age miners perhaps adopted a similar approach, but the wooden equivalents or sacks are not preserved in the mine.

Once at the surface, the miners again used cobblestone tools to further crush and grind the mine extract. At least 25 stone mortars have been recovered from deposits of spoil overlying the Tourist Cliff (Dutton, 1990, fig. 5; Lewis, 1996, p. 119). Although all were discovered in secondary contexts, similar artefacts have been found at other Bronze Age mines in Europe, such as Ross Island (Ireland) and mines in the Cabrières region of France (e.g., O'Brien, 2004, p. 356, fig. 165; Ambert, et al., 2009, fig. 3). It is therefore likely that the Great Orme examples relate primarily to activity during this period. These flattish objects have well-developed wear concavities in their upper and lower surfaces, produced by the grinding or very fine crushing of ore (Dutton, et al., 1994, p. 269; Gale, 1995, pp. 156–7). The Bronze Age miners would have placed a quantity of ore on the surface of a mortar or anvil, so that it was contained within any natural dimples or hollows in the rock (Doonan, 1994, p. 86), or perhaps even within depressions that had developed in the face through use. They would then have crushed and/or ground the ore using a stone pestle or pounder gripped either one- or two-handedly. Two 'pestle-type' cobbles, which could be positioned comfortably within the hollows in the surface of the stone mortars, were identified in the cobblestone tool assemblage (Dutton, et al., 1994, pp. 269–72).

Processing experiments using remnants of copper ore from the mine indicate that rotted dolomitic limestone is easy to crush using stone implements and can be rapidly broken down to a granular consistency (e.g., Roberts, 1998; Brocklehurst, 2001). For coarse-grained copper-bearing minerals, crushing could have achieved almost complete separation from the rotted rock matrix. Hard unaltered dolomitic rock requires more physical effort to crush and it is more difficult to fragment the rock sufficiently to release the copper mineralisation. In such cases, and when the copper minerals were more finely disseminated through the host rock matrix, the Bronze Age miners probably used a repeated sequence of crushing and grinding. It was probably important to use a stone of sufficient weight to minimise the effort needed to crush or grind the ore, while not exhausting the practitioner (Roberts, 1998, p. 8). It may also have been useful to place the mortar or anvil on a groundsheet, perhaps made of animal skin, to help retain any detached flakes or fragments of ore falling close by. Practitioners who were proficient at crushing are unlikely to have been in any danger of being struck by flying material (Doonan, 1994, p. 86). As protection from the wind and rain while crushing, grinding and sorting, the Bronze Age miners may have constructed some form of temporary shelter outside the workings, like the stake-built huts found adjacent to the prehistoric mine at Ross Island (O'Brien, 2015, p. 127). At the Great Orme mine, however, no evidence for such structures, nor primary prehistoric deposits of comminuted rock close to the workings, seems to have survived.

Once the ore had been crushed or ground sufficiently, the miners separated the black-green/blue copper-bearing minerals from the yellowish waste dolomitic rock by hand sorting – identifying and picking out the

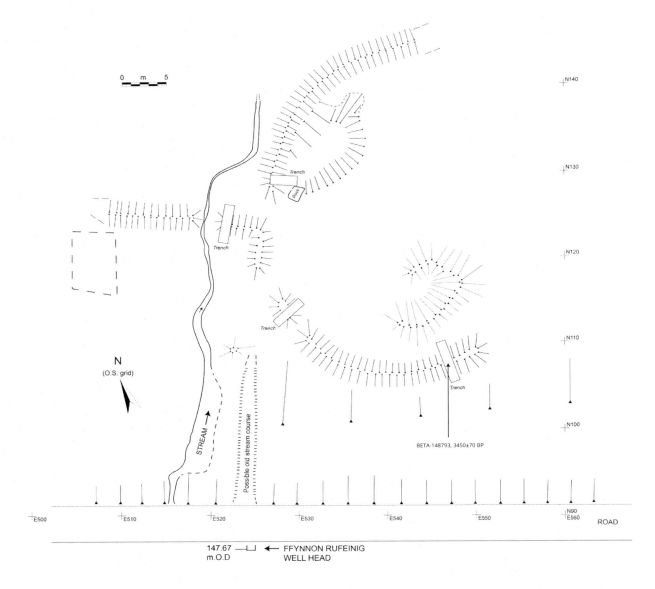

Figure 3.5: Plan of surveyed surface features at the Ffynnon Rhufeinig ore washing site. Shows locations of the excavated trenches and the bone samples taken for radiocarbon dating (with laboratory number and radiocarbon measurement) (drawing by M. Andreikos).

copper fraction by hand, or with the help of pointed bone tools, based on its colour, lustre and density (e.g., O'Brien, 2004, pp. 378–84, 2015, p. 221). It is probable that they used the spring at nearby Ffynnon Rhufeinig, less than a kilometre from the mine, to concentrate the ore by 'washing' in water. Also referred to as 'wet ore processing' or 'gravity separation', this technique uses flotation in water to separate the desired ore minerals from the waste rock by exploiting differences in their relative densities. Excavation of a spoil heap at Ffynnon Rhufeinig uncovered a complex succession of well-sorted, orangey-brown, dolomitic-limestone sand and silt layers; malachite ore fragments; finely disseminated charcoal; a single igneous rock fragment, possibly a spall from a stone hammer; and copper-stained bone fragments. These deposits, a sample of bone from which has been radiocarbon dated to 1945 to 1545 cal BC (BETA-148793; 3450±70 BP; 95% probability), are interpreted as waste from the washing of prehistoric mining spoil. Other features excavated at the site are interpreted as a system of channels and pools for ore washing. These point to repeated later phases of reworking of earlier deposits – of Bronze Age origin – at Ffynnon Rhufeinig resulting, over time, in the surface features visible at the site today (Wager and Ottaway, 2019) (Figure 3.5).

The Bronze Age workers could also have taken ore to some of the other water sources scattered across the Great Orme, such as Ffynnon Galchog (discussed in Chapter Two), as well as the undated site of Ffynnon Gaseg, for concentration by washing (Lewis, 1990a, 1996, pp. 144–9; Jones, 1994; Wager and Ottaway, 2019) (Figure 1.5). Transporting the mine extract to these locations may not have been a very arduous task, particularly if the miners first removed the bulk of the waste rock at the mine. At Bronze Age ore washing sites elsewhere in Europe, like those in the Austrian Eastern Alps at Troiboden (e.g., Stöllner, 2019; Timberlake, 2019), Bachalm/Kelchalm (e.g., Preuschen and Pittioni, 1954) and Mauk F (e.g., Schibler, *et al.*, 2011), a very simple technology of wooden sluice boxes and settling troughs was used to achieve a greater degree of enrichment of copper ores by washing than would have been possible with hand sorting alone (Wager and Ottaway, 2019, p. 26). Perhaps workers on the Great Orme used similar technology, although no evidence for this has survived.

The processing sites that presumably formerly existed at or near to the Bronze Age mine can no longer be identified. This is partly due to disturbance from pre-modern mining (such as the considerable quantities of spoil of mixed date filling the area of the Pyllau Valley between the Tourist Cliff and the Llety'r Filiast burial monument) (Figure 2.3), as well as limited surface excavation immediately beyond the mine workings. Using X-Ray Diffraction (XRD) analysis, however, it is possible to identify in the deposits of mining spoil at the various springs on the headland material from different geological areas in the mine (Timberlake and Marshall, 2019, p. 16). The results of such studies, combined with additional field investigation and scientific dating of all the potential prehistoric ore washing sites on the headland, will help to enhance our understanding of the *chaîne opératoire* of ore processing at this source and how it changed over time (Wager and Ottaway, 2019, p. 30). It is, however, clear from the evidence already available that it comprised a flexible, multi-stage sequence of active technological choices unfolding both at the mine and across the Great Orme headland. The precise steps in this sequence are likely to have varied not only over time but with the intended end use of the concentrate. The ethnographic record indicates the diversity of purposes minerals may have, even within a single sociocultural context (e.g., Sagona, 1994). As we saw in Chapter Two, ore could have been extracted to make pigments for bodily decoration, medicines or jewellery well before the customary phase of Bronze Age mining for metal began. The former practice may have continued, even as people also mined ore on the headland to make copper. Stages of very fine crushing or grinding to reduce the ore to a powder would have been essential for pigment making. For smelting, a pea-sized concentrate may have been the optimum particle size (Craddock, 1995, p. 161), although experimental smelts suggest that a more finely ground, powdered concentrate is also effective as this can then be formed into cakes using suitable binders such as fine sawdust with finely crushed charcoal (Timberlake and Prag, 2005, p. 203).

Organising the work

The physical environment of fire-setting on the N1 ore vein in the Tourist Cliff and the reconstructed *chaîne opératoire* of ore processing at Loc.007 show that the logistics of organising mining at this source could be very challenging. Due to the physical form and distribution of the workings, the Bronze Age miners must often have faced major practical difficulties when entering, working in and moving around the mine. This section considers the evidence for how they responded to the labyrinthine layout of the mine workings, roof collapse, flooding and damp, muddy conditions. We will begin by looking at how the disposal of the unwanted rock produced by ore extraction may have been organised, focusing first on mine-working in narrow ore veins. At such locations, it would have been necessary to cease mining at regular intervals, to clear accumulations of unwanted spoil away from the face being worked before it became obscured. In larger, or low but laterally extensive areas, unwanted material could have been more easily shovelled to one side so that spoil disposal did not interrupt the rhythm of extraction. Even in such locations, it would eventually have become necessary to remove the unwanted rock from a working, so that mining could proceed unimpeded. When the working was at a short distance into the hillside in, for example, the Tourist Cliff, the Bronze Age miners probably took spoil to the surface and deposited it there. The landscape around the ore deposit must gradually have been transformed by progressive accumulations of discarded rock, that grew and perhaps shifted position over time.

Further underground, unwanted rock was piled or stacked into workings where extraction had ceased, until all or most of the tunnels at a given location were completely or partially filled with spoil. This can be seen at, for example, Locs. 5 and 36 (James, 1990, fig. 2; Jowett, 2017, p. 64, fig. 8) (Figure 2.20). There are possible drystone revetment walls or 'deads-stacking' across ore veins in the Tourist Cliff, perhaps to prevent spoil spilling into the Lost Cavern (Dutton, *et al.*, 1994, p. 283, fig. 7). Two well-preserved drystone walls have been excavated at Loc.7c (Underground) (David, 1993) (Figure 3.6). Large boulders block the vertical extent of several ore veins and could have been deliberately placed to maintain access along the ore vein below while facilitating the disposal of unwanted rock above (Lewis, 1996, plate 10). The Bronze Age miners may also have strategically backfilled abandoned workings

Figure 3.6: Distinct areas of mine-working separated by bedrock ribs on the western edge of the Lost Cavern, facing southwest. A – Loc.7c (Underground), with a well-preserved drystone wall; B – Loc.7c (Surface); C – working on the West 6 ore vein; D – Loc.7b; E – an entrance to Loc.7c (Underground) (photographs author's own). (For the plan view of these locations, see Figure 2.20.)

and intentionally cleared rock debris from underground tunnels and surface entrances to aid ventilation (Lewis, 1996, p. 111). The practice of backfilling stopes and the blocking of passages and surface entrances by natural processes such as roof collapse would have altered the flow of air into and around the mine. The miners may have taken care to ensure that major routes to surface and numerous points of access from the surface into the workings were clear so that air could continue to circulate freely. This would have been particularly important when fire-setting and/or working at locations at considerable distances into the hillside, without direct connections to surface, such as Locs. 33, 34 and 35 (Figure 2.24).

As we saw in Chapter Two, the Bronze Age miners frequently had to contend with constricted passages in a complex, ramiform layout; abrupt changes in slope created by the sinuous course of the ore veins as they meandered and twisted in a horizontal and vertical plane; and steep gradients produced by the dip of the rock strata, such as the inclination of approximately 27% (15°) they would have experienced excavating Passage B at Loc.5 (Figure 2.24). While it would have been possible to crawl or even walk into some workings with direct connections to surface, access to others, such as Loc.21, would have been far more awkward, involving the negotiation of a tight squeeze and an abrupt slope (Lewis, 1996, fig. 25). Accessing areas at a considerable distance into the hillside that could only have been reached through other workings (e.g., Locs.

33, 34, 35) must have been an even more laborious and lengthy process. In such cases, moving people, ore, spoil and mining equipment must have involved considerable physical exertion and required coordination, stamina and flexibility. Consequently, the miners may sometimes have spent long periods underground at such locations, perhaps even sleeping there, despite the apparent discomfort.

How the movement of material into and out of the mine unfolded in practice probably depended upon the part of the source being accessed. Ore, spoil and equipment were perhaps moved by relay, passing from hand to hand into the mine or out to the surface. This process may have been ongoing when working in areas that were close to the surface and easy to access. At more distant and difficult to reach locations, it could have been a more episodic process. The miners could, for example, have moved the partially processed mine extract in stages. All the ore mined from the passages at a particular location was perhaps stockpiled temporarily at suitable central places, such as large underground spaces, like Loc.18 or workings at the junction between several ore veins. Such spaces may also have been used for the temporary storage of equipment, like timber or spare tools. Some of the apparently unused cobblestones found throughout the workings could be 'left over' from a potential stone tool store that was never used.

As an episode of working underground progressed, the miners would most probably have had to carry out running repairs or modifications to the tools they were using. They may even have had to make new ones. It may have been necessary to designate a person to these tasks (e.g., Timberlake and Craddock, 2013, p. 51). Although distinct areas at the mine for toolmaking and repair have not yet been identified, the more open spaces underground could have comprised arenas where miners labouring in neighbouring workings met to repair or refresh their equipment and share expertise. Such occasions may have comprised interstices in the rhythm of extraction in which the miners could perhaps eat, drink or briefly rest – a respite from the cramped conditions in the more confined mined-out areas. As we shall see, however, lighting in the workings was generally poor and limited and so it may not always have been possible to make or repair tools successfully underground. The miners may have preferred to take sufficient spare tools in their toolkit to the area where they were intending to work, to minimise the need for ongoing repairs. This is certainly likely to have been the case when they were working in the more difficult to access parts of the source, at considerable distances into the hillside.

When mining the major, vertically extensive ore channels running through the mineralised zone, the Bronze Age miners worked upwards, extracting ore overhead. This gave rise to large, steeply inclined chambers or 'stopes', as seen at Locs. 30, 31 and 34 (Lewis, 1996, App. C, 15–16) (Figure 2.24). They may have used a similar stoping technique – mining inwards and then upwards at various points on the face – when working the broad ore veins visible at the surface in the Tourist Cliff and other rocky outcrops in the Pyllau Valley. At these locations, the miners would have needed some form of support to stand on. The early/late modern miners constructed wooden platforms on stemples (horizontal timbers) placed across the ore vein (Williams, 1995, p. 37). By standing on these, they were able to extract ore at increasing heights overhead. There is no conclusive evidence for similar methods in Bronze Age mining at the Great Orme mine, although possible prehistoric stemple holes have been tentatively identified in the Lost Cavern, including in a vertical wall of dolomitic limestone on the eastern edge of a working on the E6 ore vein (David, 1997). If wooden supports were used, perhaps some of the bronze fragments mentioned earlier were from the axes and knives used to trim the timber to a precise fit while underground? Instead of using wooden supports, the Bronze Age miners could have piled up spoil to create platforms on which to stand as they worked the mineralisation above. Wooden ladders and scaffolding would, however, have been needed if lower faces of, for example, the Tourist Cliff had to be kept free of spoil to maintain access to the lower workings and when removing mineralisation much further up a rocky cliff or in a large stope. At other Bronze Age mines, miners used rock-cut foot- and hand-holds, wooden 'step-planks' and notched tree-trunk ladders to help manoeuvre themselves and materials around the workings (O'Brien, 1994, p. 171, 2015, p. 217; Timberlake, 1996, p. 60). The Great Orme miners perhaps also adopted similar methods. Rock-cut steps have been tentatively identified in two places in the sidewalls of a working on the Central Vein (David, 2000, p. 73). There are also rock-cut steps of assumed Bronze Age origin in several places along the east wall of the East 1 Vein (David, 2003, p. 96).

As the Bronze Age miners progressed working in areas where the strata capping the ore vein were unstable, extraction would have become increasingly dangerous. Numerous instances of roof collapse have been recorded in the early workings, particularly where the mineralization is confined beneath rehydrated mudstone or shale strata (Lewis, 1996, App. C). In other areas, such as the 'North-West Corner', the limestone capping the ore vein is heavily fractured and prone to collapse (David, 1998, 1999, p. 78). Despite the apparent danger, however, there is only limited evidence for the use of rock or timber roof supports at the Great Orme mine, although examples of both are known from other European prehistoric copper mines (e.g., O'Brien, 2015, p. 219). In Loc.35 at the Great Orme mine (Figure 2.24), the Bronze Age miners extracted ore from the mineral body over a lateral extent of up to 5m, only deliberately leaving in place occasional pillars or rifts of bedrock (Lewis, 1996, App. C, 16–7). They may also have used timber supports, the later decay of which could have resulted in the observed roof collapses at this location. If such structures were not regularly used and maintained during a single mining event and throughout subsequent instances of activity at the source, there would have been a serious threat of cave in at some locations.

All these hazards, practical difficulties and discomfort must have been compounded by the wet conditions underground. The Bronze Age miners would have encountered both running and standing water, due to seepage as extraction progressed along a particular ore vein and as accumulations in existing stopes. The extent to which drainage was an issue would have varied with prevailing weather conditions as well as according to aspects of the physical structure of the deposit, such as the dip of the bedding planes (O'Brien, 1994, p. 172). Water from the surrounding catchment, an area of c.260,000m², would have collected in the Pyllau Valley, in the deep depression – approximately 15–20m deep – formed by the dip of the rock strata in the mineralised zone (Lewis, 1996, p. 82). There are two possible models of the mine's hydrogeology: in one, water was in general able to drain freely away, with the level of the water table varying over time to occasionally reach depths below sea level; in the other, drainage was slower, creating a series of perched water tables (Lewis, 1996, pp. 82–3). Further hydrogeological and geotechnical studies of the mine workings are needed for a firm conclusion to be reached.

However, the evidence for Bronze Age extraction to a considerable vertical and lateral extent, over such a long period, indicates that the practical difficulties presented by the presence of water were generally not insurmountable, at least during the customary phase of mining during the second millennium BC. Discoveries by early modern miners hint that their Bronze Age counterparts sometimes constructed dams or bailing devices of plaited hazel twigs and clay to remove water from flooded workings (Lewis, 1996, p. 40). The prehistoric miners could also have employed other techniques to deal with unwanted water ingress, for which no physical evidence has survived. Like the Earlier Bronze Age miners at Copa Hill, mid-Wales, they may have fashioned and then carefully positioned wooden launders to direct the flow of water away from locations where extraction was taking place (Timberlake and Craddock, 2013, fig. 3). They could also have bailed water by hand using an animal skin receptacle to tackle localised flooding and water seepage within a discrete area of working. As well as using the materials available to them, the Bronze Age miners most probably also adopted other practical strategies in response to the groundwater conditions, such as avoiding further activity in workings with standing water or that were prone to flooding. They perhaps chose instead to work other parts of the source or even to mine during drier times of year. We will return to the issue of drainage and seasonal working in Chapter Four.

Lighting the workings

All the tasks of spoil disposal, moving around the mine, and responding to the physical conditions encountered there would have been made more challenging by the lack of illumination in the underground workings, except for those close to the surface with direct connections to the open air. The Lost Cavern and many of the workings along the ore veins in the Tourist Cliff would probably have received some natural daylight or, at night, the light from surface fires. Elsewhere underground, artificial lighting methods would have been needed when, for example, prospecting for a new ore vein to follow as mining progressed at a particular location. Fire-lighting equipment must have been a vital component of the miners' toolkit and so it is perhaps surprising that no artefacts interpreted as flint strike-a-lights have yet been found. Possible 'sooty deposits', perhaps from a form of prehistoric lighting, have been observed on some rock surfaces underground (e.g., Lewis, 1990d, p. 7). 'Nests' of charcoal fragments, elongated in form and up to 3.5cm long, in two hollows in the passage wall at Loc.7a have been tentatively identified as burnt twigs or wood shavings

(Lewis, 1996, p. 137). Branch charcoal, possibly for lighting, has also been excavated at Loc.36 (Jowett, 2017, pp. 68–9).

These finds may be the remains of tinder or short wooden spills ('lighting-sticks'), like those discovered in Bronze Age mines in the Eastern Alps, Saint-Véran (France) and on Mount Gabriel (south-west Ireland) (e.g., O'Brien, 1994, pp. 158–60, 2015, pp. 217–8; Goldenberg, 2015, abb.5). The wood species in the Great Orme examples is unidentified and so it is not possible to assess whether they fit a pattern of selective wood use specifically for lighting, such as the deliberate choice of bright-burning resinous conifer for lighting splints seen at Mount Gabriel (O'Brien, 1994, p. 158). It is possible that the Great Orme remains relate to fire-setting rather than lighting. There are no finds of oil- or fat-burning lamps of stone or clay of definite Bronze Age date from the Great Orme mine (Lewis, 1996, p. 137). They are very rare elsewhere, with only one possible example, from the Chinflon copper mining district in south-west Spain (O'Brien, 2015, p. 218).

The methods used to light the workings may have been ephemeral and left few traces (e.g., Dutton, et al., 1994, p. 279). Even in the 19th century AD, each miner on the Great Orme worked solely in the light cast by a tallow candle stuck into a ball of clay (Smith, 1988, p. 42). Whatever method of illumination was used, the practical challenges involved in working this source would have meant that maintaining a constant light source was difficult, if not often impossible. This is particularly likely to have been the case in locations at a considerable distance into the hillside, although fixed illumination may have been possible in the larger stopes. Even holding a lit spill in the mouth for any length of time would have been impractical and dangerous given the strenuous effort that must typically have been involved in mining at this source.

The Bronze Age miners may not always have used a constant source of light, but could instead have worked largely in the dark, using smooth cobbles of igneous rock as 'signposts' (Wager, 2002, p. 385). Herbert (1984, p. 59) mentions that sorting of ore could have been carried out underground in the dark in 'ancient' workings in Umkondo, Zimbabwe, for example. Much of the ore-bearing material could have been extracted from the Great Orme ore body by feel, due to the generally well-defined boundary between the rotted dolomitic limestone hosting the mineralisation and the barren unaltered dolomitic rock. Artificial illumination would have been essential at the beginning of each mining event, when choosing where to begin working in parts of the source that were beyond the reach of daylight. The miners may then have marked an access route to that location from the surface, through existing workings or natural cave passages, by positioning igneous cobblestone 'signposts' in places where a passage bifurcated, or where there were numerous possible exits from a working. These rocks would have been easy to distinguish by touch from the surrounding dolomitic wall rock due to their 'foreign' lithologies (Wager, 2002, p. 385). They would have comprised suitable waymarkers, enabling a chosen location to be reached using no, or only sporadic, illumination – particularly as the sensory deprivation the miners experienced by being underground would have heightened the perceptions of the other senses, including touch (Stöllner, 2018, p. 62). This could be another explanation for the occurrence of unused cobblestones in the mine workings. It could also explain why these items tend to have a greater degree of surface smoothness than stone finds displaying obvious wear marks (Gale, 1995, p. 150).

Contributory tasks

As the mining field of activity unfolded, the participants most probably undertook a range of contributory tasks in and around the workings and across the Great Orme and its hinterland. The *chaînes opératoires* of some of these activities, such as collecting and chopping wood and making charcoal, collecting beach cobbles, weaving baskets or making rope were, as we have seen, entangled in the task sequences of ore extraction, processing and spoil disposal. The *chaînes opératoires* of others – such as knapping flint, working leather, preparing skins, repairing clothing, constructing shelters, hunting, foraging, fishing, watching livestock, maintaining fires for warmth and light, preparing food, eating, resting and conversing – could variously have run alongside. All probably overlapped with other activity fields. All could, however, have played a part in the fulfilment of a given mining event and were perhaps considered integral to the mining field of activity by the participants.

Evidence for the content and full range of the contributory tasks carried out at the Great Orme mine is scarce. In common with many other European Bronze Age copper mines (O'Brien, 2015, pp. 241–2), no trace of a 'work camp' at the mine nor elsewhere on the Great Orme has survived. The discovery of awls in the bone tool assemblage could indicate leatherworking or basketry (James, 2011, p. 421), perhaps to repair bags or clothing. Other evidence hints at *chaînes opératoires* around food and woodland use, although the details are sketchy. As discussed, each bone from the mine – including the mining tools – would have been food at some stage in its 'lifecycle'. Some cuts of meat, such as ribs, from mainly cattle and also sheep could, however, have been intended primarily for consumption as food. We have already seen how the miners did not preferentially and intentionally select rib bones as tools. They could instead have elected to bring racks of ribs with them to the mine specifically for cooking and eating on site (James, 2011, p. 349). The elemental composition of the pig bone assemblage is statistically different from that of other species, in the number of mandibles and teeth, the ratio of rib bones to long bones and variety of body parts. Pig bones are under-represented as mining tools (James, 2011, p. 349) and are less obviously functional, due to their shape and size, than those from cattle or sheep/goat. Mining was (and still is) a very physical activity. Did the Bronze Age miners select cuts of pork, particularly protein-rich pig heads, as a preferred food source to sustain them during their labours (James, 2011, p. 373)?

The presence of hare, red and roe deer bones points to the hunting of wild animals. As at most other European Bronze Age copper mines (O'Brien, 2015, pp. 264–6), the amounts recorded are extremely small. The minimum number of individuals present in the hare assemblage from the Great Orme mine, for example, is only two, while for red and roe deer it is one and three respectively (James, 2011, pp. 302, 311). Surprisingly, given the Great Orme's coastal situation at the mouth of the Afon Conwy and the evidence for use of other littoral materials, such as beach pebbles, no fishbones and only six specimens of marine molluscs (four mussels, a clam and a winkle) have been found at the mine (James, 2011, pp. 311, 374, table 5.1). James (2011, p. 188), in her detailed study of the mine's bone assemblage, argues convincingly that the bones of fish and other small mammals are in fact missing from the site and were not overlooked during excavation and recovery. Perhaps the Bronze Age miners seldom engaged in the time-consuming tasks of fishing, foraging for shellfish or hunting game during a given mining event? The composition of the mine's faunal assemblage, however, does not necessarily reflect the prevalence of these activities at the mine nor in the broader taskspace but perhaps intentional choices not to eat fish or to dispose of the waste of hunted animals away from the source (James, 2011, pp. 387–9). The Later Bronze Age layers at nearby Pentrwyn on the headland contained edible intertidal shellfish (limpets, common winkles and mussels), indicating that shoreline products were part of local *chaînes opératoires* around food, at least during the later second millennium BC (Smith, Chapman, *et al.*, 2015, pp. 64–5). The vast quantities of edible mollusca found in Later Mesolithic and other prehistoric layers of undefined date in nearby Snail Cave (Smith, Walker, *et al.*, 2015) also attest to the long history of this practice at this coastal location (Figure 1.5). Perhaps the Bronze Age miners chose not to consume such foods at the source in preference for calorie-rich pork?

Deposition of food at the mine was probably also not simply a case of casual disposal (cf. Dutton, *et al.*, 1994, p. 277) or 'putting out the rubbish'. Such practices would have been guided by conventions governing what could be eaten and when, and how and where it was appropriate to deposit the remains. Codes dictating the 'right' way to act may have affected the deposition – and in the case of rib bones – reuse of material produced by the slaughter and butchery of animals, whether as part of food preparation, tool fabrication or other fields of activity. There is unfortunately no definite evidence for these practices from elsewhere in the local taskscape. The excavation report on the only other paleoenvironmental assemblage on the Great Orme securely dated to the Bronze Age – that from a dated Later Bronze Age layer at Pentrwyn – concludes that overall this material has little interpretive potential, due to its small size, poor preservation and fragmentary condition. Species identification was, for example, only possible for 28 of the 379 fragments recovered (Smith, Chapman, *et al.*, 2015, pp. 62–5). Faunal assemblages have only been found at a handful of presumed or actual Bronze Age sites elsewhere in north Wales, mostly in lowland locations. They are frequently small, due to the generally poor preservation of bone in the peaty soils that are widespread across this region (e.g., Lynch, 2000, p. 83).

The scale and significance of the acts of food consumption undertaken at the source were probably varied. They may in some instances have been small-scale and informal, punctuating the tempo of activity at the source. The Bronze Age miners could have chosen foodstuffs that were easily portable to take with them to eat underground. As seen earlier, some stages in the preparation of these, such as the final butchery and cooking of joints on the bone, could have been undertaken at the mine (Dutton, *et al.*, 1994, p. 277; James, 2011, p. 430). Other foods, like strips of dried meat, could have been brought ready-made to the site after processing elsewhere, leaving no archaeological trace once consumed. There may also have been larger-scale acts of collective consumption, involving all those present at the source. On such occasions, the miners could have enacted the *chaînes opératoires* of preparing and eating food and depositing the remains as moments of purposeful significance. These may have been intended to ensure or celebrate the success of the mining enterprise or to mark the beginning or end of a mining event.

We can envisage that the Bronze Age miners lit fires outside the workings for cooking, although whatever technique was used to cook meat, it does not usually seem to have resulted in burning or charring of the bone (James, 2011, p. 347). Fires would certainly also have been needed for warmth and illumination, producing at least some of the wood charcoal disseminated across the site. The species in the wood charcoal from the mine, mainly alder, oak, hazel, ash, holly and blackthorn, are those that were most likely intentionally selected for fire-setting, lighting or for cooking fires. There is also a possible wood fragment of elm (Dutton, *et al.*, 1994, p. 280; Lewis, 1996, table 1; Smith, Chapman, *et al.*, 2015, p. 66). Given the limited evidence for fire-setting at this mine, however, the gathering of branch wood for cooking fires or illumination may have been a more regular feature of the mining field

of activity than the collection of timber for fire-setting. In the Later Bronze Age layers at Pentrwyn, carbonised plant material is predominately oak with willow or poplar, while the wood charcoal fragments are from a wider range of species, including holly, yew and ash, with some heather (Smith, Chapman, *et al.*, 2015, p. 66). None of the species identified at the mine nor at Pentrwyn indicate which tree types the Bronze Age miners picked for hafting or to make other items of wooden equipment.

It is challenging to reconstruct the *chaîne opératoire* of human-woodland interaction at the mine from the wood charcoal data. The dataset is small and has such poor chronological resolution that we cannot pinpoint changes over time. Each sample was created through a complex interweaving of human activities and ecological effects (Caseldine, 1990, p. 17). It is difficult to elucidate exactly which phenomena were responsible for their characteristics and their scale or timing (Moore, 1993, p. 221). Unlike at other Bronze Age copper mines, such as Mount Gabriel (Ireland), Copa Hill (mid-Wales) and Mynydd Parys (north Wales) (e.g., Mighall, *et al.*, 2000; Timberlake, 2003b, pp. 66–8; Jenkins, *et al.*, 2021, p. 280), there are no peat bogs nearby to provide comparative environmental records that can be used to reconstruct the character of the contemporary local woodland and hence the technological choices made by the miners. Even when this information is available, it can be difficult to distinguish the local vegetation impacts caused by mining from those due to other fields of activity, such as agriculture (O'Brien, 2015, p. 277). The present-day ecology of the Great Orme is also an unreliable guide to the environment that the Bronze Age miners experienced: the Great Orme today is not a 'natural' landscape but a palimpsest of complex, multiple geomorphological events and past and present land uses (e.g., Conwy County Borough Council, 2011). Neither are the faunal assemblages from the mine and Pentrwyn straightforward indices of contemporary local environmental conditions.

The environmental samples paint a picture of oak woodland, possibly mixed with ash and an understorey of hazel and holly, as well as open heathland and poorly draining fen carr, on and around the headland (Dutton, *et al.*, 1994, p. 280; Smith, Chapman, *et al.*, 2015, pp. 66–7). Did the Bronze Age miners use whatever timber was locally available? Or did they intentionally select certain tree taxa? Mixed types of young roundwood, for example, are suitable for the "fast-blazing" fires needed for fire-setting or cooking (O'Brien, 2015, p. 236), while hardwood species like oak, which burn for long periods at high temperatures, are an effective choice of fuel for smelting. This could explain the abundance of oak charcoal at Pentrwyn (Smith, Chapman, *et al.*, 2015, p. 66). At the mine, however, the occurrence of oak in the dated Bronze Age wood charcoal samples is presumed to be the residue from fire-setting (e.g., Lewis, 1990b, p. 55). A mix of oak and hazel wood, for example, seems to have been used in the fire-setting episode at the N1 ore vein on the Tourist Cliff, as we saw earlier. This fits with the evidence from most other Welsh Bronze Age mines, such as Copa Hill, mid-Wales, where oak charcoal, found in abundance, comprised 90% of the wood fuel for fire-setting, with only small amounts of other timber (hazel roundwood) (Mighall, *et al.*, 2002, p. 1177). Hardwoods can also be crafted into items of mining equipment, holly twigs are ideal kindling and hazel can be twisted into withies for hafting cobblestone mining tools (e.g., Chapman and Chapman, 2013, p. 15; Timberlake and Craddock, 2013, p. 50). At Copa Hill mine, the Bronze Age miners not only used hazel for fire-setting but also for rope, handles, baskets and brushwood flooring (Mighall, *et al.*, 2002, p. 1177). Branches from alder carr woodland can also be used for making baskets or hurdles (Anon, 2008).

Were the Bronze Age miners actively managing trees by, for example, coppicing carr woodland (Dutton, *et al.*, 1994, p. 281)? Some species, such as hazel, thrive under regular cutting and provide an abundant renewable wood source (O'Brien, 1994, p. 162). Bronze Age miners at other European copper mines display considerable knowledge of how to make the most effective use of woodland, although conclusive evidence for deliberate coppicing is difficult to identify (McKeown, 1994, p. 279; O'Brien, 2015, p. 236). Perhaps the miners did not obtain all the wood or charcoal they needed from the surrounding locale but in some instances brought it with them from further afield – if the need for wood for a given mining event was greater than

Figure 3.7: Dr David Jenkins at the Pentrwyn smelting site shortly after its initial excavation in the late 1990s. The site's location on a narrow limestone shelf below vertical sea cliffs can be clearly seen (photograph author's own).

that locally available, or the specific types of trees they were seeking for particular purposes were unavailable locally. Or did they instead choose, when smelting, to take the processed ore to the wood? Either of these readings may be relevant to the final *chaîne opératoire* we need to unpick before moving on from the mining field of activity: smelting at the nearby site of Pentrwyn.

Making metal: character, content and scale

On the Great Orme, repeated excavation on a narrow limestone shelf below vertical sea cliffs at Pentrwyn, 1.2km to the west of the mine, uncovered a complex palimpsest of Mesolithic, Bronze Age and possibly Medieval activity (Smith and Williams, 2012; Smith, Chapman, *et al.*, 2015) (Figures 1.5, 3.7). The evidence for metallurgical activity at this site during the Later Bronze Age includes two small, adjacent, roughly circular 'pits', each about the size of a coffee mug (maximum diameter 10cm and depth 15cm), cut into a grey silty (possibly ashy) layer. An adjacent darker-grey layer yielded numerous metallurgical fragments (more than 270) and other finds of bone, charcoal, stone and flint. No crucible fragments or burnt clay was found in situ in and around the pits, but both had traces of a slight 'lip' with a notch on one side (Smith, Chapman, *et al.*, 2015, plate 9). These are interpreted as the remnants of a collar, perhaps made from fill bound with water, with a slot for a blow pipe and a tuyère, the nozzle through which air is forced into a furnace or forge (Figure 2.31).

Other evidence comprises a small, roughly conical hole, 40cm in diameter and 120cm in depth, containing charcoal-rich silt, some vitrified material and two slag fragments. The date of this feature (from sample BETA-127076) is insecure, as we saw in Chapter Two (Smith, Chapman, *et al.*, 2015, p. 55). Unfortunately too little has survived for its character and content to be reconstructed (Chapman and Chapman, 2013, p. 10). A small, charcoal-rich, V-shaped depression excavated in 1998 and interpreted as another possible smelting pit was found by later excavation to be a product of natural erosion of underlying clay deposits (Smith, Chapman, *et al.*, 2015, pp. 56–7). More than 700 small metallurgical particles, typically 1–10mm, were found in total, predominately in Later Bronze Age contexts, but also in most of the other deposits excavated across the site. These comprise copper prills and particles of black, heterogeneous, low-silica (c.5%) slag, often containing microscopic droplets of copper. There were also a few fragments of ore. Chemical and lead isotope analyses show that the slag particles are consistent with the smelting of malachite-goethite secondary carbonate oxidic ores from the mine (Smith, Chapman, *et al.*, 2015, pp. 57–70). Most of the particles are very heavily weathered; as a result only a small sample

has been investigated in detail (Williams, 2014; Smith, Chapman, *et al.*, 2015, p. 57).

The stratigraphy of this site is complex. The wide spread of the metal fragments, together with the fragmentary condition of the bone assemblage referred to earlier, indicates significant site disturbance by later processes, including erosion, small-animal burrowing and road construction (Smith, Chapman, *et al.*, 2015). However, the excavation evidence, combined with the results from replication experiments and analysis of the slag geochemistry, points strongly to the smelting at Pentrwyn of ores from the mine, using a simple low temperature, poorly reducing, charcoal-fired pit furnace process (Williams, 2014, pp. 107–8). Only small amounts of ore could have been smelted at any one time. The tiny dimensions of both pits suggest that they were dug out by an individual using one hand: "any smaller and it becomes difficult to remove the fill" (Chapman and Chapman, 2013, p. 12). Other preparatory steps in the *chaîne opératoire* would have included making a tuyère, perhaps from clay, chopped plant fibres or animal dung, and a blowpipe, probably of wood, as well as collecting kindling and making charcoal. Kindling and charcoal would then have been placed directly into the pit and set alight by blowing through the blowpipe, before sealing the pit using a piece of cut turf. Once the charcoal was red-hot, the practitioner would have inserted a small amount of crushed and concentrated ore into the pit furnace by hand. Ore does not seem to have been processed on site (Smith, Chapman, *et al.*, 2015, p. 69), but carried from the mine and/or one of the nearby springs after crushing and concentrating. Ffynnon Galchog, one of the Great Orme's water sources with possible evidence for ore washing, is only c.400m away, although as discussed in Chapter Two, this site has not yet been dated to the Bronze Age (e.g., Lewis, 1990a; Wager and Ottaway, 2019).

The practitioner could have controlled the furnace temperature and atmosphere by blowing through the blowpipe and by adding small amounts of wet charcoal and charcoal dust to create the necessary reducing atmosphere (Chapman and Chapman, 2013, p. 17). As they tended the furnace, they perhaps leant back to rest against the wall of the cliff immediately behind them (Smith, Chapman, *et al.*, 2015, p. 69). When the smelt was complete – after as little as 1.5 hours – the furnace contents could have been easily removed using a twig or wooden or bone scoop. Washing the metallurgical material in water would have helped to identify the small metallic copper droplets so that they could be picked out by hand. Pieces of slag would have been crushed to retrieve the small prills of copper trapped inside. Replication experiments by David Chapman show that the debris created by this technical sequence – charcoal fragments, ore dropped while charging the furnace, copper droplets and slag

94   COMMUNITY, TECHNOLOGY AND TRADITION

missed during initial hand sorting – litters the furnace area creating a dark, dirty working floor, very similar to the grey silty layer excavated around the Pentrwyn pits (Chapman and Chapman, 2013, pp. 17–19).

Before use, the copper prills would have to be remelted and refined to reduce iron levels and give a larger quantity of metal (Williams, 2014, p. 108). There is no surviving evidence at Pentrwyn for this next step in the copper-making *chaîne opératoire*, which presumably required some form of crucible. In this context, it is interesting to note that a "scrap of bronze" (Davies, 1973, p. 15) and successive charcoal/wood ash layers interpreted as hearths were excavated in a rock shelter at Lloches yr afr, less than 200m to the west of Pentrwyn on the Orme's north-facing cliffs (Davies, 1974, 1989, p. 97) (Figure 1.5). This evidence may point to the practice of metallurgical activity in the rock shelter, possibly remelting, alloying or casting, but it is unfortunately undated. None of the excavated evidence revealed whether the Bronze Age users of the Pentrwyn site erected permanent shelters. Indications of 'domestic' activity, in the form of animal bone, shellfish and cereal grains, point to repeated instances of people eating there while participating in smelting, rather than long-term dwelling (Smith, Chapman, *et al.*, 2015, p. 69).

This reconstructed *chaîne opératoire* indicates that the Later Bronze Age smelting process at Pentrwyn used simple technology, was relatively quick and could have been carried out by a single person. This site also appears to have been used on more than one, probably multiple occasions. The two excavated pits represent a least two smelts as there was insufficient space around them for both to be in use at the same time (Smith, Chapman, *et al.*, 2015, p. 69). Each pit could also have been used more than once. One was sealed by a discrete light-coloured deposit, possibly the remains of a turf cap replaced over the furnace at the end of the smelt (Chapman and Chapman, 2013, p. 19), perhaps in readiness for its next use. Close by are the badly damaged remains of two more possible furnace pits (Chapman and Chapman, 2013, p. 23). There may have been more as only an estimated 10–20% of the original terrace has survived after road building in the 19th century AD. Additional similar small-scale smelting episodes could have taken place on the continuation of the Pentrwyn terrace to the north, which has not yet been excavated (Chapman and Chapman, 2013, p. 3; Smith, Chapman, *et al.*, 2015, p. 69).

As we saw in Chapter Two, it is an open question whether the Pentrwyn evidence is representative of the *chaîne opératoire* of smelting during the period of customary Bronze Age mine-working on the Great Orme, as the dates associated with one of the pits (feature 111) point to its use during a single, brief episode sometime during the 10th–9th centuries cal BC, later than the modelled dated mining activity at the mine itself.

Technologically, the simple 'hole-in-the-ground' method used at this site was suited to smelting the copper ores extracted at the source throughout the Bronze Age life of the mine (Williams, 2014, pp. 108–9). While the scale of smelting is likely to have varied over this period of c.400–800 years, perhaps this very small-scale process was one of the primary methods for smelting the Great Orme ore? In this scenario, smelting on the headland was typically carried out non-intensively, on a regular and repeated basis. Unfolding alongside countless small-scale mining events, it may have been repeated many hundreds – or thousands – of times, at Pentrwyn as well as perhaps elsewhere on the headland and its hinterland, or even further afield. The archaeological traces of even very large numbers of these small pit furnaces would be ephemeral, perhaps explaining why Pentrwyn is the only site for smelting oxidised copper ores dating to the Bronze Age to have been discovered so far in the British Isles and Ireland. At Ross Island, in south-west Ireland, the only other known smelting site from this region of Bronze Age date (c.2400–1900 cal BC), a different ore type, arsenic-rich tennantite (fahlore) was smelted (O'Brien, 2004, pp. 466–72).

There is no definite trace of large, fixed smelting furnace installations of Bronze Age date on the Great Orme, although these may eventually be revealed through any future programme of extensive surface excavation of the mine surroundings in the Pyllau Valley. Closer to the summit, a geochemical survey of the Great Orme using a portable field X-Ray Fluorescence Analyser has identified a discrete copper anomaly. This may indicate a buried smelting site (Jenkins, Owen and Lewis, 2001, p. 168). Its location on a slope exposed to the strong prevailing winds would be better suited to wind-assisted, forced-draught smelting furnaces like the Early Bronze Age examples at Feinan, Jordan (e.g., Ottaway 2001, 94), rather than pit furnaces of the type found at Pentrwyn. Alternatively, as only a handful of ore could be smelted in each of the Pentrwyn furnaces at any one time, it has been suggested that this site was used primarily for trial or assay smelts, to test a handful or two of ore from the mine for its quality and/or the presence of copper (Chapman and Chapman, 2013, p. 23; Smith, Chapman, *et al.*, 2015, p. 69). Larger amounts of concentrated ore could then have been turned into metal in bulk smelts in larger furnaces somewhere else, possibly even a sea journey away via the small natural harbour close to Pigeons' Cave on the headland's north coast (Figure 3.8). The location of these potential large-scale smelting sites is not known, but it could have been close to areas with more extensive and varied tree cover than likely to have been present on the Great Orme.

The findings from Pentrwyn hint that, sometime in the early first millennium BC, people made recurring

Figure 3.8: Small natural harbour at Pigeons' Cave on the Great Orme's north coast. The eponymous hoard (HER PRN GAT4577 [accessed online 22-07-2019]) was found somewhere in the general location of this harbour and its adjacent cave (not shown) (photograph author's own).

visits there to smelt small quantities of ore from the mine, either as part of a *chaîne opératoire* of ore testing and/or as the first step in the production sequence for turning ore into metal – perhaps some of which was then being used to make tools for use in the mine. It is not possible to determine from the available evidence, however, whether all the instances of smelting at Pentrwyn occurred within a single season or whether each event took place decades – or even a century or more – apart. As there is no overlap between the modelled radiocarbon dates from the mine and those from the smelting site, we also cannot say with certainty whether the excavated evidence from Pentrwyn implies contemporary mining in the Pyllau Valley or the smelting of ore recovered by scavenging in spoil deposits of earlier date. The former scenario would suggest that, at least in the latter part of the Later Bronze Age, the act of smelting was closely embedded within the mining field of activity on the Great Orme, as a concurrent rather than 'down-the-line' activity, and so may have been undertaken by the same people. The latter model might fit better with the notion of assay smelting, to assess the copper content of the scavenged ore, but such a task could also have been undertaken at the start of a mining event, as well as during, as the miners accessed new areas of the mineral deposit.

Finally, why did the Later Bronze Age workers make the technological choice to use the terrace at Pentrwyn, more than 1km from the mine, for smelting, even if they only ever transported small amounts of ore there at any one time? Elsewhere in Europe, Bronze Age smelting evidence is often found either within or very close to the mine workings, as is the case at the Spanish copper mines of Chinflon and Cuchillares, as well as at the mines at Cabrières (France) and Ross Island (Ireland). There is no comparable excavated evidence from the Great Orme. At other sites, such as on the Mitterberg and in the Mauken Valley (Austria), the tasks of mining and smelting were spatially separated, but this choice appears to have been made for practical reasons, such as the availability of wood for fuel (O'Brien, 2015, pp. 229–30). Could the Later Bronze Age workers have selected Pentrwyn due to its east-facing position below a cliff, which provided shelter from the prevailing westerly wind (Smith, Chapman, *et al.*, 2015, p. 54)? Was it near to (currently unidentified) sources of kindling and charcoal? Or did its secluded location meet a need for privacy and secrecy during smelting? This idea will be revisited in Chapter Five.

The technology of mining: conclusions

When considered as a single assemblage, the composition of both the cobblestone and faunal tool finds from the Great Orme mine is remarkably uniform. This points to highly standardised and carefully curated methods of selecting, making, using and depositing bone and cobblestone tools, which changed little over the four to eight centuries of customary Bronze Age mining at the source (Wager, 2002, pp. 368–70; James, 2011, p. 342). A recent study of Bronze Age smithing tools has shown how tool types do not evolve in a direct evolutionary sequence from 'simple' to complex'. Instead, people adapt 'old' tools for use with new techniques, putting those tools aside or readapting them for new uses only when those techniques stop being used (Fregni, 2019, p. 41). This may help to explain the longevity of the use of cobblestone tools (as well as those of bone and perhaps also wood) at the Great Orme mine for mining and processing ore.

It has been suggested that the period post-1300 BC was characterised by a 'twilight' of low ore production following a 'golden age' of extraction (Williams and Le Carlier de Veslud, 2019, p. 1192). There is, however, no evidence from the tool finds that activity at the source was characterised by any decline in the efficacy and proficiency of mining practices during the latter part of the Bronze Age life of the mine. Rather, extracting ore from the kilometres of narrow veins threading through the mineralised zone would have required a sustained and high level of technical skill, material know-how, planning and commitment to mining. Despite its consistency

on a macro scale, as the mining field of activity on the Great Orme unfolded piecemeal over at least 400 years, there would have been myriad small-scale variations in practices from one part of the source to another and from one mining event to the next, as the miners made technological choices about how to proceed. These would have been informed not just by the physical conditions they encountered but by changing – and sometimes conflicting – ideas about the self, skill and social relations. These ideas will be the topic of Chapter Five.

Reconstructing the *chaînes opératoires* of toolmaking and repair, ore extraction and processing, spoil disposal and mine organisation reveals the ways in which the Bronze Age mining field of activity at the source was in general carefully choreographed. These tasks were carefully planned and performed on several spatial and temporal scales. They typically involved purposeful selection and use of the available materials. Some of these choices were grounded in utilitarian considerations, such as the physical properties of bone as a material, while others probably arose out of the miners' culturally situated understandings of the world they inhabited.

The geological setting of the mineral deposit, with areas of ore-bearing ground separated by hard ribs of bedrock, physically defined discrete locations of working (Figure 3.6). Within these bounded spaces, the technical sequence of spoil disposal as well as, for example, roof support and water drainage, must have been structured to facilitate ongoing extraction during a given mining event. Instances of flooding or roof collapse could on occasion have blocked routes of access, making it difficult or even physically impossible for the Bronze Age miners to re-enter certain parts of the source to resume extraction. Although the geological occurrence of the copper dictated the progress of mining to a considerable extent, transformations in the layout of the workings caused by flood or collapse events would have distorted the miners' 'mental map' of the mine. This would have led to the making of different choices about where and how to 'follow the ore', contributing to the piecemeal patterning of activity observed across the source. Tasks such as ore processing may also have unfolded in one or more spatially distinct areas, from immediately outside the mine entrances, to the nearby water source of Ffynnon Rhufeinig, as well as possibly at others on the headland.

The mining field of activity involved people working together. The physical layout of the workings would have impacted on how these interactions could have been configured. Some tasks would have involved the shared participation and collaboration of all those present at the source, while other *chaînes opératoires* were restricted to only a few or, like choices about where to place unwanted spoil, were potential arenas for conflict and social differentiation. The miners needed considerable technical skill and knowledge of the physical structure of the mineral deposit and the spatial layout of the mine to overcome the practical challenges of extracting ore from the mineralised zone. This required tremendous physical strength, stamina and flexibility, as well as – certainly from our modern perspective – the physical and perhaps also conceptual courage to venture into and work in dark, confined spaces underground, particularly in locations at significant distances from the surface (e.g., Locs. 33–35) (Figure 2.33). These ideas will all be explored in Chapter Five.

Throughout the Bronze Age, it seems probable that the miners interacted with the Great Orme's landscape and its materials to carry out the tasks involved in the mining field of activity. From possible use of the natural spring at Ffynnon Rhufeinig for ore washing, to potential smelting at secluded Pentrwyn, the steps in the metal production sequence may have been dispersed across the landscape, at least on occasion. The relevance of the Later Bronze Age smelting site at Pentrwyn to the customary period of Bronze Age mine-working is uncertain, but it certainly indicates small-scale, non-intensive activity. This may have been embedded in the mining field of activity and so could have been carried out by the miners themselves. It used simple technology that could have been easily replicated elsewhere. Whether tasks such as processing and smelting were ever centralised in a large-scale, 'permanent' 'work camp' is likely to remain an open question given that, as we have seen, the record of surface activities associated with Bronze Age activity at this mine is very poor, primarily due to the accumulations of spoil created by early/late modern mining. While it is probable that 'work camps' of varying size, form and duration existed at various times over the many centuries of mining on this headland, the best evidence for their character and configuration perhaps comes from the wider context of copper and bronze production in north Wales. We will return to this issue in the next chapter.

The *chaînes opératoires* of the mining field of activity were tied into other fields of activity in the broader taskscape occurring at other times and places, particularly those concerned with animals and the land. Cobblestones suitable for mining could, for example, have been picked up while fishing, foraging for shellfish or even collecting clay to make pottery. Decisions about how to proceed would have involved contact and communication with others, both at the mine and at the loci of interaction with these other fields of activity. It is by investigating how mining was enacted as part of this culturally and historically situated taskspace that we can begin to understand the ways in which it was not just a technical process but also a meaningful, dynamic and political "total social fact" (Mauss 1979 [1935]). The character, scale and patterning of the taskscape of north Wales from the Chalcolithic to the Earliest Iron Age is the subject of the next chapter.

4

Mining and the taskscape

The discussion so far has focused on ore extraction and related activities on the Great Orme itself during the second and early first millennia BC. As we saw in the previous chapter, the mining field of activity at this source is, however, likely to have been closely enmeshed with other fields of activity. In this chapter, the geographical scope expands to encompass the Great Orme's hinterland and the rest of north Wales, to consider the evidence for the character, scale and patterning of several other strands of the regional taskscape. The discussion opens by examining the evidence for the type and distribution of settlement and monuments in this area and how these changed over time. Here, as throughout the chapter, the period of interest is the Chalcolithic to the end of the Earliest Iron Age (c.2500–600 BC), an expanded time frame intended to situate the customary phase of dated Bronze Age activity at the Great Orme mine as part of a historical trajectory of action. It will become clear that the pattern of dwelling – in the sense of people's practical and sensory engagement with and experience of the world (Ingold, 1993b, p. 155) – on the Great Orme headland is typical of the character and scale of contemporary lifeways in north Wales. Throughout the second millennium BC, communities in this region seem to have been essentially small-scale, and the social and material conditions of life fairly scattered, while becoming gradually less dispersed by the first half of the first millennium BC, at least in some areas. The significance of this conclusion for interpreting the mining field of activity at the Great Orme mine will be considered.

As the *chaînes opératoires* of agriculture appear to have been pivotal to the practical and social dynamics of mining at the Great Orme mine, the discussion then considers this aspect of the taskscape and its entanglement with the routines of labour at this source in more detail. The patterning, character and tempo of land use across north Wales, and how these may have shifted with changes in climate towards the end of the Bronze Age, are examined. Analogy with more recent 'small-scale societies' – those composed of small, perhaps dispersed communities – reinforces the idea that tending to both animals and the land is likely to have been as fundamental to the craft and rhythm of the taskscape in north Wales throughout the second millennium BC as it was in the preceding millennium (Edmonds, 1999a, p. 16, 1999b, p. 487). Replication experiments show that small-scale hand cultivation of even only a few small fields demands relentless, labour-intensive toil (Reynolds, 2000, p. 95).

Mobility, or the movement of people, things and ideas, is the next component of the taskscape to be examined, one that is fundamental for understanding the configuration, character and scale of the networks of relations within which mine-working at the Great Orme mine took place. The discussion addresses the question of how Bronze Age mining on the headland articulated with the circulation of copper and bronze, tin and lead within and between communities across north Wales and beyond. The concluding sections of this chapter review the evidence for the content, scale and tempo of copper and bronze working and recycling in the region in the second and early first millennia BC, within the framework of the character of the wider taskscape developed throughout the chapter. It will be seen that contemporary copper and bronze working in north Wales

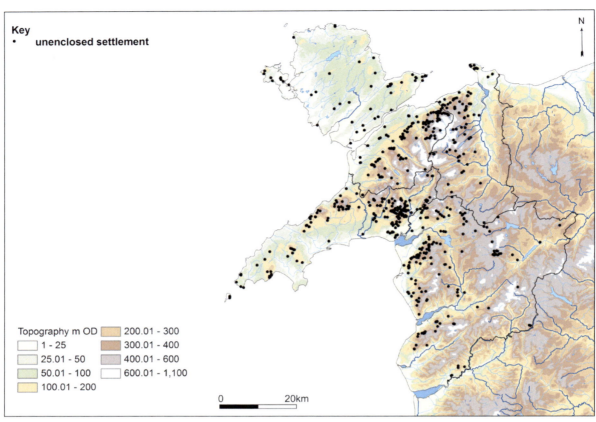

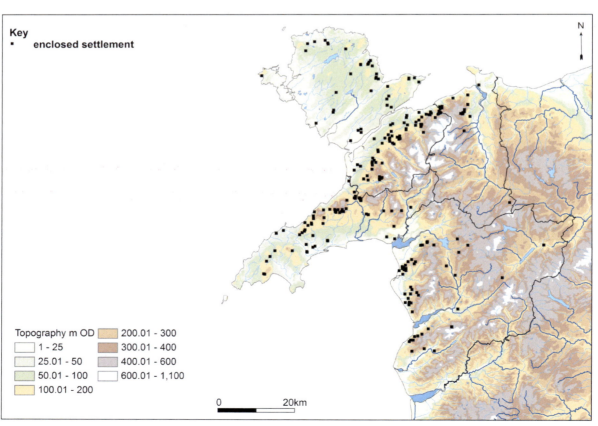

Figure 4.1: Distribution of unenclosed (top image) and enclosed roundhouse settlements in north-west Wales (after Waddington, 2013, plates 3.5, 3.6).

can be interpreted as a typically small-scale, community-organised, ideologically significant activity carried out by people for whom metalworking was only one of numerous skilled tasks (Webley, Adams and Brück, 2020). The implications of this conclusion for understanding how the prehistoric mining field of activity on the Great Orme interlocked with the other stages in the local copper and bronze production sequence will be assessed.

Settlements and monuments: the scale and patterning of community life

Both isolated and scattered single unenclosed stone-built roundhouses are a striking feature of the surviving prehistoric landscapes of north Wales. They occur in large numbers particularly in upland areas – defined as all land above the 244m contour (Silvester, 2003, p. 3) – although rarely above 500m OD. There are noticeable concentrations of these settlement types in the hilly areas on the western fringes of Snowdonia and in the northern foothills along the coast (Waddington, 2013, pp. 27, 68, 71–2) (Figure 4.1). Numerous examples have been identified in the hinterland of the Great Orme, including in the Lower Conwy Valley and the uplands of North Arllechwedd (Gwyn and Thompson, 1999). There are also a number of poorly preserved stone-built probable roundhouses on the Great Orme itself (RCAHMW, 1956, pp. 114–5; Bibby, 1984; Waddington, 2013, pp. 262–3) (Figures 1.4–1.6). Like the Great Orme examples, most such settlements in north Wales are undated. There are only a handful of dwellings of Chalcolithic/Earlier Bronze Age date or earlier. The Late Neolithic examples at Trelystan (Welsh Marches), Cefn Caer Euni (Gwynedd) and Llyn Morwynion (in the uplands of Ffestiniog on the southern fringes of Snowdonia) were all small single unenclosed

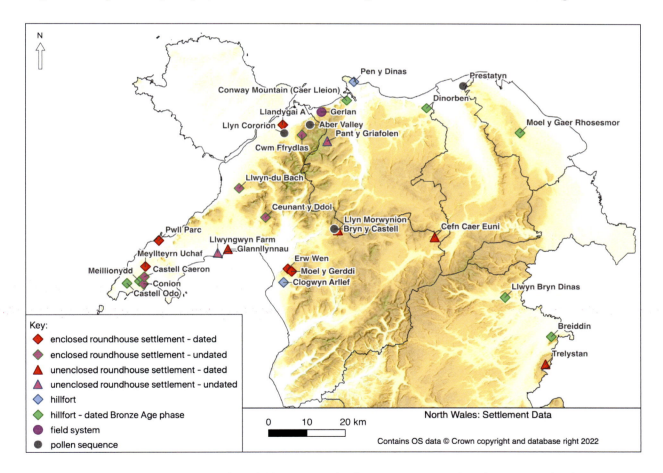

Figure 4.2: Distribution of settlements, hillforts, field systems and pollen sequences in north Wales mentioned in the text. Unenclosed roundhouse settlement – both isolated or scattered single roundhouses; enclosed roundhouse settlement – single or double-walled circular or circular concentric enclosures; hillfort types – palisaded double ringwork enclosures, single-walled enclosures or palisaded enclosures. Dated – either by scientific dating methods or association with diagnostic finds, e.g., pottery. For similar sites on the Great Orme and in its hinterland, see Figures 1.4 and 1.5.

MINING AND THE TASKSCAPE

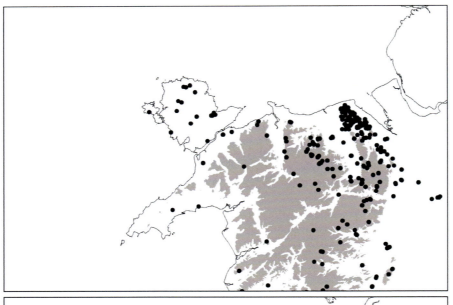
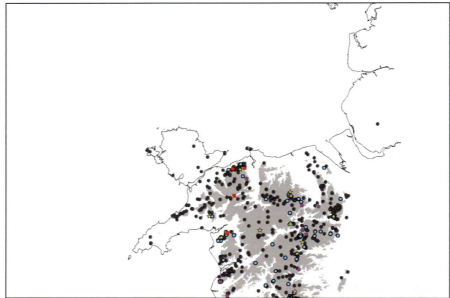

Figure 4.3: Distribution of round barrows (upper image) and burial cairns (lower image) in north Wales (land above 230m shaded). Round barrows recorded in upland areas are probably turf-covered round cairns (after Tellier, 2018, figs. 13 and 14).

Legend
- Round cairns
- Cairn circles
- Kerb circles
- Kerb cairns
- Ring cairns
- Platform cairns

stone or timber roundhouses, only c.3.5–4m in diameter (Britnell, *et al.*, 1982; Lynch, 1986; Waddington, 2013, pp. 89–90). There are also the scant remains of a small unenclosed settlement complex with a possible timber-built roundhouse at Glannllynnau in the Llŷn Peninsula. Radiocarbon dating of charred wood from a pit at this site indicates occupation potentially beginning towards the very end of the Earlier Bronze Age (BETA-204433; 1265–935 cal BC; 95% probability) (Davidson, Smith and Roberts, 2007). Pottery from another small unenclosed settlement on the Llŷn Peninsula, at Llwyngwyn Farm, possibly also points to Bronze Age occupation sometime in the last few centuries of the second millennium BC (Waddington, 2013, pp. 239–41) (Figure 4.2).

Although the roundhouse in Wales is a settlement form in use primarily in the first millennia BC/AD (Ghey, *et al.*, 2007), analogy with similar structures elsewhere in Britain suggests that small size could be characteristic of Earlier Bronze Age settlement. The numerous small single roundhouses dispersed across the uplands – such as the scattering of 28 stone dwellings, 3–5m in diameter, at Pant y Griafolen (Lower Conwy Valley, Great Orme hinterland), or the roundhouses, 3–6m across, at Cwm Ffrydlas (Ogwen Valley, north Snowdonia) and Ceunant y Ddol (Llŷn Peninsula) – could all have been inhabited from at least the mid-second millennium BC (Gwyn and Thompson, 1999, p. 32; Smith, 1999; Waddington, 2013, p. 90) (Figure 4.2). However, despite the apparently dense distribution of isolated and scattered single roundhouses at places such as Pant y Griafolen, these structures were not part of large, directly contemporary settlement conglomerations. They were instead situated within what was probably still a fairly dispersed pattern of dwelling during the Chalcolithic and Earlier Bronze Age. Their small size compared to earlier Neolithic rectangular buildings suggests a transformation in function and/or occupation by smaller, less extended family groups. At Trelystan, the hearth in each small wattle roundhouse took up much of the interior floor space, limiting both the types of activity that could have been undertaken and the number of participants (Lynch, 2000, p. 85).

The distribution of contemporary monuments, which are scattered across the region in both hilly and low-lying areas, further hints at the patterning of Earlier Bronze Age settlement (Waddington, 2013, p. 85). Architecturally, these monuments are typically earthen turf-covered round barrows, simple circular stone cairns, burial pits and cists, and structurally more complex features, including stone cairn circles, ring cairns and platform cairns (Lynch, 2000, pp. 127–8). Burial cairns of various types predominate across the interior uplands, in eastern and central Snowdonia for example, while round barrows are located mostly on the upland fringes or in the lowlands (Figure 4.3). Dense concentrations of such funerary monuments can be seen in the Vale of Clwyd, along the south-eastern edges of the Clwydian Hills and in North Arllechwedd near the Great Orme, in the hills of Penmaenmawr. There are also possible burial monuments on the Great Orme itself (RCAHMW, 1956, p. 116 no.377; Tellier, 2018, p. 26, figs. 13, 14) (Figure 4.4). A small number of sites containing multiple cremation deposits have been uncovered in the lowlands, such as on the island of Anglesey at Bedd Branwen and Treiorwerth, and in the Welsh Marches at Trelystan (e.g., Tellier, 2018, pp. 28–33) (Figure 4.5).

Upland burial structures often occur in clusters of other monuments of similar age, like standing stones, stone circles and embanked stone circles. In the hinterland of the Great Orme, for example, the burial mounds on Penmaenmawr headland are part of a remarkable prehistoric landscape of tightly grouped features including an unusually impressive embanked stone circle (known as the Druid's Circle), cairns, stone circles and standing stones. Nearby, in the hills of the Lower Conwy Valley, there is a complex of Neolithic and Bronze Age monuments including the Maen-y-Bardd burial chamber, a stone circle, cairn and standing stones (Anon, 1998, pp. 82–3, 114–5; Gwyn and Thompson, 1999, p. 36). Concentrations of stone circles occur in the mountain mass of Snowdonia (e.g., Lynch, 1984, 1986). Other monument types, like Late Neolithic/Chalcolithic henges, smaller banked ditches, free-standing timber rings and cursus monuments, such as the Sarn y Bryn Caled cursus running through the Severn Valley south of Welshpool, are usually situated on flat land in valley bottoms (Lynch, 2000, pp. 129–31, fig. 3.18) (Figure 4.5).

There are challenges in interpreting the scale and intensity of the Chalcolithic and Earlier Bronze Age fields of activity producing these different feature types. Their surviving distribution could be misleading, due to subsequent ad hoc removal of stone by local communities; the effects of recent intensive agriculture, settlement and other land uses; and original construction in timber (Hughes, 2003, p. 118; Ghey, et al., 2007). Such factors could explain the limited evidence today for burial and settlement in river valleys and along the low-lying plateaux fringing the coast of Anglesey and the Llŷn Peninsula (Waddington, 2013, p. 68). A mound containing cremated remains, formerly located close to the foot of the Great Orme, was levelled during construction in Llandudno in the late 19th or early 20th century AD (Roberts, 1909). The relatively dense concentrations of burial monuments noticeable in, for example, the Vale of Clwyd could likewise be a product of research and preservation biases. Both burial monuments and roundhouses are hence likely to have been even more widely dispersed than they currently appear – although Waddington (2013, p. 68) notes that the very limited evidence for prehistoric settlement above 500m AOD is most likely representative of past dwelling patterns.

Interpreting the archaeology of the Great Orme described in the opening chapter exemplifies these challenges (Figure 1.5). Although most of these features are not securely dated, the Great Orme was clearly not considered a marginal landscape in the past. Given its long history of land use, there could previously have been many more sites dotting the headland, now destroyed. For example, most of the recorded sites and finds on the Great Orme lie outside areas of enclosed farmland. In aerial photographs of the Parc Farm taken just after the Second World War, extensive earlier field systems can be seen within its perimeter. These are much less evident in later photographs, most likely due to the impact of recent intensive agriculture and land improvement (Hopewell, 2013b, p. 187). Geoenvironmental evidence also shows that serious floods and accompanying massive landslides have occurred periodically on the headland over many millennia, most recently in 1993 (Wood and Campbell, 1995). The erosion caused by such events may have affected the survival and visibility of the archaeological material connected to the mine, including any nearby encampments and other activity areas for metal production.

Viewed from a different perspective, however, the Great Orme's surviving archaeology is still very rich and varied. The intensity of land use on the headland in both the past and the present is also little different to that experienced elsewhere in lower-lying areas of north

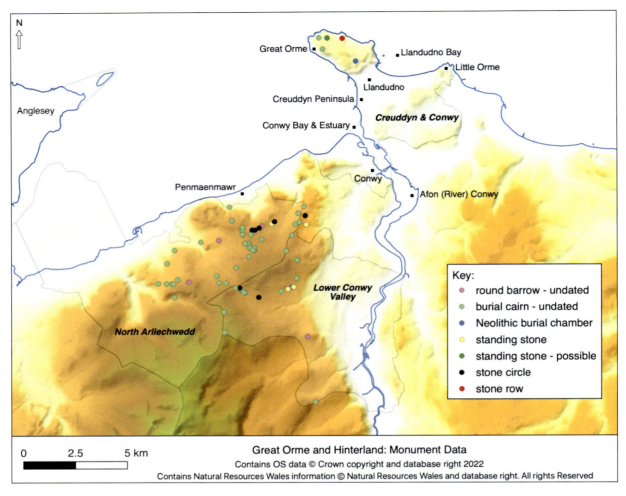
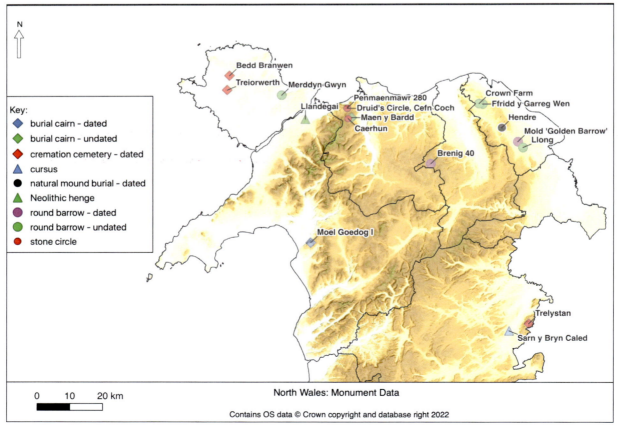

Figure 4.4: Distribution of funerary and other monuments on the Great Orme and in its hinterland. Grey-shaded areas (labelled in italics) represent the hinterland areas defined in the Landscapes of Historic Interest in Wales, Part 2.1 (Anon, 1998). Round barrow – earthen turf-covered mound or cairn; burial cairn – round cairn, cairn circle, kerb circle, kerb cairn, ring cairn, platform cairn; stone circle – includes embanked stone circles.

Figure 4.5: Distribution of funerary and other monuments in north Wales mentioned in the text. Round barrow – earthen turf-covered mound or cairn; burial cairn – round cairn, cairn circle, kerb circle, kerb cairn, ring cairn, platform cairn; stone circle – includes embanked stone circles. Dated – either by scientific dating methods or association with diagnostic finds. For approximate location of burial cairns at Llanllechid and Llanelian, and the Darowen round barrow, see Figure 1.2. For similar sites on the Great Orme and in its hinterland, see Figure 4.4.

Wales. Erosion and practices such as land improvement and ploughing can destroy sites, but they also play an important role in the recovery of items such as metalwork. Elsewhere in north Wales, developer-funded excavation in advance of infrastructure improvements, for example, is contributing to the recovery of increasing numbers of timber dwellings (Waddington, 2013, pp. 27, 77), albeit still within an overall pattern of relatively sparse occupation. The archaeology of the Great Orme has, since at least the late 18th century AD, received a great deal of attention from travellers, antiquarians and archaeologists. This interest has been intensive if not detailed and systematic. If the range and density of prehistoric sites and finds on the headland were originally much greater than is indicated by surviving examples, some evidence for this would by now most likely have been identified. The caveat to this is, of course, if such features were previously present in areas that became the focus for subsequent mining and spoil accumulation, either later in the Bronze Age or during the early/late modern period. In this case, it is probable that such sites will never be found. It is, however, also possible to suggest from the preceding observations that the absence of evidence for extensive large-scale settlement or production sites on the headland and in its hinterland is due to the actual absence of these features. To move the argument forward, this is the perspective adopted over the following chapters: that the Great Orme's surviving archaeology is probably reasonably representative of the character, scale and patterning of activity around the copper source not only during the Chalcolithic and Earlier Bronze Age, but throughout the second and into the early first millennium BC. It is certainly noteworthy that the prehistoric character of the Great Orme complements that seen in its hinterland

and throughout the rest of north Wales (Conwy County Borough Council, 2011, p. 35) (e.g., Figures 1.6, 4.4).

Taken together, the evidence from the wider region points to the use of a broad range of landscapes across north Wales in the Chalcolithic and Earlier Bronze Age, in a pattern of relatively dispersed fields of activity. The taskscape was, however, slowly transforming during this period. During the first half of the second millennium BC, people gradually built fewer burial monuments and other types of circular monument. The small, c.4m-wide stone circle known as Penmaenmawr 280 (HER PRN GAT544 [accessed online 18-06-2022]), in the hills of Penmaenmawr in the Great Orme's hinterland, is the only circular enclosure monument in Wales that has been dated to or after the 17th century cal BC (Figure 4.5). This is by association with a nearby firepit with which it is assumed to be contemporary (NPL-12; 3080±145 BP; 1670–925 cal BC; 95% probability) (Tellier, 2018, pp. 44–5). Another indication of a gradual transformation in patterns of dwelling towards the end of the second millennium BC, from unenclosed to enclosed settlements, comes from Mellteyrn Uchaf on the Llŷn Peninsula (Ward and Smith, 2001). Sometime around 1200 BC – and at the same time as the customary Bronze Age phase of mine-working on the Great Orme was ongoing – people constructed and inhabited a roundhouse encircled by a large double-ditched enclosure on a slight hill promontory overlooking a river valley, on freely drained brown earth soils. This is the earliest securely dated enclosed settlement in north Wales (Waddington, 2013, pp. 91, 231–32).

The evidence for settlement in this region is, however, still relatively limited into the Later Bronze Age (Waddington, 2013, p. 5). By this period the architecture of settlements had largely replaced that of monuments as a permanent feature of the landscape (Barrett, 1999, p. 497). People were no longer building Earlier Bronze Age monument types, such as stone circles, embanked stone circles and ring cairns (Tellier, 2018, p. 121). They only rarely chose to construct new burial mounds, like those at Llanllechid near Bangor and Llanelian further east towards Colwyn Bay (Davies and Lynch, 2000, pp. 211–12) (Figure 1.2). Although communities in Later Bronze Age north Wales continued to know and use a broad range of landscapes, a new pattern of inhabitation emerged, with less dispersed settlement reorientated away from the mountainous interior to the foothills and low-lying areas, including river valleys, edging the coast. The character of settlement was also changing, with construction of circular and circular concentric enclosures and the earliest hillforts, as part of a trend for larger, more nucleated and monumental dwellings that becomes more pronounced throughout the first millennium BC (Waddington, 2013, pp. 60, 85). In some cases, settlements reused and modified earlier henge monuments, such as Late Neolithic Henge A at Llandegai, which became the site of a settlement occupied sometime between the 9th to the 6th century cal BC at the Later Bronze Age/Earliest Iron Age transition (Figure 4.5). At other sites, like the possible Later Bronze Age enclosure at Pwll Parc on the Llŷn Peninsula, the form and layout of these emerging settlements appear similar to earlier henge architecture (Waddington, 2013, p. 92) (Figure 4.2). Other examples, such as the bivallate enclosure at Llwyn-du Bach in a coastal setting in north-west Gwynedd (Waddington, 2013, p. 49), continued the tradition of building double-ditched concentric enclosures first seen at Mellteyrn Uchaf from around 1200 BC. Smaller single-walled circular concentric enclosures, like those at Erw Wen and Moel y Gerddi on the edge of the uplands in the Rhinog Mountains, western Gwynedd, also emerge at the beginning of the first millennium BC (Waddington, 2013, pp. 248–50), at the very end of or even post-dating the customary Bronze Age phase of mine-working at the Great Orme mine: the period when people regularly and repeatedly worked there. As we saw in Chapter Two, this most probably ended shortly before 900 cal BC.

At both Moel y Gerddi and Erw Wen, an earlier timber version was consolidated by rebuilding in stone (Kelly, 1988, pp. 130–2). Other indications of monumentality include the formal entranceways found in Later Bronze Age enclosed settlements. The multiple roundhouse sequences identified by excavation at such sites indicate that they were a focus for dwelling for long periods of time, throughout the first millennium BC. Although the communities occupying these structures are still relatively small-scale, larger numbers of people could potentially have come together to take part in the fields of activity occurring within these enclosures than in the Earlier Bronze Age. Although only one or two roundhouses were typically enclosed, their size increases over time, from c.4–6m in diameter for the roundhouse at Mellteyrn Uchaf to c.9–10m for the central roundhouses at Moel y Gerddi and Erw Wen (Waddington, 2013, p. 91). Waddington (2013, p. 92) suggests that up to 80 people could have been seated comfortably in the 10m-diameter roundhouse at Llandygai (Figure 4.2). Such sites were presumably both inhabited and used for large gatherings.

The appearance of hillforts in prominent raised settings at the beginning of the first millennium BC primarily in north-east Wales, the Welsh Marches and on the Llŷn Peninsula also shows that the taskscape was becoming less dispersed, at least in some parts of the region. Later Bronze Age hillforts are known from other parts of Britain but, as Waddington (2013, p. 17) points out, the clustering of sites observed in north Wales is unusual. The earliest examples, such as the small (no more than 1ha.) palisaded double ringwork enclosures at Castell Odo and Meillionydd, both on the Llŷn Peninsula,

were built sometime in the 9th/8th century cal BC, at the Later Bronze Age/Earliest Iron Age transition (Figure 4.2). They were morphologically like the circular and circular concentric enclosed settlements that were appearing in the landscape at the same time, although somewhat larger: the hillfort enclosure at Meillionydd, for example, was c.100m across and contained several roundhouses. Single-walled forts of similar size in other areas, such as Clogwyn Arleff, south-west Gwynedd, could also be of similar date (Waddington, 2013, pp. 88, 93).

Some of the earliest hillforts, often in timber, would go on to become large, monumental, stone-built features of the Middle Iron Age landscape in north-east Wales and the Welsh Marches: Moel y Gaer Rhosesmor, Flintshire; Llwyn Bryn Dinas, a hilltop commanding a spectacular view over the Tanat Valley in northern Powys; and at the Breiddin in the Welsh Marches (e.g., Davies and Lynch, 2000, p. 150; Waddington, 2013, p. 17). Closer to the Great Orme, to the east of the Creuddyn Peninsula at Dinorben, construction began on a timber palisade and bank (Figure 4.2). The date of this earliest phase of construction is uncertain. While erection of an initial timber palisade may have begun as early as the 9th/8th century BC (e.g., Waddington, 2013, p. 94), Guilbert (2018) has recently argued that the dating evidence is too imprecise for any activity earlier than c.800 BC to be identified. There is no clear association between the date of deposition of a Later Bronze Age hoard of horse harness fittings outside the western ramparts (the Parc-y-meirch hoard) (HER PRN CPAT101999 [accessed online 12-03-20]), probably sometime in the 9th century BC, and construction of the palisade. Unfortunately, as the hillfort has now been destroyed by quarrying, no better evidence is likely to be forthcoming to resolve this issue (Figure 4.6).

Some three centuries or so after the modelled probable end in the mid-10th century cal BC (at 68% probability) of the customary Bronze Age phase of mining at the Great Orme mine, people in the hinterland of the headland started building the earliest phases of a hillfort on Conwy Mountain, also known as Conway Mountain, Castell Caer Lleion or Caer Seion, on a coastal ridge at the mouth of the Conwy Estuary (Smith, *et al.*, 2012, pp. 8–13; Waddington, 2013, p. 40) (Figure 4.2). Even closer to the mine, activity at Pen y Dinas hillfort on the Great Orme has recently been placed in the 4th/3rd century BC, in the Middle Iron Age (BETA-254961; 2270±40 BP; 400–205 cal BC; 95% probability) (Smith, *et al.*, 2012, p. 8) (Figures 1.5, 4.7).

A ceramic fragment from this site, which had previously been tentatively identified as a sherd of Middle Bronze Age to Middle Iron Age pottery (Wager, 2002, p. 227), has recently been reidentified as a piece of fired daub, of probable Middle Iron Age date (Smith and Jacques, 2009, table 1, find number 38). The significance

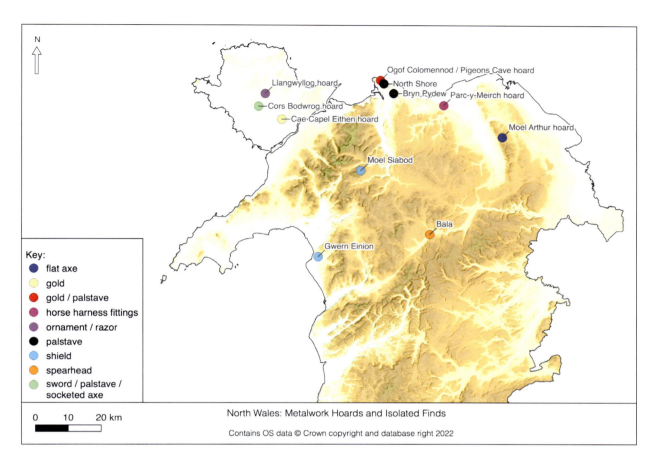

Figure 4.6: Distribution of hoards and isolated metalwork finds in north Wales mentioned in the text. For the distribution of similar finds on and around the Great Orme, see Figures 1.4 and 1.5.

of this hillfort's location close to the copper mine will be returned to in Chapter Five.

To conclude, the monument and settlement evidence indicates a gradual transformation in the scale and patterning of the taskscape from the Chalcolithic to the end of the Earliest Iron Age in north Wales. The transition from the Earlier to the Later Bronze in this region does not represent a 'watershed' marked by abrupt changes in ways of dwelling, although there are evident distinctive – and sometimes subtle – regional variations. The Great Orme and its hinterland, for example, were not a focus for the development of distinct building traditions at the end of the second millennium BC, unlike the western tip of the Llŷn Peninsula, western Anglesey and Ardudwy in south-west Gwynedd, where the construction of enclosures seems to have been an important field of activity, occupying time, effort and materials (Waddington, 2013, pp. 81, 85–6). Neither the beginning nor the end of the customary Bronze Age phase of mining-working at the Great Orme mine corresponds to a sudden or major change in the scale and patterning of community life. Of the known Later Bronze Age/Earliest Iron Age hillforts and enclosed settlements in north Wales, only the hillforts at Conwy Mountain and Dinorben were constructed in the hinterland of the mine. (Although Dinorben is outside the 'hinterland' area defined in this study, being further east, its near location to the Creuddyn Peninsula along the north Wales coast indicates that it should also be considered part of the headland's 'local' context.) There is no chronological overlap between the customary phase of copper mining at this source and the development of these structures, nor is the Great Orme headland, with the mine as a prominent landscape feature, a focus for hillfort building activity. Rather, the mining field of activity on the Great Orme was one of the many strands of the taskscape that created the material and social conditions out of which enclosed settlements and hillforts emerged in the latter part of the second millennium BC. This gradual process of transformation eventually led to a clustering of medium and large hillforts along the north Wales coastline by the Middle Iron Age.

MINING AND THE TASKSCAPE 107

Figure 4.7: Pen y Dinas hillfort, Great Orme, view facing south towards the hillfort (centre of image) (photograph author's own).

Tending animals and the land: issues of tempo and temporality

What do we know about the landscapes in which the settlements and monuments discussed so far were situated? It is probable that forest cover in north Wales reached its maximum extent c.4500 BC. While the landscapes of the highest mountain ridges and plateaux with naturally thin soils would always have been fairly bare, valley bottoms and hill slopes were most likely dominated by open oak and hazel woodland, perhaps up to a height of at least 750m OD where the latter were well drained (Smith, 1999, p. 46; Lynch, 2000, p. 81). The soils under this deciduous forest would have been fertile and suitable for arable and pastoral farming, although prone to acidification, deterioration and eventual peat growth in areas of acid parent rock. Scattered grassy clearings would have punctuated the woodland canopy, due to the nature of forest growth and the intentional clearing and maintenance of open ground and the forest margins (Barrett, 1999, p. 496; Barnatt and Smith, 2004, pp. 15–16). Environmental studies from areas of both lower and higher altitude across north Wales point to expanding woodland clearance and the emergence of substantial openings in the forest cover from the Later Mesolithic onwards (e.g., Warrilow, et al., 1986, p. 56; Caseldine, 1990, pp. 35–6; Musson, et al., 1991, p. 173; Mighall and Chambers, 1995, pp. 318–9; Smith, 1999, p. 46; Watkins, Scourse and Allen, 2007). Throughout the second and into the first millennium BC, the general regional picture obtained from pollen records is of an ongoing reduction in tree cover, accompanied by increased soil acidification and the beginnings of valley mires, which eventually created an open landscape. Due to the topographic and vegetational diversity of north Wales, however, there are local variations in the timing and nature of the impacts causing these changes (Mighall and Chambers, 1995, p. 319; Watkins, Scourse and Allen, 2007, p. 171).

The decrease in tree pollen accompanied by a rise in arable grasses associated with the Earlier Bronze Age enclosed roundhouse at the lowland site of Meyllteyrn Uchaf on the Llŷn Peninsula (95m OD), for example, indicates that this settlement was built in what was already an open grassy landscape (Waddington, 2013, p. 66). At Bryn y Castell, in the uplands of southern Snowdonia (c.364m OD) (Figure 4.2), almost complete removal of the tree cover seems to have occurred somewhat later, in the Later Bronze Age (GrN-17580; 2720±50 BP; 940–800 cal BC; 95% probability), by which time the presence of charcoal and cereal pollen suggests that people were practising both arable and pastoral agriculture in the locality (Mighall and Chambers, 1995, pp. 317–8). Closer to the Great Orme, most palaeoenvironmental records are either from high-altitude sites or deal with sea level changes (e.g., Walker, 1978; Bedlington, 1994). There are only a few local pollen sequences from sites that are both contemporary with Bronze Age mining on the headland and from broadly comparable environmental settings. A similar Bronze Age sequence to that from Bryn y Castell is, however, evident around Llyn Cororion, a lake on a low-lying coastal plateau (the Arfon Platform) approximately 20 miles to the west of the Great Orme with one of the most complete, well-dated and detailed Holocene vegetational records in lowland north Wales (Watkins, Scourse and Allen, 2007, p. 179). Pollen and plant macrofossil evidence from an environmental sampling pit at Melyd Avenue, Prestatyn, on the north Wales coast approximately 20 miles to the east, indicates an episode of clearance in the latter part of the second millennium BC, corresponding to an increase in both pastoral activity and crop cultivation at that location (Bell, *et al.*, 2007, pp. 305–6). In the lower reaches of the Aber Valley of north-west Snowdonia, an upland area (c.420m OD) with a concentration of roundhouses and relict field systems of probable Bronze Age date, tree taxa begin to decline between 2000 and 1000 BC, although this change occurred at different times and intensities throughout the valley (Woodbridge, *et al.*, 2012, p. 91).

Removing trees and maintaining open spaces were clearly activities that Earlier Bronze Age communities across north Wales engaged in, whatever the exact timing and tempo of these practices at a given location. Wood clearance would also have been integral to other fields of activity, such as stock keeping or planting cereals. The traces of cereals – mainly barley but also emmer wheat and oats – and the remains of animals, including cattle and sheep, at a small number of sites across the region show that growing crops and rearing animals were also part of the contemporary taskscape (Waddington, 2013, p. 66). The frequent juxtaposition in the uplands of unenclosed scattered and single roundhouses with numerous field clearance cairns and small irregular curvilinear fields, such as characterise the fieldscape around Gerlan, North Arllechwedd in the Great Orme hinterland, demonstrates that even isolated settlements were part of a wider agricultural landscape (Gwyn and Thompson, 1999, p. 25; Lynch, 2000, p. 91; Waddington, 2013, pp. 69–70) (Figure 4.2). Most of these relict field systems are undated

and have not been surveyed fully. At least some, like the roundhouses with which they are associated, are likely to have been created during the Earlier Bronze Age – and even earlier. The small single unenclosed Late Neolithic roundhouse at Llyn Morwynion, for example, was accompanied by a field enclosure. Individual roundhouses were associated with garden plots and paddocks with clearly delimited boundaries and may have been situated in upland pastures (Smith, 1999, pp. 41–2; Waddington, 2013, pp. 66–70, 90). Although our own perception of the hills and mountains of north Wales is that they have always been agriculturally marginal (cf. Roberts, 2007, p. 104; e.g., Williams and Le Carlier de Veslud, 2019, p. 1189), suited only to non-intensive sheep grazing, tending to animals and the land during the Chalcolithic and Earlier Bronze Age is likely to have involved a wide range of practices, even in the uplands.

The content of these activities is, however, difficult to describe at anything other than a macro scale. Interpreting the paleoenvironmental record for north Wales presents similar challenges to those encountered when using like data from the Great Orme to reconstruct the *chaînes opératoires* of making and repairing bone tools and other contributory tasks at the mine (Chapter Three). Much of the cereal evidence, for example, comes from burials or other contexts unassociated with settlement. It could point to local crop cultivation, or to the storage or consumption of cereals brought in from elsewhere (e.g., Lynch, 2000, p. 83). Even pollen and plant microfossils recovered from roundhouse sites, like the traces of wheat and barley identified at Meyllteyrn Uchaf, may not indicate onsite cultivation (Caseldine in Ward and Smith, 2001, pp. 25–31). The faunal data likewise do not fully represent contemporary local animal resources and subsistence strategies (James, 2011, p. 393), such as the relative proportion of animal to arable farming at a particular settlement. Nor do the terms used here to describe different working practices, such as 'arable farming', 'pastoralism' and 'mixed farming', capture fully the richly varied ways in which people would have engaged in agriculture across north Wales during the second millennium BC. Environmental evidence suggestive of 'clearance' or 'grazing' can refer to a diverse and complex range of historically situated practices. For example, people may have hewn down trees to clear forest or combined the felling and burning of trees with repeated animal grazing (Moore, 1993, pp. 220–1). The pollen record for an episode of woodland clearance does not reveal whether it was purposeful or opportunistic, nor is it always possible to determine whether a fire identified from the environmental evidence had a human or natural cause (Watkins, Scourse and Allen, 2007, pp. 178–9). In some times and places animals could have been encouraged to roam in search of fodder at a distance from an enclosure or they may have been kept close to the farmstead, perhaps in pens (Briggs, 1985, p. 306). The high values of barley recorded in pollen sequences on the Berwyn Mountains, on the Powys/Denbighshire border (Caseldine, 1990, p. 57) (Figure 1.2), could indicate cultivation of this crop for animal fodder somewhere in the vicinity, as part of a mixed farming regime – although we cannot always be certain that such evidence represents in-situ crop cultivation rather than cereals brought in from elsewhere (Manley, 1985, p. 342).

Cattle and sheep may sometimes have been seasonally relocated from a permanent dwelling in the lowlands in the winter to rough pasture around a temporary dwelling in the uplands in the summer. Known as transhumance, the seasonal relocation of cattle and sheep between lowland and upland areas was practised in Wales from the 10th into the 18th centuries AD (Allen, et al., 1979, p. 3; Manley, 1985, p. 343; Chambers, Kelly and Price, 1988, p. 345). This historic form of land use arose from particular socio-economic conditions and had a diverse pattern and structure, which differed from place to place. There is no universal model of historic transhumance in Wales that can be used to explain all prehistoric farming practices across this region. We should not assume that the latter were always organised predominately around a pastoral routine. Rather, the movement of livestock was probably only one of several chaînes opératoires in the farming field of activity for a given community and would have varied in form from region to region (Kelly, 1982, pp. 885–6; Briggs, 1985, pp. 302–5; Manley, 1985, p. 343). Enhanced phosphate levels at the Meyllteyrn Uchaf roundhouse and the later enclosed settlement of Moel y Gerddi, for example, point to animal corralling all year round (Waddington, 2013, p. 92).

Due to these ambiguities in the evidence, the technical sequence of bone and wooden toolmaking at the Great Orme mine (Chapter Three) cannot be related directly to particular farming practices and routines at specific sites or locales in the contemporary taskscape, neither in the immediate hinterland nor further afield. It is probable, however, that the timing of the mining field of activity was embedded in the agricultural cycle. This would itself have been entangled in the tempo of the seasons and other non-human phenomena such as the cycle of the moon and tides and the lifecycles of plants and domestic animals (Ingold, 1995, p. 9). There is historical and ethnographic evidence from Europe and Africa for mining during slack periods in the agricultural routine (e.g., Blanchard, 1978, pp. 2–3; Herbert, 1984, p. 69; Childs, 1998, p. 127). During these lulls in an otherwise relentless farming regime – which in European contexts occurred in winter – members of a community could potentially participate in other fields of activity such as mining. There is tentative faunal and tree-ring evidence for mining in winter at the Bronze Age copper mines of Ross Island and Mount Gabriel, both in

Ireland (Figure 3.3). Winter mining has also been proposed for the high altitude (2400–2600m AOD) Bronze Age copper mine at Saint-Véran, south-east France, where freezing conditions would have reduced water seepage into the deeper workings (O'Brien, 2015, pp. 118, 267). The widely spaced regular growth rings in some of the samples of oak charcoal recovered from the Great Orme mine may hint at winter activity: suggesting fast, young growth, they point to the use of coppice stools and pollards. These are often cut in winter once the sap has fallen (Edmonds, 1999a, p. 24), but as they can also be cut at other times of year and then stockpiled prior to use, this evidence does not rule out mining in other seasons (Jenkins and Lewis, 1991, p. 155). As seen earlier, the extent of the mine suggests that the Bronze Age miners perhaps typically chose to access this source during drier times of year, to minimise water levels in the workings.

Although we cannot say with certainty when mining habitually occurred during the customary Bronze Age mining phase, the environmental evidence hints that it was most likely episodic rather than continuous, with numerous, repeated extraction events. Procuring wood for each mining episode, whether the miners obtained this material locally or from further afield, would have resulted in recurring, small-scale woodland clearances. These would have contributed to the gradual opening of the landscape due to agriculture. Deforestation on the headland and in its hinterland due to Bronze Age mining is, however, likely to have been less rapid and extensive than previous studies have suggested (e.g., Jenkins and Lewis, 1991, p. 158; Dutton, et al., 1994, p. 255; Wood and Campbell, 1995, p. 17). Only relatively small amounts of wood may customarily have been needed over the course of a mining event, for the various chaînes opératoires discussed in Chapter Three. Episodic working would also have allowed some vegetation recovery between each mining episode, as would species selection and other woodland husbandry practices. Detailed palaeoenvironmental investigations of the impact of Bronze Age mining at other sites in Britain and Ireland, such as Copa Hill (mid-Wales) and Mount Gabriel (south-west Ireland), reveal a similar picture (e.g., Mighall and Chambers, 1993; Mighall, et al., 2000).

As we saw in Chapter Three, the generally careful and consistent choreography of the mining field of activity points to knowledge of extraction technologies and working practices persisting between each instance of activity at the mine. James (2011, p. 343) suggests that this indicates that each episode of extraction was made in living memory of previous visits. Over the Bronze Age life of the mine, the intervals between occurrences probably varied. Even if undertaken seasonally, it may not always have been a yearly or regular event. The pace of working may also sometimes have risen or fallen. We have seen how the rhythm of working underground was not dependent on natural daylight. Consequently, its timing may not have been tied into the cycle of day-night and so may sometimes have been going on continuously for the duration of a given mining event – excepting those occasions when fire-setting underground meant that some or all of the workings had to be vacated while full of smoke.

Whatever its exact timing and pace, the way in which Bronze Age communities understood the passing of time is likely to have differed significantly from modern Western perceptions of temporality. In non- and pre-industrial societies, time is essentially social and task orientated, i.e., it does not exist as an objective entity but is intrinsic to the performance of tasks and the rhythms of non-human phenomena (Ingold, 1995, pp. 7–9). Inherent in the daily patterning of people's routine activities, time can only be measured by reference to how long something else takes. Similarly, to explain when an event occurred, it must be related to another activity that occurred concurrently. The Bronze Age miners may therefore have understood the passing of time at the source in relation to their experience of mining there. Each mining event perhaps ceased when, for example, a certain quantity of ore had been obtained, recurring roof collapses impeded extraction or crops were ripe for harvesting. Its duration could also have been measured and understood in relation to the resonance of cycles such as the ebb and flow of the tides or the ageing of the moon. The appropriate time to begin and end a mining event may likewise have been determined in relation to the occurrence of other habitual events in the taskscape, such as the harvest or a particular ceremony. Other fields of activity that could have influenced the tempo and temporality of Bronze Age mining at this source – not only those related to farming – will be discussed in the next chapter.

Land division, climate change and new ways of being in the world

Earlier practices revolving around tending to animals and the land appear to have continued throughout the Later Bronze Age and Earliest Iron Age in north Wales (e.g., Chambers and Price, 1988; Watkins, 1991, pp. 253–5; Waddington, 2013, p. 67). They would have remained pivotal to many people's routine experience. The enclosed roundhouses at Moel y Gerddi and Erw Wen, for example, were built in what was already an open grassy upland landscape by the 9th century cal BC, a consequence of ongoing woodland clearance at higher altitudes (Waddington, 2013, p. 66) (Figure 4.2). The traces of arable weeds, such as corn spurrey, and several varieties of cereal at these settlements could indicate local crop cultivation, while the layout of their enclosures points to stock keeping (Chambers and Price, 1988, pp. 99–100; Kelly, 1988, p. 141; Caseldine, 1990, p. 69; Waddington, 2013, p. 66). The latter part of the second millennium BC in north Wales

was, however, marked by the development of terraced fields, typically rectangular and accompanying enclosed settlement in the lowlands (with terraced cultivation landscapes up to 300m OD in north-west Wales). Although their precise chronology and relationship to roundhouse settlement in north Wales is essentially unknown, some examples, such as the terraced fields associated with activity at Erw Wen, are certainly Later Bronze Age in date (Waddington, 2013, pp. 71, 118). Heavily weathered traces of small narrow linear fields with parallel land divisions are still faintly visible on the Great Orme. Part of the extensive, multi-phase evidence for agriculture across the headland, which includes field systems, terraces and banks, these relict fields may indicate pre-Medieval and possibly even later-prehistoric spade cultivation (Aris, 1996, p. 91; Hopewell, 2013a, p. 11). Today, most of the Great Orme is used for sheep grazing (Conwy County Borough Council, 2011, p. 4). Clearly, people have certainly not always viewed this area as agriculturally unfavourable (cf. Williams and Le Carlier de Veslud, 2019, p. 1189).

The outcome of sustained, systematic cultivation, these fields signalled the emergence of new forms of land division and use in north Wales from the latter part of the second millennium BC, continuing throughout later prehistory. Appearing somewhat earlier elsewhere in Britain, these practices both expressed and reproduced a gradual transformation in people's relationship to the land and particular places (e.g., Barrett, 1994, chap. 6, 1999, pp. 497–8; Waddington, 2013, p. 66; Brück, 2019, chap. 5). They have, however, also been interpreted as evidence of agricultural intensification, particularly in southern and eastern Britain. One argument for the peak in mining on the Great Orme that some researchers have identified from the changing chemical and isotopic signature of Bronze Age metalwork is that this intensification created a demand for bronze axes and other tools around 1600 BC, which in turn drove up the mine's productivity over the following few centuries (Williams and Le Carlier de Veslud, 2019, p. 1192). A link between the increased construction of bounded fields and more intensive farming in a given area is, however, difficult to substantiate. On Dartmoor, south-west England, for example, there is no compelling evidence for any significant pressure on land availability in the years before the appearance of coaxial field systems. Detailed examination of the layout and structure of prehistoric land boundaries across Britain also shows that these features were typically not the outcome of single enclosure events or phases, but the products of lengthy, episodic and gradual periods of construction, alteration and maintenance (Brück, 2019, pp. 188–94). Rather than emerging out of any pre-planned 'grand design', they were most probably unintended consequences of local routines of practice, as they unfolded over quotidian,

MINING AND THE TASKSCAPE   111

seasonal or more extended time scales (Barrett, 1999, p. 498). Their appearance in the landscape therefore does not represent a discrete horizon of marked agricultural intensification that could then in turn have triggered a sudden steep increase in the demand for copper ore from the Great Orme mine.

Between the 13th and 12th century BC and again in the 9th century BC, marked climate deterioration led to cooler, wetter summers and more severe winters across Britain, accompanied by increased storminess, a relative rise in sea level and rising water tables in many areas (Brown, 2008, cited in Waddington, 2013, p. 13). In north Wales, this most probably produced localised flooding of the fertile, low-lying coastal plains. The flat open landscape around the Afon Conwy in the Great Orme's hinterland, for example, is a flood plain (Anon, 1998, p. 81; Gwyn and Thompson, 1999, fig. 17). From c.800–650 BC, there is a significant rise in alluvial deposits across Britain and rivers like the Afon Conwy are likely to have silted up and shifted course considerably (Waddington, 2013, p. 13). Palaeoenvironmental data also seem to point to deteriorating soil quality and the increasing development of peat and heath from the Later Bronze Age throughout the Earliest Iron Age in upland and lowland areas. Pollen recovered from the enclosures at Moel y Gerddi and Erw Wen indicates that, during the mid-first millennium BC, the soils at these sites became wetter and more acidic and that local heath vegetation increased, perhaps due to repeated small-scale short-lived intensive episodes of clearance and grazing in a worsening climate (Caseldine, 1990, p. 69; Waddington, 2013, p. 66).

Mining on the Great Orme continued during the initial period of climatic deterioration in the 13th/12th century BC (although the dated evidence from the mine – at 68% probability – indicates that extraction had most probably ceased before the second phase of deteriorating climate from 800–409 BC). The miners, like people engaged in other fields of activity elsewhere in the taskscape, would have encountered the practical consequences of the changing climate, such as quantities of water and spoil blocking access to workings due to flooding; more frequent poor harvests; or the closing of traditional routes, perhaps on a seasonal basis, as fords became impassable or coastal seas increasingly treacherous. Communities are, however, unlikely to have had any awareness that these events were part of what we now know to have been a long-term climatic trend, instead perceiving them only at the scale of their everyday experience (Barrett, 1989b, p. 114; Edmonds, 1995, p. 157).

The effects of changing climate would not have been felt at the same time and in the same way everywhere across north Wales. People's responses would also have varied from place to place (Waddington, 2013, pp. 14, 66). How the communities mining on the Great Orme chose to inhabit these changing conditions – the technological choices they may have made to, for example, mine only at a particular time of year, start extracting ore from a different part of the source, pool communal resources or switch between forms of arable cultivation and animal husbandry – were not simply adaptive responses to environmental stimuli and constraints, such as soil quality, drainage, aspect, tree cover, hours of sunshine or amount of rainfall. Such strategies, like the routines of working at the mine discussed in Chapter Three, arose through a complex ongoing interchange between people, their perceived environmental conditions (the plants, animals, rocks and minerals, soils and other land attributes they encountered as they went about the routine business of their lives) and their socially constructed understandings of place and appropriate ways of 'going on' (Barrett, 1988, p. 31, 1989b, p. 114, 1999, pp. 498–9; Edmonds, 1999a, pp. 20–1, 36–7). From these interactions and understandings emerged new types of action and ways of being in the world, resulting in the shifts in the pattern and character of dwelling that slowly transformed the taskscape of north Wales during the second and early first millennia BC. These include the reorientation of activity away from some upland areas to the coastal fringe; the appearance of increasingly monumental, enclosed settlements, hillforts and terraced fields; and the changing *chaînes opératoires* of burial and commemoration, which led to people ceasing construction of funerary and other monuments in areas of high ground by the Later Bronze Age. Although increasing rainfall and cooler temperatures may eventually have inhibited year-round use of the highest peaks and mountain ranges and made it more difficult to grow crops at higher altitudes, there is no clear evidence that communities in this region abandoned areas of high land directly due to climate changes. Some form of agricultural activity continued in many upland areas (Manley, 1990; Davies and Lynch, 2000, p. 145; Woodbridge, *et al.*, 2012; Waddington, 2013, pp. 13–14). Like any other set of technological choices, the tasks of tending to animals and the land were concerned, consciously or otherwise, as much with the making and upkeep of human relations as with the production of food for survival (Barrett, 1989b, 1999).

It would likewise be too reductive to link the cessation of Bronze Age mining on the Great Orme directly to the effects of a worsening climate. Recurrent, perhaps seasonal, flooding could gradually have led to extraction in some of the deeper workings becoming untenable, even though as we saw in Chapter Three, it is probable that the miners were familiar with a range of techniques for draining and redirecting incoming water. It is also unlikely that all parts of the source would have been flooded at the same time, or to the same extent throughout the year. The end of Bronze Age activity at the source could, of course, have been brought about by the eventual complete removal of the accessible, easily smelted malachite-goethite copper ores the miners were seeking (e.g., Lewis, 1996, p. 170). Even in this case, however, the changing material conditions – the gradual depletion of the sought-after ore; increasing water ingress; the harder, more difficult to work dolomitic rock encountered at depth – would have been understood and responded to over time in the context of changing practices in the wider taskscape. These would have included the increased prevalence of bronze recycling over the course of the Bronze Age rather than the use of metal derived from a single mine, as well as a gradual transformation in the social relevance of bronze and bronze working, both points we will return to in the next chapter.

The movement of people in the north Wales taskscape

Throughout the second and early first millennia BC, the taskscape for communities in north Wales involved a degree of mobility over a variety of terrain. During the Chalcolithic and Earlier Bronze Age, this could have included residential mobility, by at least some of the population at certain times. The scattered pattern of short-lived land clearances recorded in the pollen record of Lake Cororion on the coastal plain to the north of Snowdonia, for example, perhaps represents the shifting, small-scale land-husbandry practices of a semi-nomadic community involved in pastoral farming around the lake during the Chalcolithic (Watkins, 1991, p. 249). By the Earliest Iron Age, this temporary form of dwelling had probably become less common. The evidence for repeated rebuilding of successive roundhouses on the same site, as seen at Moel y Gerddi and Erw Wen, points to permanence and the long-term inhabitation of specific places in the landscape, in some cases over several centuries (Barrett, 1999, p. 497; Waddington, 2013, p. 15). People would, however, have continued to move around the north Wales landscape as they participated in community interactions like marriage or the *chaînes opératoires* of tending to animals and the land, visiting burial grounds and other monuments, or sourcing raw materials. The uplands of north Wales are criss-crossed by numerous prehistoric trackways, such as the examples of assumed Bronze Age date running across the upland moorland of Waun Llanfair (North Arllechwedd) in the hinterland of the Great Orme. A track running over the hills from the Conwy Valley to the coast, which has probable Bronze Age origins, was until the 18th century AD the only route travelling westwards from the Great Orme that avoided a hazardous voyage by sea around the treacherous coast of Penmaenmawr (Gwyn and Thompson, 1999, pp. 4, 36; Lynch, 2000, p. 95; Roberts, 2007) (Figure 1.6).

While that stretch of coastline may have been best avoided, finds of planking from sewn-plank paddle boats show that this form of transport was in use in the Bronze Age. There are at least 10 such examples from Wales and England with dates spanning the second millennium BC. One is a timber of Early Bronze Age date found in north-west Wales on the Llŷn Peninsula near Porth Neigwl (Figure 1.2). Elsewhere in Wales, some of the fragments of boat planking excavated at Caldicot and Goldcliff on the Welsh side of the Severn Estuary have been scientifically dated to c.1200–1100 BC (Van de Noort, 2006, p. 273, fig. 3) (Figure 1.1), at the same time as there was also Bronze Age activity at the Great Orme mine. Such discoveries probably represent only a tiny sample of the total number of such boats constructed and used throughout the Bronze Age in Britain (Clark, 2004; Van de Noort, 2006, pp. 273–5; Smith, *et al.*, 2017). These vessels would certainly have been capable of making journeys across and around Conwy Bay and along the lower reaches of the Afon Conwy. People probably also used log and skin boats or coracles on the smaller inland rivers (Lynch, 2000, p. 96; Van de Noort, 2006, p. 274). There are examples of Bronze Age log boats from Carpow, south-east Scotland, and from Must Farm, eastern England (Strachan, 2010; Must Farm, 2021).

Evidence that the technology existed to attempt longer sea journeys during the Bronze Age, and even earlier, comes from the trace element composition of the very first copper objects found in north Wales and elsewhere in Britain. Items such as the three flat axes found in the Chalcolithic/Earlier Bronze Age hoard from Moel Arthur, north-east Wales, which chemical analysis indicates were made using copper from ore mined c.2400–1900 BC at Ross Island in south-western Ireland (Lynch, 2000, p. 97; O'Brien, 2004; Bray and Pollard, 2012, pp. 856–8) (HER PRN CPAT102280 [accessed online 21-06-2022]) (Figure 4.6), demonstrate that travel across the Irish Sea between Britain and Ireland was possible even in the latter part of the third millennium BC. The 'Amesbury Archer', an individual living sometime between 2470–2240 cal BC (OxA-13541; 3895±32 BP; 95% probability) and buried at Amesbury in Wiltshire, southern England, whose origins were in Europe, possibly in the central Alpine region, is further evidence for sea crossings between continental Europe and the British Isles during the same period (Fitzpatrick, 2009, pp. 176–7) (Figure 1.1). At least some of the Bronze Age communities in north Wales could have made longer journeys by boat along the coast or across the sea to Ireland or the Isle of Man, but it is not possible to establish how routine this was, nor whether every community took part.

More locally, at least some people on occasion moved between the valleys or the coast and the uplands of north Wales. The turves used to construct some of the Brenig Valley burial mounds on Mynydd Hiraethog, Denbighshire, contained weed pollen suggesting that they were brought to the site from lower down the valley (Caseldine, 1990, p. 57; Lynch, 2000, p. 83) (Figure 4.5). Mineralogical analysis of soil adhering to cremated bone from Moel Goedog Circle I in south-west Gwynedd likewise indicates an origin in a lower-lying region, perhaps in the vicinity of Llanbedr or Dyffryn Ardudwy (Lynch, 1984, p. 28) (Figures 1.2, 4.5). The scale of the distances involved could typically have been measured by journey durations of several days (Edmonds, 1999b, 489) – and longer, in some cases and for some people, as the continental origin of the Amesbury Archer hints. Although our limited understanding of the spatial patterning of contemporary settlement, monument and agricultural activity makes this difficult to assess accurately, strontium and oxygen stable isotope analysis of tooth enamel from burials elsewhere in Britain presents a similar picture, at least for the Chalcolithic and the beginning of the Earlier Bronze Age. These results show that people – both men and women – could be highly mobile both locally and across longer distances, but that relatively few individuals journeyed as far as the Amesbury Archer (e.g., Booth, et al., 2021, pp. 7–8).

The Great Orme was part of these patterns of activity. During the Bronze Age, it may have been almost an island, separated from the mainland by a narrow strip of marsh. This topographic setting does not seem to have restricted or inhibited the movement of people to or away from the mine. The striking black-green/blue mineralisation scarring the rock exposures along the north-western edge of the Pyllau Valley was most likely first encountered by people moving onto and around the headland during routine fields of activity, such as tending to animals and the land, or retrieving minerals for pigment making, rather than being 'discovered' by prospectors specifically seeking sources of ore to smelt into copper metal (cf. Timberlake and Marshall, 2018, pp. 427–8). This context of routine mobility, together with an awareness of the unusual rocks and minerals to be found in the Pyllau Valley discussed in Chapter Two, is likely to have created the practical and conceptual conditions out of which copper mining for smelting on the Great Orme during the Bronze Age first began. The island character of the headland could even have enhanced the perceived qualities of the ore from this source.

Networks and the circulation of copper, tin and lead

The probable origins of amber, jet, faience, flint, pottery, gold, copper and tin objects found in north Wales show that the fields of activity taking place on the Great Orme and across its hinterland from the Chalcolithic into the Earliest Iron Age were also variously part of a taskscape comprising networks of connections stretching in different directions. These links extended between this region and the Welsh borderland, the North of England, Scotland and Ireland, and into continental Europe as far as the Baltic (e.g., Lynch, 1983, p. 18, 2000, pp. 99–121; MacGregor, 2012,

p. 102). The use of Ross Island ore from south-west Ireland to make the earliest copper metalwork found in north Wales shows that the products of copper mining were moving considerable distances from the beginning of the use of copper metal in Britain and Ireland in the Chalcolithic. This phenomenon continued throughout the Bronze Age. From around the 22nd century BC onwards, tin was increasingly alloyed with copper to make bronze. Combined chemical and lead isotope analyses have recently demonstrated that copper ore from the Great Orme was used to make bronze objects, particularly shield-pattern palstaves. This is a distinctive form of metal axehead with developed (pronounced) flanges and a stop-ridge. In widespread use from the mid-second millennium BC onwards, finds of this tool are distributed across Wales, lowland England and the Continent as far afield as Brittany, The Netherlands and Sweden (Adkins, Adkins and Leitch, 2008; Williams and Le Carlier de Veslud, 2019, p. 1185, fig. 4). Two palstaves have been found on the Great Orme: a looped example assigned to the Later Bronze Age, which was recovered with two gold lock-rings and a socketed punch or awl in the Ogof Colomennod or Pigeons' Cave hoard (HER PRN GAT4577 [accessed online 22-07-2019]) (Savory, 1958, pp. 14–15); and a badly corroded shield-pattern palstave from the North Shore beach at Llandudno (HER PRN GAT4690 [accessed online 22-07-2019]). There is a further palstave find from the Creuddyn Peninsula, at Bryn Pydew (HER PRN GAT7889 [accessed online 22-07-2019]) (Figures 1.4, 1.5). The significance of these finds in the vicinity of the Great Orme mine is considered later in this chapter.

Evidence for the long-distance movement of tin metal by the Later Bronze Age comes from the presence of an object of tin – unalloyed to copper – described as an 'oval mount', in the Llangwyllog hoard, Anglesey (Appendix 6, Figure 4.6). The only sources of tin in Britain are in the South-West (West Devon and Cornwall), with Brittany the nearest continental source (Figure 1.1). The tin used to make bronze in north Wales presumably came from either, or perhaps both, of these sources (Rohl and Needham, 1998, pp. 14–15; Barber, 2003, p. 97; Kienlin, 2013, p. 420; Timberlake, 2017). It has only very recently become possible, using a combination of geochemical parameters, to begin to distinguish tin from either area (e.g., Berger, et al., 2022, p. 6). Nonetheless, even though the archaeological field evidence for Bronze Age tin extraction at the South-West tin field, one of the largest in Europe, is still relatively insubstantial (Timberlake, 2017), it is reasonably certain that this was the main source of tin used with Great Orme copper to make bronze. The Leverhulme-funded 'Ancient Tin' project (University of Durham, 2021) should throw more light on the connections between Wales and Devon/Cornwall during this period.

From the mid-second millennium BC onwards, lead was routinely added to tin bronzes to reduce metal viscosity and lower the liquid temperature during casting (Northover, 1982, p. 91). In contrast to the movements of both copper and tin, the circulation of lead could represent networks of connections performed on more local scales. There is a major lead ore field in north Wales with some undated single finds of stone mining tools at locations without any associated copper ores (Timberlake, 2017). In mid-Wales, Earlier Bronze Age lead mining on a very small-scale has been identified at the Comet Lode Opencast, Copa Hill, although it is unclear whether the miners there were preferentially extracting the lead ore or were removing it as a by-product of the working of copper or even silver ores (Timberlake, 2003b, p. 100, 2017).

How did the Bronze Age mining field of activity on the Great Orme fit into these networks of contacts, which were in effect the interface between the *chaînes opératoires* of mining and the making and consuming of things made of metal? As we saw in Chapter Three, the relevance of the Later Bronze Age smelting site at Pentrwyn to the customary period of Bronze Age mine-working on the Great Orme is uncertain and so there is no direct evidence for the form in which the mined extract was leaving the mine. No other possible Bronze Age sites for smelting oxidised copper ores have yet been identified in Britain or Ireland. Consequently, in north Wales – as elsewhere in Britain – most of the products of copper mining have been recovered in a finished or semi-finished state, the outcomes of stages three, four and five of the copper/bronze metalwork production sequence (melting and casting, smelting and finishing, repair and modification). As discussed in the opening chapter, such objects, which occur in a wide variety of shapes, are present in the landscape in burials, as hoards (collections of more than one item) or as isolated finds (objects unaccompanied when found). The latter may in some cases be the only items recovered from disturbed hoards. Following Needham (1989, p. 55) and Barrett and Needham (1988, p. 129), in this study isolated finds and hoards are treated as a single category of material as both may be the outcome of purposeful acts of deposition.

There is a small number of finds of copper/bronze from north Wales that appear to have been deposited before being cast or worked into a definite or final object form (although none are from the Great Orme's hinterland) (Appendix 6). These include scarce and indeterminate metal fragments from Earlier Bronze Age burials, such as the possible rivet from a round barrow at Porth Dafarch and two copper/bronze fragments from another barrow at Ty'n-y-pwll, both on Anglesey; and four separate finds of whole and fragmented bun ingots ('cakes') of copper or bronze. Only the ingot from Trefnant, in north-east Wales, was recovered as an isolated find; the others are part of hoards of items most probably deposited in the Later Bronze Age. Most of these ingots are from the well-known

Guilsfield hoard, in the south-eastern part of the study area (Davies and Lynch, 2000, fig. 4.14) (Figure 4.8).

Bronze Age finds made mainly of tin or lead are extremely rare in north Wales (and, again, none are from the Great Orme mine's hinterland). The Llangwyllog hoard already mentioned contains the only example of unalloyed tin from this region. There are very few such finds from elsewhere in Britain, even close to tin sources in the South-West. There are similarly few discoveries in Britain of lead of Bronze Age date (Timberlake, 2017) and only one from north Wales: the Later Bronze Age Llangollen Community hoard from the north-east of the region. This comprised three lead alloy objects: an unusual socketed object, part of a socketed axe and a fragment of metal (Figure 4.8).

The find context of all these metal objects is evidence of the practices around their final deposition, plus the outcomes of any subsequent, post-depositional taphonomic processes. Even when scientific provenancing techniques highlight the probable source of the ore used to make the metal in a particular item, there is no intrinsic link between where the thing was found and where it was made (e.g., Barrett and Needham, 1988, p. 29; Bradley, 1998, pp. 33–4; Radivojević, *et al.*, 2019). The geographic distribution of deposited metal objects does not in itself indicate how far the ores mined at a specific source were travelling, nor is it necessarily a direct reflection of the supply routes along which ores, or metal, or cast and/or worked items of metalwork travelled (cf. Williams and Le Carlier de Veslud, 2019, fig. 8). Unworked and/or unfinished or only partially finished finds of copper/bronze, tin or lead may, however, provide insights into the forms in which these metals circulated.

Two collections of metalwork found on the seabed in near-coastal locations at Salcombe and Langdon Bay, along the south coast of England, perhaps provide rare examples of the products of mining deposited in the act of being moved from one place to another (Needham, Parham and Frieman, 2013; Wang, Strekopytov and Roberts, 2018; Radivojević, *et al.*, 2019) (Figure 1.1). Careful analysis of each assemblage and its find location suggests that both deposits were created by shipwreck, although no traces of either boat have survived. The assemblages have both been dated typologically to 1300–1150 BC. A single object from Salcombe can be assigned to 1000–800 BC. Given that the Salcombe evidence is actually from two sites, 400m apart, it could indicate two distinct events occurring more than a century apart, or a single event involving 'old' curated objects (Wang, Strekopytov and Roberts, 2018, p. 103). Both the Langdon Bay and Salcombe metalwork collections were very large. More than 360 mainly unleaded tin-bronze objects in a wide variety of forms were recovered from Langdon Bay, including many complete tools (chisels, palstaves and other axe types) and numerous broken swords and rapiers (Needham, Parham and Frieman, 2013,

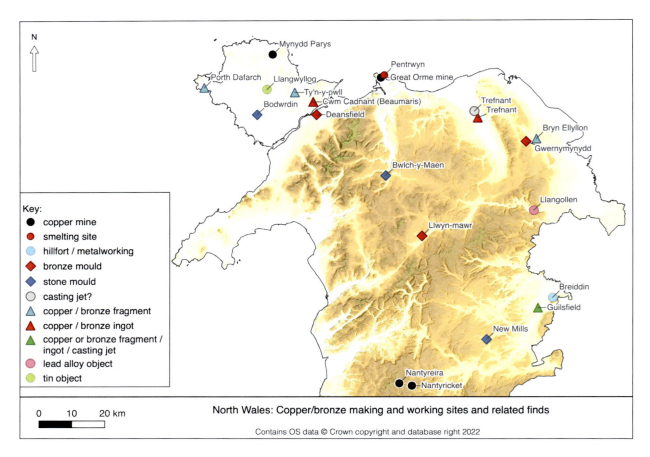

Figure 4.8: Evidence from north Wales related to the making and working of copper and bronze. Finds recorded in Appendix 6 and not shown here: copper/bronze ingot, Abergele Community hoard (Conwy); bronze bivalve mould, Conwy Community (Conwy); mould ingate/casting jet, Flintshire; possible casting jet, Trearddur (Anglesey). For the distribution of all metal production sites on the Great Orme, see Figure 1.5.

p. 109, table 5.6). The 400 objects from Salcombe were of gold, tin, copper or bronze. They included 280 copper or bronze ingots and 40 bun ingots of tin, the largest discovery of metal in this form from north-west Europe (Wang, Strekopytov and Roberts, 2018, p. 102).

The Salcombe assemblage is the best direct evidence to date that, during the later second millennium BC, people were moving the metals copper and tin in ingot form. It is one of several recent studies demonstrating that what were previously presumed to be bronze ingots found in hoards on land are in fact made of unalloyed copper (Wang, Strekopytov and Roberts, 2018, p. 112). Three bun ingot fragments in the Later Bronze Age Cwm Cadnant Community hoard, Anglesey, for example, have recently been identified as copper (Appendix 6, Figure 4.8). This may also be the case for some of the ingots in the Guilsfield hoard: three are of bronze (Northover, 1982, p. 74) and the metal composition of the remainder is not yet known. The Salcombe and Langdon Bay evidence together hints that, at least from the later centuries of the Earlier Bronze Age, the products of copper and tin mining participated in networks of circulation not only as (semi-)finished objects but at various stages of becoming and transformation, from the 'raw' metal onwards. Perhaps people regularly moved the output from the Great Orme mine as ingots of unalloyed copper metal before it was further transformed into axes, spears, chisels and other objects by remelting, alloying, casting, working and polishing? As we shall see later in this chapter, the products of mining, in whatever form, would have moved within the context of a dispersed metalwork production field of activity, at least in north Wales.

There is, however, so far, no compositional link in published provenance data between Great Orme ore and any ingots found in north Wales or elsewhere in Britain and Europe. The latest research into the lead isotope ratios of the metal in the Salcombe copper ingots suggests ore sources in Sardinia and/or southern Spain as the origin for the copper used, rather than a British mine, but it has not yet been possible to determine exactly how and where the unalloyed copper in these ingots was made, despite detailed metallographic and compositional analyses (Berger, *et al.*, 2022, p. 23; Wang, Strekopytov and Roberts, 2018, p. 115). The latest research suggests that they may not all have shared the same production history (Berger, *et al.*, 2022, p. 12). There are also no land finds from Britain of copper or bronze ingots that can be dated to the Earlier Bronze Age. Due to their very simple shape, the ingots found in north Wales, such as the isolated ingot find from Trefnant, are, like all examples from elsewhere in Britain, dated to the Later Bronze Age by comparison with other similar ingots, or by typological association with any accompanying metalwork (Wang, Strekopytov and Roberts, 2018, pp. 103–4) (Appendix 6, Figure 4.8). The copper ingots in hoards such as that from Guilsfield perhaps point to a change in depositional practice over time, as people chose not only to move metal in this form during the later second and early first millennia BC, but also to deposit it in the ground in this way.

The Langdon Bay and Salcombe assemblages appear to be frozen in time, perhaps at the interface between the 'producer' and 'consumer'. While they clearly show that people were moving things, they do not, however, demonstrate the character of the interactions that eventually resulted in their deposition on the seabed. The specific mechanisms by which objects circulated – like the routes along which they travelled – cannot simply be 'read' from artefact distribution patterns. This would assume a direct relationship between the structure of a deposit and the character of the relations that produced it, when very different practices can result in deposits that appear archaeologically similar (Barrett and Needham, 1988, p. 138). The occurrence of Great Orme copper in a bronze shield-pattern palstave found in Sweden (Williams and Le Carlier de Veslud, 2019) likewise does not necessarily indicate the existence of direct connections between Bronze Age communities in north Wales and those elsewhere. Nor does it show that people – whether individuals or groups – routinely journeyed between the Great Orme and the Baltic. Within the fluid patterns of mobility discussed earlier, in which direct travel over such long distances could have been relatively rare (e.g., Booth, *et al.*, 2021, p. 18), things of copper or bronze, or their constituent raw materials, may customarily have moved through the north Wales taskscape and beyond in short steps, passing directly hand to hand over even long distances (Barrett, 1985, p. 102; Edmonds, 1995, p. 56). This most probably occurred within some type of exchange relationship, which is the subject of the next section.

Exchange and the relations between people and things

The ways in which Bronze Age communities in north Wales chose to move a thing would have related to its meaning and value. These concepts are not inherent to things, nor are they passive or static. Nor do people simply 'give' meaning to things. Rather, the meaning and significance of a thing arose dynamically out the ways in which people interacted with it in the practice of their everyday lives. For many things, its value would have been created by the act of exchange. The value of others was created by their identification with a particular person or persons, or symbolic association with specific local cosmologies and fictive or actual genealogies or events. When such an inalienable object was exchanged as a 'gift', these associations would also have been transferred from the giver to the receiver, creating a bond that lasted beyond the moment of receipt. The receiver was placed in a relationship of debt and obligation to the giver, whose political power and authority, personal identity and worth were in turn reinforced and enhanced (e.g., Bradley and Edmonds, 1993, pp. 12–13; Edmonds, 1995, pp. 55–60; Barrett, 2012, p. 12; Brück, 2019, p. 227).

In this way, the circulation of things made of copper, gold, tin, amber, flint, pottery and jet through the north Wales taskscape and beyond was a social and political act. Like the practice of any technology, it was as much about the classification of people and the establishment and display of differences in social standing, the reproduction of political power and the creation and maintenance of social relations, as it was about the acquisition or sharing of things. Through specific socially and historically contingent practices of exchange, communities in north Wales – particularly those along the northern coastal fringe, including on the Great Orme and its hinterland – were situated throughout the third to mid-first millennia BC within geographically extensive networks of relations of affiliation, obligation and reciprocity. The exact form of these exchange relations, as well as the networks within which they occurred, are unknown but they are likely to have been complex and very varied (Barrett, 1985). In local settings, the outputs from the Great Orme mine could have been exchanged as inalienable gifts. As the mine's products were moved through widening spheres of social relations, they may have become increasingly detached from their web of symbolic, personal and political associations, until eventually notions of debt and reciprocity had no meaning and their demands little power. They were then perhaps exchanged by barter or even trade as a largely alienable commodity. The Great Orme's products therefore most probably moved within multiple overlapping spheres of circulation, of which 'trade' was only one amongst numerous *chaînes* of social interaction (Radivojević, *et al.*, 2019).

This distinction between an inalienable gift and an alienable commodity is likely to have been fluid and ill-defined, as the value of the Great Orme's products shifted depending on the social relationship between giver and receiver (Edmonds, 1995, p. 59, 1999a, pp. 126–8). Brück

(2015, pp. 51–2), for example, highlights the ways in which, even in contemporary Western societies, commodity transactions are facilitated by personalised relations between business partners. The relations through which people chose to exchange the mine's output would not, however, have been dictated directly by its relative abundance in or between different communities. The social distance at which the things made of Great Orme ore became inalienable commodities may likewise not always have mapped onto spatial distance. If, for example, the *chaîne opératoire* of smelting was closely integrated into the mining field of activity, as may have been the case on the Great Orme for at least some of the Later Bronze Age, (semi-)processed ore may not have been moved very far from the mine before it was turned into metal. Even this geographic proximity to the source may not, however, have precluded its inclusion in local exchange networks.

The Llangwyllog hoard from Anglesey contained, alongside the tin object and items of bronze, beads of jet and amber (Appendix 6, Figure 4.8). Like the tin, these latter materials must have travelled considerable distances before they were finally deposited on Anglesey (the amber from the Baltic and the jet from Whitby in north-east England) (Figure 1.1). They may not, however, have all travelled through the same spheres of value and interaction. Nor does their deposition together in north Wales demonstrate that they were directly exchanged for Great Orme metal. The results of geochemical and isotopic analyses of copper and bronze metalwork indicate that patterns of copper ore/metal/object circulation around Earlier Bronze Age Europe were more complex than unidirectional movement outwards from specific ore sources (Radivojević, *et al.*, 2019). The output from the Great Orme mine is similarly unlikely to have flowed in one or more fixed directions outwards from the source, with another type of thing, such as amber, moving in the opposite direction in exchange (cf. Williams and Le Carlier de Veslud, 2019, p. 1194). Even the orientation of the exchange networks in which the presumed shipwrecks at Salcombe and Langdon Bay participated is unclear, as the precise points of departure or destination are unknown in each case (Wang, Strekopytov and Roberts, 2018, p. 115). For either assemblage, the voyages represented could have been a one-off trip or, equally, one of a series of regular events.

Compositional analysis of Earlier Bronze Age copper and bronze artefacts from Britain and Ireland has shown that these often had complex life histories, circulating through numerous processes of transformation – smelting, melting, alloying, casting, smithing, reworking, remelting, mixing with metal from other sources, recasting, etc – from the mined ore to the point of final deposition over, in some cases, many centuries (Bray and Pollard, 2012; Bray, *et al.*, 2015). We should assume that this was typically the case

for the products of the Great Orme mine. Even the palstave made of Great Orme metal found near Pigeons' Cave on the headland itself could, before its final deposition, have been moved in multiple directions, involving the entanglement of many different people in communities at various intra- and inter-regional scales (Figure 1.5). Its recorded find location, behind a large rock somewhere close to the Great Orme's only natural harbour, may be significant in this regard, although it is doubtful whether it was originally deposited exactly where found. The fact that it was recovered with an old break to the butt end (Savory, 1958, pp. 56–7) also hints at a less simple life history than its find proximity to the mine would suggest.

The obligations underpinning the gift exchanges in which the products from the Great Orme mine participated most probably included not only the movement of, for example, gold, jet, amber and tin but of animals, foodstuffs, textiles, baskets and other things made of organic materials; people (both living and fragments of the dead) and ideas, as well as the provision of hospitality and labour (Bradley and Edmonds, 1993, p. 12; Brück, 2007). The copper and bronze metalwork that has survived for us to recover may not even always have been the most important component of these interactions (Edmonds, 1999a, p. 128). Brück (2019, pp. 104–5), for example, has suggested that cloth played a significant role in Bronze Age exchange networks, alongside bronze. Knowledge of where minerals, including tin and lead ores, occurred in the landscape could also have travelled in a similar way as a valued item of exchange.

There are likely to have been changes over time in the orientation, scale and configuration of the networks of relations within which communities in north Wales were embedded. This is illustrated by the ceramic evidence from the region. The restricted range of form, limited distribution and poor quality of the pottery used in north Wales during the Later Bronze Age compared to the Chalcolithic/Earlier Bronze Age points to an increased emphasis on local production and the technological choice to use readily available raw materials (Davies and Lynch, 2000, p. 199; Williams and Jenkins, 2019, pp. 32–3). This suggests that, by the early first millennium BC, communities in this region had ceased to include things made from ceramics and their component raw materials in the exchange relations through which things of gold and amber continued to be moved. Pots and clay were no longer part of the same "regime of value" (Edmonds, 1999a, p. 128). The exchange relations through which pigment made from Great Orme ore perhaps moved in the Chalcolithic most probably differed in character, purpose, scale and orientation to those in which metal from this source was circulating by the Later Bronze Age. The latter relations would, however, have emerged slowly from one generation to the next through the gradual reworking of such existing traditions of action in the north Wales taskspace.

From the Chalcolithic into the Earliest Iron Age, the exchange networks in which the products from the Great Orme mine participated were, like other contemporary fields of activity such as farming, reproduced by local instances of action (Barrett, 1989b, p. 114). In common with their perception of large-scale forces such as climate change, Bronze Age miners on the Great Orme would have been unaware of the full geographic extent of the web of exchange relations of which they were a part. There was most probably no coordinated long-distance flow of metal from the Great Orme, under some form of centralised control (cf. e.g., Earle, *et al.* 2015). The circulation of the outputs from the Great Orme mine instead occurred through an interlinked series of dynamic local traditions of kinship, affiliation, identity and political authority, the boundaries of which were diffuse and which played out across a range of spatial and temporal scales (Edmonds, 1998, p. 252). This is not to say that people in these essentially small-scale communities did not control and manipulate the circulation of copper ore, metal or objects (as well as things of other materials) to their own advantage, but that this would have unfolded in the context of local relations, understandings and perceived needs.

Moulds and the *chaîne opératoire* of copper and bronze casting

Whatever the structure of the relations by which the raw materials to make bronze circulated in the taskscape, transforming them into a finished object such as an axe, bracelet or sword required several technical steps: remelting, refining, alloying, casting, smithing and finishing (the metalwork production stages three and four outlined in Chapter One). By the end of the Bronze Age, people were also working sheet bronze to create elaborate objects, like the two bronze shields found in bogs in north Wales, one on Moel Siabod mountain (HER PRN GAT3760 [Accessed online 12-03-20]), Snowdonia, and the other at Gwern Einion (HER PRN GAT4785 [Accessed online 12-03-20]), close to Harlech (Davies and Lynch, 2000, p. 187) (Figure 4.6). No Bronze Age evidence for any of these production steps – such as melting hearths, moulds, crucible sherds and other refractories, metalworking tools and casting waste – has been identified on the Great Orme or in its hinterland. Elsewhere in north Wales, finds of casting paraphernalia offer the best evidence for copper and bronze production. Three stone moulds, all recovered singly; two bronze moulds found together in the Deansfield hoard discovered close to the coast at Bangor; the single find on the same stretch of coast of a matching pair of bronze valves for a palstave mould; and a possible isolated find of a mould ingate or sprue from Flintshire

can all be assigned to the Earlier Bronze Age, a period that in this study covers both the traditional Early and Middle Bronze Age periods (Appendix 6, Figure 4.8).

Despite its small size, this mould assemblage illustrates the gradual changes in casting technology taking place during the first half of the second millennium BC. People initially cast a relatively restricted range of forms using 'open' one-piece moulds, like the stone mould for flat axes found in a field while ploughing at Bwlch-y-Maen, Conwy. Two-piece (bivalve) moulds, like the example from Bodwrdin, Anglesey, appear in Britain around 1700–1500 BC (Figure 4.8). With this kind of mould, it became possible to cast complex socketed objects, as well as a wider range of three-dimensional forms (Lynch, 2000, p. 102; Barber, 2003, p. 118). Carved from a single stone block, the Bodwrdin mould had matrices on all four sides for casting three different things: a long, thin bar, perhaps for an armlet or bracelet; double-looped socketed chisels; and end-looped socketed spearheads (Barber, 2003, pp. 118–9). Whoever cast weapons in this mould seems to have also used it to make tools and ornaments, hinting at the making of a variety of different object types by the same hands (Webley, Adams and Brück, 2020, p. 51). From c.1500 BC, two-piece moulds made of bronze first appear, typically for casting axes and a restricted range of other tools (Webley and Adams, 2016, pp. 324–5), like the two intact examples from the Deansfield hoard for making two different forms of palstave. Although bronze moulds were probably mostly used for direct casting of copper or bronze, they could also have been used for casting other materials, such as lead or even wax to create patterns for investment casting or for making bivalve clay moulds. There are, however, no adhesions of lead on the inner faces of the Deansfield moulds (Webley and Adams, 2016, pp. 331–2). The lead objects in the Later Bronze Age Llangollen Community hoard, from north-east Wales, seem to have been cast either in unfinished or hastily or provisionally formed moulds of an unknown material that were made specifically for casting these items (Appendix 6, Figure 4.8).

The moulds from north Wales suggest that casting copper or bronze was a component of the Earlier Bronze Age taskscape of this region. However, the context in which this took place is unclear. As with items of deposited metalwork, the findspots of these moulds may not be where they were made and used. The sources of the stone – sandstone, in the case of the Bodwrdin and New Mills moulds; quartz-grit of possible north Wales origin for the Bwlch-y-Maen mould (Barber, 2003, p. 118; Webley, Adams and Brück, 2020, p. 238) – are not known (Appendix 6). Lithological examination of similar early one-part stone moulds from other areas of Britain has shown that they were typically made of material local to the place where they were deposited (Webley, Adams and Brück, 2020, p. 52). This could point to the use of these pieces of metalworking equipment generally close to where they were made. Both the Bwlch-y-Maen and New Mills moulds were, however, found at the top of steep slopes overlooking river valleys, with no contemporary dwellings or monuments nearby, hinting that they had been intentionally deposited at these selected locations (Webley, Adams and Brück, 2020, pp. 53–4) (Figure 4.8). The Bwlch-y-Maen mould (illustrated in Britton, 1963, fig. 8) was weighty but still transportable and so could potentially have travelled during its use-life. This is even more likely for the smaller, more compact Bodwrdin mould (Britton, 1963, fig. 16). The bivalve moulds in use towards the end of the first half of the second millennium BC often moved considerable distances from their original stone source (Webley and Adams, 2016, p. 332; Webley, Adams and Brück, 2020, p. 74). Moulds like the Bodwrdin example could have circulated by exchange, passing through the networks of connections stretching between Bronze Age communities. Alternatively, the mobility that is likely to have been a part of Bronze Age life for some individuals may have become more of a feature of copper and bronze working in north Wales, with the smaller, more compact bivalve stone moulds perhaps also travelling with the movement of people (Webley and Adams, 2016, p. 333). The social status of these individuals or groups, and whether they were directly involved in metalworking, is however unclear (Webley, Adams and Brück, 2020, p. 123).

Bronze moulds may also have moved along similar paths, although they seem typically to be found in the same geographic distribution area as the object forms they could be used to make. Finds of bronze moulds are relatively rare, with only around 55 known for the entire Bronze Age in Britain, suggesting that such items were usually melted down into new forms when their use-life as moulds came to an end (Webley and Adams, 2016, pp. 324, 333). The Deansfield moulds had for some reason been treated in a different way. They were found with an unlooped trident-patterned palstave that could not have been made in either mould (HER PRN GAT2304 [Accessed online 12-03-20]) (Appendix 6). The things in this hoard therefore seem to have been deliberately selected for deposition together. They are not a snapshot of the tools and products from a casting event held at that location. The Deansfield moulds hint that metalworking equipment, like metal axes and other objects, could have complex life histories. This is also illustrated by the Bodwrdin stone mould, one face of which has a matrix for a unique form of double-looped socketed chisel. Not only are there no known metal equivalents for this type of chisel, but there are no finds of socketed chisels of any kind from Britain before the Later Bronze Age, several centuries later than the currency of the end-looped socketed spearheads represented by the matrices on two of the other faces. Could this mould have been carefully curated by the community that used it, the different matrices being carved at different times (Barber, 2003, p. 119)? A single example of an end-looped socketed spearhead that was perhaps cast in the Bodwrdin mould has been found at Bala (HER PRN GAT3207 [Accessed online 17-03-20]), approximately 90km south-east of the mould's findspot on Anglesey (Figures 4.6, 4.8). While this could signify circulation of this object away from the locale where it was formed, both the spearhead and its mould could have moved through many spheres of interaction before final deposition.

It is difficult to assess from the small size of the mould assemblage how widespread copper and bronze casting was in this region and at what scale it unfolded. The six stone and bronze moulds found could not have been used to produce all the things of metal being made in this region during the Earlier Bronze Age. Communities in north Wales probably also used bivalve clay moulds for casting from at least the mid-second millennium BC onwards, although no examples of this date have been found there. The earliest clay moulds in Britain to which a date can be assigned, from a midden at Grimes Graves flint mine in eastern England, are unlikely to be older than c.1300 BC (Barber, 2003, p. 118; Healy, et al., 2018, p. 295). Shallow fired-clay moulds appear to have been used to cast the copper ingots of broadly similar date from the Salcombe seabed assemblage (Wang, Strekopytov and Roberts, 2018, p. 114). Casting in sand moulds is also a possibility, a practice that would have left few field traces. The irregular shapes of the tin ingots from Salcombe have been interpreted as evidence that they were cast in either sand or earth (Ottaway and Seibel, 1998; Wang, Strekopytov and Roberts, 2018, p. 114). Stone and bronze moulds were reusable and probably were used multiple times, perhaps up to several times a day (Barber, 2003, p. 118; Webley and Adams, 2016, p. 333). In this way, Bronze Age practitioners in north Wales could have produced large numbers of things of the same design, with a degree of standardisation, although this appears to have become more difficult to achieve when casting longer, larger items like swords and rapiers (Barber, 2003, p. 122).

Bronze melting and casting at the Breiddin hillfort

The only in-situ evidence for bronze melting and casting in north Wales comes from Later Bronze Age contexts at the Breiddin hillfort in the Welsh Marches (Musson, et al., 1991) (Figures 4.2, 4.8). During this period, the Breiddin comprised a low timber-framed earth and stone bank enclosing an area of some 28ha, within which there were timber roundhouses and midden deposits (Waddington, 2013, p. 15). Clay mould and crucible fragments, minute pieces of copper alloy melting slag, pits, 'furnaces' and 'working-hollows' were excavated in the southern part of the interior. A timber building or enclosure, c.6m in diameter, may have been associated with this group of furnaces/hearths, samples of charcoal from which were dated to around the 9th century cal BC. Finds of a crucible fragment and copper alloy melting slag from another area in the enclosure suggest bronze casting was taking place at more than one location at the hillfort (Musson, *et al.*, 1991, pp. 149, 178; Adams, Brück and Webley, 2017; Webley, Adams and Brück, 2020, figs. 4.13–14.14).

The size of the enclosure at the Breiddin indicates that it was a focus for activities involving large groups of people, but the scale of the metalworking field of activity there, and whether the evidence represents a single instance of bronze casting or several recurring events, are unclear. Clay moulds could only be used once and had to be smashed to release the casting. Only five moulds (for a socketed object and a possible sword blade, spearhead, pin shank and socketed axe) have been tentatively identified from the c.50 small fragments found (Needham and Bridgford, 2013, p. 74, table 3.7; Adams, Brück and Webley, 2017). The smashed fragments tend not to survive well unless buried (intentionally or incidentally) soon after use (Needham and Bridgford, 2013, p. 74) – perhaps explaining why no such moulds of Earlier Bronze Age date have yet been found in north Wales. The mould assemblage from the Breiddin may consequently not be fully indicative of the number of castings carried out there. It is also uncertain whether the bronze finds from Later Bronze Age contexts at the site, which included a complete socketed axe, a knife, and sword and spearhead fragments, were made in situ (Musson, *et al.*, 1991, fig. 56 [nos 138–159]).

The Breiddin is the only hillfort in Wales with evidence for Later Bronze Age copper metalworking (Adams, Brück and Webley, 2017). Traces of copper working at other hillforts in north Wales, such as Llanymynech Hill and Llwyn Bryn Dinas (both in Powys), are of Middle Iron Age date. Copper working at Pen y Dinas hillfort on the Great Orme during this later period is also a possibility, although there is no direct evidence for this (Smith, *et al.*, 2012, pp. 3–8; Waddington, 2013, p. 102) (Figure 1.5). The variety of finds from the Breiddin – pottery, tools, weapons and 'personal ornaments' including bronze pins (Musson, *et al.*, 1991, sections 7.3, 7.9) – indicate that people were participating in a range of production and consumption activities on this hilltop at the beginning of the first millennium BC. These related to bronze metalworking and farming, feasts and gatherings (Waddington, 2013, pp. 96–7). Like other contemporary hillforts in north Wales, the Breiddin enclosure does not seem to have been the focus for any one specific field of activity. Occupation may have been intermittent, perhaps seasonal (Waddington, 2013, p. 18), or scheduled according to the tempo of other fields of activity or non-human phenomena. The Breiddin was therefore most probably not a permanent production centre for bronze casting on a large scale, nor for specialist manufacture of a single object type. The small number of moulds recovered does, however, indicate that a wide range of objects (swords, spearheads, axes and pins) was being produced on site. This in turn points to the presence of a range of different skills sets, suggesting that the Breiddin was a significant place for the making of metalwork. The casting event(s) that took place in this hilltop enclosure are unlikely to have occurred spontaneously. Each stage in the casting *chaîne opératoire* would have required longer-term planning and community collaboration. The making of a clay sword mould, for example, involved a sequence of steps that had to be carried out over several weeks (Needham and Bridgford, 2013, pp. 65–6; Webley, Adams and Brück, 2020, p. 32). In preparation for casting, as well as making moulds, people would have had to collect wood and make charcoal, build a furnace or hearth, and form the crucibles from suitable clay sources (Barber, 2003, pp. 119–20). A source of metal was obviously also required and, of course, the practical skills and technical expertise needed to undertake the bronze melting and casting *chaîne opératoire*.

During the Later Bronze Age in north Wales, bronze melting and casting perhaps on occasion took place within the group (Stig Sørensen, 1996, p. 49). Analogy with the evidence from Later Bronze Age enclosed roundhouse settlement from elsewhere in Britain and Ireland suggests that similar sites in north Wales, like those at Llandygai and Pwll Parc, could have been loci for metal making and working during the late second and early first millennia BC, alongside the earliest hillforts such as the Breiddin (Waddington, 2013, p. 92; Webley, Adams and Brück, 2020, p. 123) (Figure 4.2). Could this activity have developed from and consolidated an existing tradition of metalworking at unenclosed roundhouse settlement in this region during the Earlier Bronze Age? Occasional finds of casting and smithing debris from settlements of this period elsewhere in Britain indicate that this may have been the case (Webley, Adams and Brück, 2020, p. 65). The Breiddin assemblage of casting waste and broken moulds and crucibles is typical of the character of bronze metalworking debris excavated at these other contemporary settlements, in that there are no bronze moulds (Webley and Adams, 2016, p. 334). The two bronze moulds from north Wales with a Later Bronze Age deposition date were recovered as an isolated find from Llwyn-mawr, near Llyn Tegid, Bala, and in the Gwernymynydd hoard found on Hafod Mountain near Mold, north-east Wales (Appendix 6, Figure 4.8). As with their Earlier Bronze Age counterparts, we do not know exactly where casting using this equipment took place. The absence of bronze moulds on settlement sites like the Breiddin could suggest either that they were never used and/or that they were excluded from deposition in such settings – although they may, of course, have been routinely recycled after use. Clay moulds, in contrast, particularly for weapons, were often included in deliberate deposits placed to mark significant stages in the lifecycle of buildings or enclosures, such as foundation or moving away (although it is unclear whether they were used in this way at the Breiddin) (Webley and Adams, 2016, p. 334; Webley, Adams and Brück, 2020, p. 185).

The Llwyn-mawr and Gwernymynydd bronze moulds are typical of other examples of similar date in Britain in that they are both for making axes (Webley and Adams, 2016, p. 324). Despite the availability of bronze as a possible mould material, Later Bronze Age communities in Britain seem to have chosen clay and stone moulds over ones of bronze to cast not only tools but also objects such as weapons, razors and ornaments. Only clay moulds were selected to make swords (Webley and Adams, 2016, p. 333; Webley, Adams and Brück, 2020, p. 185). These preferences can be seen in the variety of object forms represented by the clay mould fragments from the Breiddin. Altogether, the mould evidence points to the existence of social and spatial conventions concerning the materials and equipment involved in copper and bronze working during the early first millennium BC. It is noteworthy that no stone moulds assigned to the Later Bronze Age have been discovered in north Wales, when they comprise the bulk of the Earlier Bronze Age mould assemblage. As this cannot be explained by lack of suitable local sources of stone (Webley, Adams and Brück, 2020, pp. 73–4), it may point to changes in the technological choices underpinning mould selection in this region over time. These were perhaps linked to changing perceptions of the value of stone (e.g., Edmonds, 1995). The choice of mould medium could have revolved around socio-cultural perceptions of the differing sensory effects afforded by its material properties. Smashing a clay mould to release the freshly formed object, for example, would have been a moment of drama (Webley, Adams and Brück, 2020, p. 186). The spectacle afforded by the *chaîne opératoire* of melting and casting at the Breiddin would also have been enhanced if it was carried out at night, when it would have been easier to discern the flame colours and hence whether the right temperature for pouring had been reached (Barber, 2003, p. 122). Ethnoarchaeological studies of metalworking show that the performance of this field of activity, as well as the materials used, could be as important as the product itself for some communities (Webley and Adams, 2016, p. 333). This element of performance could explain why bronze casting took place at the Breiddin hillfort, a site that was perhaps an 'arena' for periodic communal gatherings involving acts of exchange or trade, ceremony or veneration, and display (Waddington, 2013, p. 97).

Copper and bronze recycling

The results of chemical and isotopic characterisation studies indicate that the recycling of copper and bronze by intentionally melting together different objects was a feature of the metalworking field of activity in Britain from at least 1600 BC onwards (Bray, *et al.*, 2015, p. 203; Wang, Strekopytov and Roberts, 2018, p. 103). It also took place in earlier periods, when it typically involved the remelting of a single object to create a new item (Bray, 2012; Webley, Adams and Brück, 2020, p. 45). Many metalwork provenancing studies, including interpretations of the Great Orme mine's ore 'signature' (e.g., Williams and Le Carlier de Veslud, 2019, p. 1183), consider the chemical and isotopic composition of Bronze Age metalwork to derive principally from the original geological ore source. Alternatively, it has been suggested that it is in fact the culmination of repeated instances of remelting, metal mixing and smithing (e.g., Bray, *et al.*, 2015; Radivojević, *et al.*, 2019). If the latter is indeed the case, then increases or decreases identified over time in the proportion of metalwork finds consistent with Great Orme ore may not be the direct outcome of changes in productivity at the mine, but due to variations in the technical sequence of bronze making and working. These could have included the increased frequency of recycling and greater metal mixing.

The results of recent trace element and compositional analysis of the bronze objects found in the Salcombe assemblage indicate that these items were produced by the mixing or recycling of different copper-tin alloys, whether from ingots or 'scrap' metal, rather than the alloying of raw copper and tin. It is also possible that copper ingots from the same collection (which, as we have seen, were made of metal possibly sourced from southern Europe) had been added as a third component to the metal mix (Berger, *et al.*, 2022, p. 23). This demonstrates how complex the copper-metal make up of an individual item could be. The researchers describe their results as:

> "the most tangible evidence to date for the mixing/ recycling of bronze in the Middle Bronze Age" (Berger, et al., 2022, p. 23).

With more such studies, it may become increasing difficult to tie metalwork definitively to a single specific source and from there to draw a straight line to peaks in mining activity.

A bucket containing a collection of broken and used bronze objects, recently discovered at the exceptionally well-preserved Later Bronze Age roundhouse settlement at Must Farm, Cambridgeshire (Wiseman, 2018, pp. 46–7; Knight, *et al.*, 2019) (Figure 1.1), is compelling evidence that, by the end of the second millennium BC, the *chaînes opératoires* of people's everyday lives involved gathering metal, presumably for remelting or reworking into new forms – although there is no evidence for either practice taking place onsite at this settlement (P. Northover 2021, personal communication, 2 November). Like the ingots from the seabed assemblages at Salcombe and Langdon Bay, the finds from this site are therefore not purposeful, curated deposits, but a glimpse of routine fields of activity in action (Knight, *et al.*, 2019, p. 661). Later Bronze Age communities appear to have considered 'finished' things of bronze or copper as much a source of metal for remelting as were copper ingots (Bray, *et al.*, 2015, p. 203).

At the Breiddin, no copper or tin ore, nor ingots of raw copper, tin or lead were found during excavation. There was also no trace of in-situ copper smelting (Musson, *et al.*, 1991, p. 149), an activity perhaps more likely to have been undertaken at or close to a mine source. As in the Earlier Bronze Age, it is unclear how the *chaîne opératoire* of bronze melting and casting at this hillfort was entangled with the other steps of metal making and working, and with contemporary networks of circulation, across the taskscape. The occurrence of casting at the Breiddin was very broadly contemporary with smelting at Pentrwyn on the Great Orme and so perhaps activity at the Great Orme mine during the Later Bronze Age (whether ore mining or scavenging; see Chapter Three) was part of the same metal making and working field of activity? No other copper mines have been found in Wales nor elsewhere in Britain and Ireland that are contemporary with metalworking at this hillfort. The copper being used there could therefore have originated at the Great Orme mine or else from a Continental ore source. The *chaîne opératoire* of casting at the Breiddin is, however, perhaps more likely to have involved recycling of bronze, using metal of multiple origins rather than from a single mine. This may have been added directly to the crucible as 'finished' whole items of metalwork, or as fragments after heating and then hammering to break up a sword, spearhead, axe or sheet metal into pieces small enough to fit into the melting pot (Musson, *et al.*, 1991, fig. 60, nos 255–9; Bray, *et al.*, 2015, p. 203; Knight, 2019, p. 268). Later Bronze Age smiths seem to have been able to manipulate and control the alloy properties of both recycled and 'raw' metal to produce the distinctive and specific proportions of tin-copper identified in bronze swords, shields and cauldrons in Britain from c.1300 BC onwards (Wang, Strekopytov and Roberts, 2018, p. 115). Bronze of similar alloy composition was used to make both tools and weapons, hinting that the making of these different object types was organised along the same lines – and perhaps even undertaken by the same people – at least at the casting stage (Radivojević, *et al.*, 2019; Webley, Adams and Brück, 2020, pp. 122–3).

Less than 5km from the Breiddin hillfort, the Guilsfield hoard of more than 120 objects, including the copper or bronze ingots previously discussed, a faulty spearhead casting and mould ingate, numerous broken weapons and 22 pieces of sheet metal, was discovered in the mid-19th century AD by workman digging in a field (Davies and Lynch, 2000, p. 182; Musson, *et al.*,

1991, p. 178) (Appendix 6, Figure 4.8). This assemblage has been interpreted as an archetypal Later Bronze Age 'scrap' or 'founder's' hoard: a large and varied collection of whole and deliberately broken metal objects, stockpiled temporarily for recycling, which the smith, for some reason, did not retrieve (e.g., Musson, *et al.*, 1991, p. 178; Bradley, 1998, p. 12; Davies and Lynch, 2000, p. 179). It is one of a series of such hoards assigned only to the Later Bronze Age in Britain (Webley and Adams, 2016, p. 334). The notion that such collections of objects were compiled as stocks of metal for practical use has its origins in longstanding assumptions about the deposition of bronze metalwork in wet places compared to in dry ground. The former is considered to indicate ritual intent and the latter to be evidence for strictly utilitarian activity, relating to the practical organisation of metal production. As many previous researchers have noted, this is, however, an artificial dichotomy. If the placing of metalwork in lakes or rivers (see Chapter One) can be interpreted as a meaning-filled practice integral to the way in which Later Bronze Age communities made and reworked ideas about the self and society, there is no inherent reason why dry-land deposits were not also made for similar reasons – particularly given that, as we have seen, cosmological beliefs are integral to the practice of all, even the most mundane, fields of activity (e.g., Bradley, 1998, pp. xviii–xix; Barber, 2003, pp. 44–53; Brück, 2015, 2016; Webley, Adams and Brück, 2020, pp. 118–9).

The link between the Guilsfield hoard and metalworking at the Breiddin is, therefore, ambiguous. There is no evidence to relate the hoard's deposition directly to the *chaîne opératoire* of metalworking at this hilltop enclosure, nor to demonstrate that it was part of the organisation of metal supply to the hillfort. The two events may not even be contemporary, as some of the swords in the Guilsfield assemblage have been dated from their form to at least a century before bronze casting at the Breiddin (Davies and Lynch, 2000, pp. 180–2). Evidence for the differential treatment of different objects in other dry-land 'scrap' hoards like that from Guilsfield – with some things being heated before deformation and breaking, while others were deposited in their as-used state – suggests that, rather than being a product of single instances of deposition, these assemblages were created as people deliberately selected, modified, arranged, added and removed pieces over perhaps long periods of time (Knight, 2019, p. 269; Webley, Adams and Brück, 2020, pp. 119–20). If contemporary with the Breiddin, the Guilsfield hoard could be a further indication that the curation of 'old' things (as suggested above in relation to the Deansfield stone mould) was part of the Later Bronze Age taskscape. The compiling, placing and curating of such a large collection of metalwork was clearly a significant act. Webley, Adams and Brück (2020, p. 120) suggest that it may have been considered a fundamental part of the metalworking process, essential to ensuring its conceptual success, even if the assembled copper and bronze was not all, or ever, intended for future recycling. The deposition of the Guilsfield hoard so close to the Breiddin is therefore unlikely to be coincidental. It perhaps represents the culmination of practices carried out primarily to evoke cosmological support for or in some other way to reference the bronze casting activities taking place at the nearby hilltop enclosure (after Webley, Adams and Brück, 2020, p. 120).

The character, scale and tempo of copper and bronze working

From the casting and metal recycling evidence, it can be concluded that copper and bronze working in north Wales was, throughout the Bronze Age, similar in character, scale and tempo to the other contemporary local fields of activity – dwelling, and tending to animals and the land – discussed earlier in this chapter. Despite the planning, careful choreography and degree of crafting skill involved in, for example, episodes of casting at the Breiddin, north Wales bronze smiths, like those elsewhere in Britain, do not seem to have been full-time economic specialists, producing for and supported by a network of consumers. Instead, they seem to have been members of the same communities that consumed the metalwork they produced (Webley, Adams and Brück, 2020, pp. 187–88). Nor does there seem to have been a formally organised, centrally controlled production 'industry' in or beyond the region to receive the outputs of the Great Orme mine, even during its customary phase of Bronze Age activity (cf. Williams and Le Carlier de Veslud, 2019; Webley, Adams and Brück, 2020, pp. 65, 125, 184).

Rather, the *chaînes opératoires* of copper remelting, refining, alloying, casting, smithing and finishing appear to have been carried out on a small-scale and dispersed across the taskspace, perhaps in ways that were entangled with the distribution of contemporary settlement and routine patterns of dwelling and mobility (Webley, Adams and Brück, 2020, p. 182). The metalworking residues from the Breiddin hillfort could be evidence for the metal making and working field of activity in this region becoming gradually less dispersed and more centralised throughout the second and into the early first millennia BC, as hillforts like this one, and larger enclosed roundhouse settlements, emerged. In Britain and Ireland as a whole, for example, there is tentative evidence for a shift in the organisation of casting and smithing, from tasks that took place in distinct spatial and/or social contexts to ones that, by around the middle of the second millennium BC onwards, were interlinked and so probably undertaken by the same practitioners making a variety of different object types. Definite workshops for casting or smithing, however, remain scarce, despite the dramatic increase in excavated sites in Britain with evidence for copper/bronze metalworking in the last 25 years, a consequence of the expansion in developer-funded excavation over this period (Webley, Adams and Brück, 2020, pp. 12, 66–7, 122, 182–4). Within this context, the casting and working of metal from the Great Orme mine may not have been undertaken by the same people as those who mined and then in turn perhaps also smelted it (Webley, Adams and Brück, 2020, p. 122). Wherever the actual smelting took place, primary smelted metal may have circulated in ingot form to locations that were both socially and geographically distant from the mine and its associated smelting sites. If the smelting evidence from Pentrwyn is representative of Bronze Age production on the Great Orme – and, as already discussed, there are good reasons why it may not be, not least its very late date relative to the mine's chronology – it may also indicate that, even by the early first millennium BC, the practice of copper smelting continued to be both small-scale and technologically simple, without any permanent, large-scale supporting infrastructure.

Like the regional traditions of building observed in Bronze Age north Wales, there would probably have been distinctive local differences in the metal making and working field of activity (Webley, Adams and Brück, 2020, p. 182). The apparent discontinuation of the use of stone moulds from the Earlier into the Later Bronze Age in north Wales points to a distinct regional practice there compared to elsewhere in Britain, but the production evidence is too sparse to identify with certainty other inter- or intra-regional patterns. Regionality has been interpreted from the results of the analysis of the form; chemical, isotopic and alloy composition; and metallographic structure of the far more ubiquitous finds of copper and bronze metalwork. Trends in these data have been used to develop a model of distinctive regional 'industries' or 'traditions' of metalworking across Bronze Age Britain, in each of which smiths produced particular forms of copper and bronze objects using specific metal compositions or 'types' (Northover, 1980a, 1980b). In this model, smiths in north Wales working in the 'Acton Park' tradition, for example, around the mid-second millennium BC, are thought to have alloyed large quantities of metal originating from the Great Orme mine with tin and lead to create a tin bronze with exceptional casting qualities, toughness and hardness, which they worked into a new innovative design of axehead, the palstave (e.g., Lynch, 2000, pp. 103–5).

The analytical data may point to commonalities and differences in, for example, the amount of tin and lead alloyed to copper, but they could also indicate variations in the trace element composition of the ore used. The model, however, conflates the distribution of items of deposited metalwork with the organisation of production, by identifying the centre of each 'industry' from the spatial patterning of metal finds. Instead, as we saw in a preceding section, many individual metal objects had complex life histories, travelling through a dispersed *chaîne* of production and then numerous spheres of circulation and consumption on a range of spatial and temporal scales before finally being deposited. The metalwork available now is also only a partial sample of the things that Bronze Age communities were making and depositing, as the double-looped socketed chisel matrix on the Bodwrdin mould indicates (Appendix 6).

Peter Bray (2019, pp. 128–9) has recently drawn attention to some important characteristics of the analytical data available for Earlier Bronze Age copper/bronze artefacts from Britain: what he calls their 'clumpiness'. The same suite of techniques, e.g., compositional analysis, hardness testing and metallography, tends to be used in conjunction, but rarely with, for example, radiocarbon dating of associated deposits. This has led to a tendency to come up with the same types of explanations for the material. These data therefore provide only a partial glimpse of the full life histories of individual or groups of objects and how these overlapped across the taskscape. Perhaps the use of prosopography, a methodology for comparing, synthesising and statistically analysing all known information on a topic, no matter how fragmentary (Bray, 2019, pp. 129–31), may eventually help to write a narrative of the Bronze Age metal making and working field of activity that fully integrates both the analytical and production data from north Wales, as well as the rest of Britain and Europe.

The extent to which individual Bronze Age communities in north Wales would have identified with or understood themselves to be part of a regional 'Acton Park' metalworking tradition is also very much open to question, given the character of casting and smithing described earlier. Some researchers have suggested a direct link between a peak in productivity at the Great Orme mine in the middle centuries of the second millennium BC and the development of the 'Acton Park industry' in north Wales, as we first saw in Chapter Two (e.g., Williams and Le Carlier de Veslud, 2019, p. 1192). This model presupposes that a high ore output created the conditions for innovation in axe design by large groups of smiths, at the same time as the development of the palstave and its role in the intensification of agriculture drove increased ore production. This reasoning does not, however, explain what may have been the initial stimulus for the proposed production increase at the mine. As we have seen, there is overall no evidence for a corresponding sharp intensification in the scale and organisation of metalworking in Britain at any time during the Bronze Age, nor for the routine practice of contemporary bronze smithing in north Wales on a large scale. The model of the metalworking field of activity presented here (and supported by other recent studies, e.g., Webley, Adams and Brück, 2020) as small in scale and community organised accords well with the conclusions about the scale and character of mining on the Great Orme reached in Chapter Three.

In this chapter, the character, content and scale of the copper and bronze making and working field of activity has been considered separately to the ways in which objects made of these materials may have moved. This separation is, of course, artificial as the production, circulation and use of things are all part of a unified cycle of interaction. In Bronze Age north Wales, as we saw earlier, the value of things made of metal was created through the character of the exchange relations of which they were a part. The tasks of metal making and working during the second and early first millennia BC would also have contributed to the ways in which a thing of copper and bronze was valued and consumed by Bronze Age communities (Barrett, 1985; Barrett and Needham, 1988, pp. 127–8). We will now return once again to the Great Orme mine to consider what role mining there during the Bronze Age may have played in the reproduction of the social relevance of copper and bronze.

5

Mining and the construction of community

In 2005, Prof. Alice Roberts (University of Birmingham) visited the Great Orme mine while making a television programme for the BBC (David, 2005, p. 146). She examined the only human remains known from this site: a lower jaw (mandible) and part of a collarbone (clavicle), plus a few loose teeth (Lewis, 1996, p. 126). These few finds are a reminder that a social prehistory of mining at this – or any other – mine is as much a narrative about people and the creation, reproduction and reworking of relations between them as it is about the technological sequence of copper ore extraction, processing and smelting. The preceding chapters have examined the material conditions of mining on the Great Orme and the sequence of the mine's development over time (Chapter Two); the content of the mining field of activity at this site (Chapter Three); and its wider practical and social context (Chapter Four). This chapter considers Bronze Age activity at the Great Orme mine from the perspective that it was a "total social fact" (Mauss, 1979 [1935]). The evidence for the scale and form of the social relations characterising the labour force, and its age and gender make up, will be assessed. We will explore in detail the archaeological evidence for the concept of personhood in Bronze Age north Wales and the implications of this for the role participation in mining on the Great Orme played in the making and (re)working of both individual and group identities. Undertaking the *chaînes opératoires* of mining not only reworked ideas about the self but also the worldviews of the participants, and the evidence for the various ways in which this may have occurred will be reviewed. The chapter closes with an appraisal of the nature of the intersecting relationship between mining, politics and power at the Great Orme mine, including the question of whether this source was ever 'owned' or access to it controlled, in prehistory. We will begin by considering in more detail the evidence for the structure of social life in Bronze Age north Wales, returning again to the human remains from the mine as our starting point.

The structure of social life in Bronze Age north Wales

Both the jaw and the collarbone found at the Great Orme mine have very similar calendrical radiocarbon dates (OxA-14308; 3344±27 BP; 1730–1535 cal BC; 95% probability, and OxA-14309; 3362±28 BP; 1740–1540 cal BC; 95% probability) (Appendix 2), a time frame that corresponds to the modelled start of Bronze Age activity at the mine. As we saw in Chapter Two, a date overlap does not necessarily indicate exact contemporaneity. Finds of human remains from European prehistoric mines are generally very scarce (O'Brien, 2015, pp. 260–62), however, and so it seems reasonable to conclude that both the jaw and the collarbone are from the same individual, possibly an adult male in his early 20s (Lewis, 1996, p. 126; James, 2011, p. 127). The results of recent oxygen and strontium stable isotope analysis of the tooth enamel from the mandible indicate someone who was local to the Great Orme (David, 2005, p. 146). The rugosity of the muscle and ligament attachments of the collarbone points to repeated powerful shoulder actions (R. Schulting 2001, personal communication). This, together with the fact that the remains were found at the mine, suggests that this individual was directly involved in mining sometime during the beginning of Bronze Age activity there. Both bone fragments were, however, found out of context (Appendix 3) and so why and how they came to be deposited in the workings are unclear. In common with the handful of human remains discovered at other prehistoric copper mines in Europe (O'Brien, 2015, p. 260), accidental death is one explanation for the presence of this individual, or they may instead have been intentionally selected for burial at the source, a point returned to later.

These remains do not tell us whether the miners were always, or all, local throughout the at least 400 years of Bronze Age extraction at the Great Orme mine. Apart from the human skeletal fragments, other traces of the miners themselves are lacking. There are no contemporary adjacent settlements nor burials that have been linked to the mine. Various human remains have been discovered in several caves on and around the Great Orme, and are certainly prehistoric in date (e.g., Chamberlain, 2014; Smith, Walker, *et al.*, 2015, p. 114). Only two examples – skeletal fragments from an adult and a child found in North Face Cave (Ogof Rhiwledyn) on the Little Orme – have been securely dated to the second millennium BC (OxA-16568; 2979±35 BP; 1375–1055 cal BC; 95% probability, and SUERC-62072; 3065±36 BP; 1415–1225 cal BC; 95% probability) (Schulting, 2020, p. 189) (Figure 1.4). While contemporary to the mine-working and found only c.5km from the mine, there is no evidence directly linking these individuals to this activity.

The evidence from the mine itself, such as the origin of the rock types forming the cobblestone tool assemblage, are the product of complex and varied procurement and use strategies, not a straightforward index of the miners' 'provenance'. The coastal and riverine setting of the Great Orme would have been accessible to groups coming from across north-western and central England, south-western Scotland and the Isle of Man, as well as from elsewhere in Wales. Its topographic situation raises the possibility that, within the pattern of mobility discussed in Chapter Four, communities from these different locations may have come together on the headland to participate in the mining process. The ore and/or metal outputs from this activity were then perhaps similarly dispersed in many different directions. The close entanglement between the *chaîne opératoire* of Bronze Age mining at the Great Orme mine and other fields of activity in the contemporary taskscape means that, to understand in more detail who the miners may have been, it is necessary to look first at the scale and structure of the broader social landscape. This would have defined the organisation of the labour force for each mining event in terms of the relations between participants, their age and gender (Wager, 2009, p. 111). One way to investigate this social landscape is from the contemporary settlement and burial evidence.

The isolated and scattered single unenclosed roundhouses in Earlier Bronze Age north Wales seen in the previous chapter point to a society structured around small-scale, intimate kin-based relations, with the members of an extended family living in individual farmsteads or two or three dwellings spread across a hillside. Each such community would have been tied into socially and geographically broader crosscutting networks of affiliation, alliance and reciprocity (and their associated relations of obligation and debt). The configuration of these relations and their geographic location may have shifted from one generation to the next, as each roundhouse was perhaps abandoned on the death of its owners. This is certainly a possible explanation for why so many of the Earlier Bronze Age settlements excavated in southern Britain are single-phase sites (Brück, 1999a, p. 149).

Single inhumation burials and their associated grave goods (Tellier, 2018, pp. 112–4), like the Beaker and two copper or bronze knives accompanying the inhumation burial in a barrow at Darowen, Powys (Tellier, 2018, p. 161), have been interpreted as the burials of high-status individuals: important chiefs or community leaders (Figure 1.2). This is seen as evidence for the emergence of institutionalised structures of authority within a vertical social hierarchy. The dead do not bury themselves (e.g., Barrett, 1988, p. 31), however, and so while grave goods may sometimes have belonged to the deceased in life, they could have belonged to and been collected together by the mourners. As a result, the items selected for burial did not serve only to reflect the wealth or status of the deceased, but to represent perceived symbolic or idealised aspects of the dead individual's roles and identity. They may have been part of the process by which the social relations between the living were renegotiated following a death. There is also evidence of things being made specifically for burial or as part of the mortuary rite. Objects such as knives and awls could have been used to prepare the funerary feast or sew the burial shroud before being interred (Tellier, 2018, p. 113; Brück, 2019, chap. 3). Grave goods cannot therefore be 'read' as straightforward indices of the 'wealth', role and identities, or social position of the deceased person (e.g., Barnatt and Smith, 2004, p. 36; Brück and Jones, 2018, pp. 237–8; Brück, 2019, pp. 88–9). Rather, the specific biographies of these things and their deliberate placing around the body, together with the other rituals and architecture of death, were a means for the remaining members of a given community to rework and renew the fabric of social relations structuring their lives. They provided the living with a way of dealing with the rupture and instability inherent in death and the concerns it may have provoked, such as issues of inheritance (Barrett, 1988, p. 31, 1989b, p. 124; Brück, 2019, chap. 3).

It is highly likely that not everyone in a position of authority in Earlier Bronze Age north Wales received formal burial in a burial mound or cairn. Only a few members of a given community, probably fewer than one individual per generation, were treated in this way on death (Lynch, 2000, p. 121; Barnatt and Smith, 2004, p. 36; Caswell and Roberts, 2018, p. 15). They may have been chosen not because they were important chiefs or tribal leaders but because they were perceived, through other aspects of their social identity, such as their relative position in kinship networks, to best represent a community as a whole and the interpersonal, familial and genealogical ties materialised through the burial rite (Barnatt and Smith, 2004, p. 36; Brück, 2019, p. 65). This may have drawn on aspects of their identity or role when living, like their importance to intergroup alliances, or the possession of shamanistic powers (Barnatt and Smith, 2004, p. 36). The spectacular embossed gold cape found wrapped around a skeleton in the 'Golden Barrow', Mold, for example, appears to have been attached to a lining and so worn in life (Appendix 6, Figure 4.5). As its form would have significantly restricted movement of the arms and upper body and hence the wearer's bodily engagement with the world, it is interpreted as an item of regalia perhaps reserved for a shaman or 'priest' (Lynch, 2000, pp. 102, 112; Anon, 2013). While its presence in the grave no more makes the specific individual wrapped in it a shaman than the accompanying urn makes them a potter (after Barber, 2003, pp. 125–8), it certainly seems to speak to the particular moments of ceremony or the places when it was worn, or of the identities of those associated with those events. These qualities were, however, then redefined by the mourners during the mortuary rite. The cape itself had been cut and folded at some point before deposition, implying an act of symbolic decommissioning, perhaps to mark the death or to contain the power of a ritually dangerous object (after Brück, 2019, p. 86). This splendid item may also have been curated as an heirloom for some time before being placed in the grave, perhaps to stress family or genealogical relations.

People often reopened and enlarged burial mounds and cairns after initial construction to insert new burials of both burnt and unburnt bodies (Tellier, 2018, pp. 111–14). At Merddyn Gwyn on Anglesey, for example, a small round cairn holding a central inhumation burial was enlarged to take an urned cremation accompanied by a ceramic Bowl (Figure 4.5). Both were placed just outside the original kerb. Two further urned cremations were subsequently dug into the top of the enlarged mound (Lynch, 2000, pp. 122, 126). At Llong, near Mold, the small cairn covering the central primary inhumation burial was later enlarged asymmetrically to the west and two or three unaccompanied cremation burials added to the mound (Lynch, 1983, p. 17). This reuse typically occurred over periods of 100 to 200 years (Tellier, 2018, p. 111), so for perhaps no more than five generations before a burial

monument was finally abandoned. This suggests that burial of both burnt and unburnt bodies in many cases involved interaction of the living with known deceased individuals from the same lineage or community.

aDNA analysis, however, from elsewhere in Britain shows that the make up of each 'family' or kinship group did not always depend on direct lines of biological descent but on relations that were complex and variable. Close relatives were not always buried in the same cemetery, which could sometimes contain the remains of individuals placed in proximity but with no shared genetic ancestry, such as the four biologically unrelated pairs of individuals interred at Amesbury Down cemetery, Wiltshire (Booth, *et al.*, 2021, pp. 8–13). As Booth, *et al.* (2021, p. 8) point out, ethnographic studies of kinship structures reveal how these are socially mediated rather than a predetermined outcome of biological relatedness. Paternity, for example, may not be assigned at birth but may have to be earned, and so may not match the genetic reality. In Bronze Age north Wales, ties of kinship could have been made and sustained through the practice of various fields of activity such as, for example, people living or labouring together, a point we will return to later. However, stable isotope analysis of human remains is revealing how even genetic relatives, who on death were buried together, did not necessarily dwell in the same area when living. aDNA evidence also indicates regional variability in kinship structures, with a possible system based on matrilineal descent identified at Trumpington Meadows cemetery, Cambridge, compared to paternal links as the key organising principle at Amesbury Down (Booth, *et al.*, 2021, p. 14).

The importance of small-scale familial ties, localised lines of authority and close connections and cooperation within and between kinship groups – however these were constituted and understood – appears to have increased throughout the Bronze Age (Brück, 2019, pp. 159–62). The feasting and other communal displays of conspicuous consumption and reciprocity taking place at the larger enclosures that emerged in some parts of north Wales around the turn of the first millennium BC would have served to strengthen the social and cosmological order. At the same time, the boundaries of these sites, whether ditches or banks of timber, earth or stone, and their frequently impressive topographic locations physically marked these areas off as distinct moments in the taskscape. While this would have signalled and reinforced social differences between individuals, particular families or kinship groups – between those hosting and those attending an event, for example – there is still no convincing evidence for the emergence of chiefdoms or other formalised vertical social hierarchies in the Late Bronze and Earliest Iron Age (Waddington, 2013, pp. 97, 115; Brück, 2019, pp. 155–8). Overall, the contemporary burial and settlement record of Bronze Age north

Wales points to a small-scale, communally organised, socially and geographically dispersed society, whose largely horizontal social structure emphasised ties of actual, socially constituted or mythologised lineage and genealogical ancestry. It appears to be typical of the form of social organisation characterising life elsewhere in Britain during the same period (e.g., Brück, 2019).

The scale and character of social relations at the mine

The concepts of kinship, alliance and obligation that were created by exchange and structured social life in north Wales throughout the Bronze Age would have been important principles shaping the composition of the workforce at the Great Orme mine. Ethnographic studies illustrate how the configuration of various forms of social relations plays a role in determining who is involved in the mining field of activity. Among the Toro of Uganda, for example, prospection for iron ore was limited to men, typically those who were older, married and experienced in iron production. The only younger men allowed to participate were their sons and nephews, subordinate male relations over whom they could exercise control in line with the prevailing social hierarchy. Once ore had been discovered, the news travelled rapidly and lots of people, mainly members of the discoverer's family and clan, would gather at the source to begin mining. Ironworkers from other clans and villages were also involved, but only those who were considered close kin after undergoing a blood-brother ritual with the discoverer (Childs, 1998, pp. 127–9). At the Great Orme mine, mining could have been undertaken by people between whom ties of close kinship existed and who identified themselves as members of the same community. As we saw earlier, however, the composition of these groupings is likely to have been more fluid and complex than 'family' units of closely genetically related individuals (cf. Williams and Le Carlier de Veslud, 2019, p. 1192). The Great Orme workforce may also have been structured according to other forms of relations, crosscutting notions of kinship and arising out of, for example, acts of reciprocity. In this case, mining could have been undertaken by people who belonged to different communities, but who were nonetheless linked together through ties of obligation or intergroup alliances.

Examination of the layout and form of the prehistoric mine and the physical structure of the source suggests that both readings are relevant to our understanding of the character of the labour force there over the Bronze Age life of the mine. The restricted size of many of the prehistoric workings, such as those at Loc.21, means that few people would have been able to work in a particular passage at any one time. The entrances at the surface to specific networks of underground working are often clustered together, in discrete parts of the source. The areas of working focused around and adjacent to Loc.7c (Surface), for example, appear to be bounded entities, defined by upstanding ribs and folds of barren bedrock (Figure 3.6). Five entrances into Loc.17 occur in proximity in the Tourist Cliff. The small and apparently isolated outcrop of rock in the 'North-West Corner' is riddled with passages. This spatial patterning suggests that, in each case, we are seeing distinct areas of working. It points to mining by small groups of people who were closely affiliated. They could have been close kin – people who typically resided and laboured together – together with more distant relations, perhaps those with obligations to discharge.

These groupings are likely to have reproduced those occurring at other places in the local taskscape throughout the Bronze Age. The small size of many stone circles in north Wales points to their use by individual communities. The 4m diameter of the stone circle known as Penmaenmawr 280 (HER PRN GAT544 [accessed online 18-06-2022]), in the Penmaenmawr hills close to the Great Orme (Figure 4.5), certainly evokes the practice of more intimate rites and a concern with expressing and maintaining immediate ties, rather than those stretching over extended spatial and social distances (e.g., Edmonds, 1999a, pp. 148–9). The layout of single and scattered unenclosed roundhouses and their associated fields, like those in the Great Orme hinterland, also points (if contemporary) to the cooperation of a few small 'family' units in tending to animals and the land. As discussed in the opening chapter, the absolute number of individuals making up the mine workforce for any given mining event may never have been large – and it was certainly never 'industrial' in scale. The piecemeal patterning of mine-working identified in Chapter Two could be most representative of a scenario in which only a limited number of stopes and passages, and their associated entrances, were being worked at once. The spatial arrangement of the mine workings would, however, have provided opportunities for relatively larger groups of people to participate than was possible in confined passages. There are low but laterally extensive mined-out areas at Locs. 10 and 35 (Figure 2.16), for example, as well as huge chambers such as the Lost Cavern and Loc.18. Mining in such spaces was perhaps more likely to have involved the members of more than one community working together, possibly on a seasonal basis, some of whom did not habitually dwell or labour alongside one other.

There are numerous ethnographic examples highlighting that even small-scale societies without centrally organised power structures can organise the labour for large-scale communal tasks (Kienlin, 2013, p. 425). The scale of this group activity at the mine similarly evokes that at the earliest hillforts of the Later Bronze Age. There is, however, no dated evidence securely tying mining in the Lost Cavern and Loc.18, where larger numbers of people could have taken part, to particular periods in the Great Orme mine's development. Larger-scale collaborative working was a component of the north Wales taskscape even during earlier periods. The dimensions of some circular monuments, like the Late Neolithic henge monument (Henge B) at Llandegai near Bangor, for example, indicate that they were places where a large number of people gathered, emphasising the extensive ties that existed between communities (Edmonds, 1999b, p. 486) (Figure 4.5). Their construction would also have required large numbers of people, most probably drawn from more than one 'family' or community. Such tasks helped to create the historical conditions for copper mining at the Great Orme mine, which may have unfolded at a scale greater than that of residence and close kinship from perhaps even its very earliest phases. Each mining event would therefore have been an occasion when the members of different – and perhaps socially and geographically dispersed – communities came together.

Each rock outcrop in the Pyllau Valley may have been a focus for mining by people who, while sharing strong links with each other, were less closely connected to those working at another point in the source. Of course, it is currently unknown whether activity at each outcrop was in fact exactly contemporary, but an explanation for the piecemeal patterning of extraction evident in the mineralisation accessed through the Tourist Cliff is that each mining event typically comprised several small groups labouring at several different locales across the mineral deposit. Hence, differentials between people, such as who could work with whom, could be partly responsible for the character of working at this mine. The convoluted form and layout of the Great Orme mine points to the accumulation over a long period of the knowledge and technical expertise needed to extract ore from this source at considerable distances from the surface. This, together with the generally consistent choreography of the mining *chaîne opératoire*, hints that the Great Orme mine was, like some burial monuments and other significant places in the contemporary taskscape, such as the nearby monument complex on Penmaenmawr (Figure 4.5), visited and worked by the same communities for generations, an issue that will be returned to. This does not, however, mean that the structure of the social relations of the labour force at the mine was always the same. Rather, it would have been messy and flexible, as the configuration of kinship ties was, for example, transformed by events in the broader taskscape, such as a marriage or a death. This fluidity in the social make up of the groups mining at the Great Orme mine could even have influenced the form in which the 'products' from each mining event were dispersed from the source. For some individuals or social groupings, a single batch of ore could, for example, have been processed to produce more than one type of concentrate, composed of varying combinations of mineral and gangue, for differing purposes. For others, it first had to be transformed into copper metal by smelting.

The age and gender structure of the labour force

In her analysis of recent iron ore mining by the Toro as a fundamentally social activity, Childs (1998, p. 129) identifies a clear gendered division of labour, with only men allowed to participate in digging out the ore. Understanding the gender and age make up of the Bronze Age labour force at the Great Orme mine is far more challenging. As we saw earlier, the only human remains found at the site are possibly from an adult male, while the very narrow width of some of the smallest workings – only 30cm wide in places at Loc.21, for instance, with more restricted examples known throughout the mine (Lewis, 1996, p. 105, App. C, 11) – could point to the use of children to extract ore underground (e.g., Lewis, 1996, p. 105). Other explanations are also possible: some of the very narrow tunnels could in fact be natural cave passages, part of the underground karst system, or the miners could have excavated inwards at arm's length from either end of some of the tightest passages (Wager, 2009, p. 112). Some of the cobblestone mining tools are small and irregularly shaped, rather than ovoid. One explanation for this (others were discussed in Chapter Three) is that individuals chose stones of a size and shape to suit their own grip (Dutton, *et al.*, 1994, p. 269), pointing to the participation of women or younger age groups at the source. In contrast, most of the bone tools appear to fit adult and not child-sized hands. The rib tools are an exception to this (James, 2011, p. 180), although it is difficult to interpret this finding in age or gender terms as these particular tools were not preferentially selected but used opportunistically when to hand. Many of the tasks making up the total mining field of activity, such as collecting wood and cobblestones, preparing food, processing ore, repairing tools and tending fires, could have been carried out by anyone at the source (e.g., Williams and Le Carlier de Veslud, 2019, p. 1192). This group could have included younger and older people as well as those with mild physical or cognitive disabilities.

Together, these points hint that the labour force for any given mining event at the Great Orme mine during customary Bronze Age mine-working there could have had a broad and variable age and gender structure, involving adult men and women, and children. A similar community profile has been identified from osteoarchaeological analysis of burials in a cemetery associated with Early Iron Age salt mining at Hallstatt, Austria. Differences in the muscle groups developed in biological women and men at that site reveal the gendering of different tasks associated with this activity, with men making repeated striking movements (presumably to extract the salt) while women regularly lifted, carried or pulled heavy loads, an essential task both underground and at the surface. The skeletons of children, including those from 'wealthy' graves, also had anatomical changes likely to have been caused by habitually carrying heavy loads on their heads, suggesting that they too were involved in the movement of the salt after extraction (Kern, *et al.*, 2016, p. 140; Pany-Kucera, Kern and Reschreiter, 2019). The Hallstatt example confirms that not only men were involved in hard physical labour in prehistory. This point is reinforced by the remains of a biological female found in a rock fissure on the Little Orme, a headland immediately to the east of the Great Orme. Radiocarbon dated to the Earlier Neolithic (Beta-87306; 4720±50 BP; 3635–3370 cal BC; 95% probability), this woman's skeleton showed a pattern of degeneration consistent with using her head to carry heavy loads (Gregory, *et al.*, 2000, p. 7).

At the Great Orme mine, unlike at Hallstatt, we cannot identify with any certainty which specific tasks were assigned to men, women, boys or girls. The age and gender profile of activities like extracting ore, making tools or hauling waste rock would have referenced that of other *chaînes opératoires* in the contemporary taskscape, such as agriculture, where there was overlap in the materials, tools or techniques used (e.g., Stig Sørensen, 1996). Examples of such loci of interconnection can be observed ethnographically. In sub-Saharan Africa, carrying tasks are typically assigned to women, whatever the context, and so it is almost always they who transport iron ore away from the source after mining (Herbert, 1993, p. 28, 1998, p. 149). Unfortunately, we do not know who in Bronze Age north Wales used bronze axes, bone or cobblestone tools, for example, nor who was responsible for digging fields, grinding corn or making things from clay, such as cooking pots, crucibles and moulds (Brück, 2019, pp. 159–60; Webley, Adams and Brück, 2020, p. 62). We cannot read the gendering of everyday activities from grave goods, which, for the reasons already discussed, are most likely to reflect idealised identities, rather than the gendering of roles amongst the living (Tellier, 2018, p. 112). Similarly, while the layout of settlements and individual roundhouses points to the spatial division of tasks, it is not known whether or in what ways these were associated with different age and gender categories in Bronze Age Britain (Brück, 2019, p. 144). Whether men, women or children were responsible for carrying out specific stages in the *chaîne opératoire* of mining at the Great Orme mine, participation in (or exclusion from) these various tasks would have been caught up in the reproduction of 'personhood': what it meant to be a person in Bronze Age communities, including how people understood concepts of self-identity such as age and gender (Harris and Cipolla, 2017, p. 52). Such ideas, constituted through the performance of activities in the wider taskscape, would have been restated and reworked as people engaged in mining. Mortuary rites provide the best window into concepts of personhood in Bronze Age north Wales.

Concepts of personhood in Bronze Age north Wales

During the latter third and early second millennia BC in Wales, disposing of the dead was a complex and protracted field of activity, in which bones were selected and removed from both burnt and unburnt bodies for curation, circulation, display and deposition in new bodily configurations elsewhere. Deliberate manipulation of the human body, both before and after deposition, was a key feature of the *chaîne opératoire* of burial (Tellier, 2018; Brück, 2019, p. 62). The low recorded weight of cremation deposits indicates that, while there is evidence from pyre sites for the almost complete collection of human bone, not all was then placed in the grave. Increasingly throughout the second millennium BC, cremation burial comprised only a token handful of bone fragments. The cremated remains from the Moel Goedog 1 burial cairn in south-west Gwynedd, which had been moved to this location from an initial burial site (Lynch, 1984, p. 28), as well as those disturbed soon after deposition in one of the burial mounds at Trelystan 1 (Welsh Marches), show that people also removed bone fragments from graves after the initial internment (Figure 4.5). Cremated bone often seems to have been curated for a time before deposition (Tellier, 2018, pp. 114–7, 122; Brück, 2019, pp. 32–4). The mingling of the burnt remains of several people in the same burial was also a relatively common feature of the mortuary *chaîne opératoire*, although it occurred less frequently as the second millennium BC progressed, with 90% of cremation deposits in Wales after c.1700 BC containing fragments of only one individual (Tellier, 2018, pp. 114–7).

Even in inhumation burials, unburnt bodies were rarely deposited intact and whole. Although the generally poor preservation of bone in north Wales could account for most skeletons recovered in that region being incomplete, burial evidence from elsewhere in Britain indicates that adding, removing and rearranging body parts in the grave was a common practice during this period (Tellier, 2018, p. 113; Brück, 2019, p. 62). In some instances, bodies were disarticulated by dismemberment or excarnation, before being inserted into existing burial monuments, as at the Ffridd y Garreg Wen and Crown Farm burial mounds in north-east Wales. On other occasions, they were placed into pits, like the disarticulated remains of at least four individuals found in a natural mound at Hendre, also in north-east Wales (Tellier, 2018, pp. 111–3) (Figure 4.5). Evidence from elsewhere in Britain points to the treatment of unburnt bodies in a variety of ways before burial, including mummification and smoking (e.g., Booth, Chamberlain and Pearson, 2015; Brück, 2019, pp. 24–7).

Mortuary practices in north Wales may have involved similar technologies (Tellier, 2018, p. 117).

So what does the fact that it was "ideologically acceptable" to separate and disperse the human body (Brück, 2019, p. 62) reveal about concepts of the self during the Earlier Bronze Age? Joanna Brück and others argue that it points to a relational understanding of personhood, in which the self was seen to be composed of unbounded, non-unified, separable parts, which were capable of being transformed and combined in new ways over the course of a person's life (e.g., Fowler, 2001; Brück, 2002, 2006a, 2006b, 2019; Harris, 2020, p. 129). In this 'dividual' concept of personal identity, which was very different to the characteristically modern Western understanding of the person as a bounded, homogenous, independent individual (Harris and Cipolla, 2017, pp. 61–6; Brück, 2019, p. 62), part of the self transcended the confines of the physical body, to be constantly created, maintained and reworked through a person's relationships with other people, places and things. Notions of personhood constructed in this way were therefore shifting and unstable, contingent on context and the demands of others.

One way that these ideas were reproduced was through the placing of objects in the grave. In Welsh inhumation burials, tools (bronze and flint knives, daggers and flakes, bronze awls) and items of personal ornament (jet/lignite buttons and necklaces, beads of amber, bone and stone) are more commonly found alongside adults than individuals who died before reaching full skeletal maturity, i.e., 'non-adults' or those younger than 18 years old (Tellier, 2018, pp. 50, 113). These items may have been chosen by the mourners to express ideas of adult identity due to aspects of their materiality such as colour. This perhaps symbolised personal qualities or relationships which may have been reserved for adults, like specific kinship ties created through inter- and intra-group marriage. Alternatively, objects like amber beads may have been intended to evoke fields of action that only adults could participate in, such as gift giving and its associations with connections and ideas relating to boundaries and the movement between worlds (Brück, 2019, pp. 86–89). Age distinctions were also recognised in Earlier Bronze Age cremation rites in Wales, with the burnt bodies of non-adults more likely than those of adults to be interred with a ceramic Food Vessel (Tellier, 2018, pp. 115–6). Brück (2019, p. 50) has suggested that the presence of pottery vessels with cremation burials was a 'second skin' intended to replicate or replace the relationship between the recognisable, whole body of the living individual and its disaggregated, indistinguishable burnt remains. Perhaps it was more important to make this connection when burying younger people rather than adults?

The ways men and women were treated in death reflected their different roles in kinship structures and the reproduction of ideational gendered identities within a sociocentric conception of society (Brück, 2009). Biological males are overrepresented in inhumation burials in Wales (Tellier, 2018, p. 112), for example, indicating that male gendered identities or lines of descent occupied a significant place in Chalcolithic and Earlier Bronze Age genealogical networks and ancestor/founder cosmologies. This picture is reinforced by the results of recent biogenetic studies of inhumation burials elsewhere in Britain, which demonstrate that paternal relations and patrilineal lineages were important factors structuring social identify during this period. Maternal connections were, however, sometimes also expressed through the burial rite and biological females were on occasion allocated a prominent central position in the grave monument, highlighting that some women may have had significant status and played a role in ancestral lineages (Booth, *et al.*, 2021, pp. 13–14). In contrast to inhumation deposits, early cremation burials in Wales show no biological sex selection (Tellier, 2018, p. 115). The act of cremation fragmented the human body, creating inalienable tokens of human bone – 'ancestral relics' – that could be distributed to the mourners for curation, gift exchange and deposition elsewhere in the taskscape. In this way, the living were able to make, rework and restate social and political identities for both themselves and the dead (Brück, 2019, p. 62). Elsewhere in Britain, there is a bias towards biological females in Earlier Bronze Age token cremation deposits. This perhaps points to the important role female gendered identities played in making and sustaining relationships created through different types of marriage arrangements both within and beyond the immediate kinship group (Brück, 2009, 2019, p. 38). In Wales, no demographic data are available from token cremations with low bone weights (Tellier, 2018, p. 115). The absence of biological sex selection in more substantial cremation burials suggests that, in this region, both men and women were considered appropriate for symbolising and evoking the relations expressed through the creation, manipulation and deposition of burnt bone.

The fragmentation and dispersal of the human body after death continued to be an acceptable practice in Wales throughout the second and into the early first millennia BC. Like elsewhere in Britain, there was a shift from the practice of both inhumation and, more commonly, cremation burial at the beginning of the Earlier Bronze Age to cremation only from around 1700 BC (Tellier, 2018, table 47). From this period onwards, no patterns in age or biological sex can be identified (Tellier, 2018, p. 117). Although this may be due to the small size of the dataset from Wales – there are only 50 cremation burials known in Wales dating from c.1700 to 1200 BC, compared to 117 in the preceding 500 years (Tellier, 2018, table 47) – it is typical of cremation burials throughout Britain from the 17th to 12th centuries BC. These are characterised by the deposition of both sexes and all age categories, usually with few or no grave goods other than pottery vessels (Caswell and Roberts, 2018, p. 13). There are no inhumation burials in Wales with radiocarbon dates later than the 18th century BC, although they do occur occasionally elsewhere in Britain (Tellier, 2018, p. 117; Brück, 2019, p. 42). The 'Golden Barrow' inhumation burial at Mold, which contained the spectacular gold cape discussed earlier and which is dated by associated grave goods thought to have been produced between 1900–1600 BC, is a rare example of the later burial of an unburnt body (Lynch, 2000, p. 102; MacGregor, 2012, pp. 101–2; Anon, 2013) (Appendix 6, Figure 4.5). As the skeletal remains were unfortunately dispersed soon after discovery in 1833, there is no information on the age or biological sex of the interred individual. It has been suggested that the narrow dimensions of the gold cape would only have fitted an adult woman or an adolescent male (MacGregor, 2012), which indicates an association between either of these identities, rather than adult male constructs of personhood, and the ceremonial or ritual places and events signalled by the presence of this object in the grave. By the Earliest Iron Age, the mortuary *chaîne opératoire* of cremation and, most probably, excarnation, resulted in very extensive fragmentation and scattering of the human body (Brück, 2019, chap. 2). Notions of personhood for communities in this region were clearly still characterised in this later period by a relational conception of identity, in which there were many different components to the self, each constantly being remade through relations with particular people, places and events (Brück, 2006a, 2019).

From the mid-second millennium BC onwards, the increasingly monumental architecture of settlement and the fields of activity associated with dwelling seem to have become important areas for the reproduction and reworking of ideas about society and the self, including concepts of age and gender (Brück, 2019, pp. 116–7, 134). Using the settlement record from southern Britain, Brück (e.g., 1999a, 2019, p. 161) has shown how roundhouses were not passive backdrops to human activity but had 'lifecycles' that were closely entwined with those of their inhabitants, such that people's identity and understanding of their place in the world were constantly being defined and reworked in response to changes in the character, scale and structure of the household and its positioning in the social universe. This interrelationship played out at both a practical and metaphorical level. Sequences of building, rebuilding and repair, like those identified in the early first millennium BC at both Moel y Gerddi and Erw Wen, could have been undertaken as members of the community living there were born, married and died (Brück, 1999a, p. 145, 2019, p. 117; Waddington, 2013, pp. 91–2). Roundhouses, like people, could also have been 'born' and 'died'. The occupants' relationship to the form of a settlement, and the social values and ideas –

including concepts of personhood – materialised through its structures and layout, seem to have been very different to those reproduced by dwelling in the 21st-century AD Western world (Brück, 2019, pp. 125, 159–60). In the same way, as we saw in Chapters Two to Four, Bronze Age copper mining on the Great Orme was different in character, content and scale to modern, industrial mining enterprises and so its materiality and practice would have facilitated the making and reworking of relations of age and gender that were distinct from our own. In this context, it should be noted that, despite similarities between, for example, the scale of prehistoric extraction on the Great Orme and that recorded in some ethnohistorical descriptions of iron or copper mining, it is deeply problematic to apply the age and gender roles described in such studies onto the past, for the reasons discussed in Chapter One (Wager, 2009). Rather, such accounts and the relations they describe, which were themselves the products of specific historical situations, prompt us to consider the practical and ideational understandings of self and society that were created and transformed as gendered subjects participated in mining during the Bronze Age.

Mining and the (re)making of identities

Concepts of age and gender appear to have been recognised and considered important for the relational construction of self in north Wales during the Bronze Age, albeit in ways that are difficult to grasp fully. What can be said is that, across the taskscape, the wielding of tools of stone, bone, wood or bronze and the techniques involved in their making and use would have been associated with particular age/gender categories. These gendered and age-related identities would have been reproduced and expressed as people used and encountered these tools, materials and techniques during mining on the Great Orme. Age and gender categories are, however, not fixed according to a single ideological definition of, for example, childhood or masculinity in a given society, but are inextricably entangled with each other and with other personal attributes in ways that change over the course of a person's life (e.g., Herbert, 1993, p. 19; Sofaer Derevenski, 1994; Kamp, 2001; Wager, 2009). As such, it was not whether an individual was a gendered woman or man that primarily dictated which tasks at the Great Orme mine they could participate in or were excluded from, nor only ideologies of gender that were reproduced in this way, but the interplay of gender with other constructs. At the mine, we can suggest that these included: age; skill and experience; physical strength, size and (dis)abilities; and relative positioning in networks of obligation and debt. The dynamics of the relationship between adults and children, for example, as they worked at the source would have played out at the intersection between conceptualisations of age, gender, maturity and skill in contemporary society.

Age and identity

As already discussed, the burial evidence indicates that, like adults, younger age groups received careful treatment on death and hence were perceived as having value in their own right in Bronze Age communities in Wales. This could imply that children participated in mining at the source as dynamic practical and social agents, and that the relationship between individuals who were more and less physically mature was not characterised by the 'use', or exploitation, of the latter for mining in spaces that the former could not physically occupy (cf. e.g., Lynch, 2000, p. 97). The possibility that younger people were actively involved in extraction at considerable distances underground hints that they were able to acquire identities expressive of the possession of technical skill and know-how in the various mining *chaînes opératoires* and that such identities were not reserved for adults. Age does not seem to have been a principal variable determining who did the unskilled tasks at the source.

The concept of the 'child' is, of course, socio-historically situated (Sofaer Derevenski, 1997, p. 193). Understandings of this construct and its chronological boundaries in Bronze Age north Wales are likely to have differed significantly from our own (Wager, 2009). They most probably also varied with gender and other attributes considered socially important. Hence, an individual young person's experience of mining on the Great Orme, in terms of the tasks in which they could participate and the status this afforded, is likely to have been transformed from one mining episode to the next, as they achieved various states of biological and social maturity. It may also have differed from that of other children of different engendered ages at the source.

Skills and experience

Nash's (1993) account of two 20th-century AD tin miners in the Bolivian highlands positioning a timber support underground highlights how age and other factors can interact to define the relations between people:

"The two worked in perfect co-ordination with few words spoken, accustomed to respond to the needs of the task without asking questions. ... Although the two men interchange roles, each doing the same acts, Manuel, as the master carpenter takes the initiative, sets the pace, and makes the judgements in each of the acts they undertake. It is a perfect blending of cooperative and managed operation, with the two men responsive to the total needs of the task. At the same time, the older master carpenter is accorded the authority to make judgements and issue commands which the younger man accepts, deferring to the authority without losing his own sense of initiative" (Nash, 1993, p. 176).

In this example, age correlates to experience and superior practical know-how and so is the determining factor in the configuration of the relationship between the two men, but the agency of both contributes to the task in hand. As we saw in Chapter Three, many steps in the sequence of mining on the Great Orme required technical skills and knowledge: for example, to locate and follow the desired ore throughout the mineralised zone; to organise working at a distance into the hillside; and to extract ore on occasion by fire-setting. Replication experiments illustrate that the latter demanded considerable empirical experience of fuel selection and ventilation control to be successful (e.g., Lewis, 1990b; Timberlake, 1990a). Ore processing experiments show that, while crushing, grinding, washing and hand sorting are technologically simple activities, a sound embodied understanding of the relative material properties – colour, lustre, hardness and density – of all the rocks and minerals in the mined extract would have been required to perform them effectively to produce the desired concentrate. A 'trained eye' would have been needed to ensure that this mainly comprised fragments of the desired copper ore and not any other unwanted minerals or rock types (e.g., Doonan, 1994; Ottaway, 2001, p. 92; Wager and Ottaway, 2019). Similar aptitudes were also essential when selecting and making mining tools and were even more important when it came to the smelting of ore.

Such practical knowledge – or 'savoir-faire' (Bamforth and Finlay, 2008, p. 3) – would have built on proficiencies gained through participation in other fields of activity elsewhere in the taskspace, which were then developed through practice and by active engagement in the *chaînes opératoires* of mining at the source. As some people learnt the best way of undertaking a particular task at the mine by, for example, observing and imitating others, and/or being instructed and supervised, others with greater technical competency and experience directed, guided or monitored their actions. These processes of knowledge transmission were one way in which mining was constantly caught up in making, reworking and embodying identities arising from different levels of practical proficiency, from novice to skilled practitioner. They would have been important for the initiation of children and young adults as miners, thereby ensuring transmission of practical mining 'know-how' from one generation to the next. The deftness and dexterity with which someone worked and their bodily gestures as they moved around the workings would all also have tacitly signalled to observers disparate access to certain categories of knowledge and/or technical skills. While there was clearly a temporal dimension to the piecemeal patterning of extraction across the mineralised zone (as discussed in Chapter Two), it is possible that some instances of contemporary mine-working of different parts of the source could reflect activity by miners with varying technical aptitude and experience. For example, perhaps the apparent occurrence of mining at both greater and smaller distances into the hillside from the Tourist Cliff represents extraction by people who had more – or more limited – practical knowledge than others? Experimentation and increased familiarity with and experience of this source and its products could have led to mining of different areas of mineralisation at different times. Additional scientific dating in more areas of the mine would enable this possibility to be investigated.

It has been suggested that Bronze mining-working on the Great Orme was carried out by occupational or 'economic' specialists: groups or individuals whose only or primary occupation was mining, and whose other practical needs were met through the exchange (or trade) of ore or metal for resources such as food (e.g., Williams and Le Carlier de Veslud, 2019, p. 1189). This model is familiar from descriptions of ethnohistoric mining communities. While such accounts can certainly provide useful insights (and have been used widely throughout this volume), they may not be universally applicable. Rather, as Edmonds (1995, p. 68) has pointed out, if we accept that the Great Orme mine could have been worked episodically and on a relatively small-scale, as one of the many routines making up the contemporary taskscape, there is then no need to draw on a model of full-time occupational specialists to explain the character of the prehistoric labour force. Skill – what Kuijpers (2018, p. 552) calls 'material specialisation' – is also distinct from 'economic' specialisation. Although the two are often linked, the former does not inevitably lead to the latter, and it is possible to become a highly skilled practitioner without any associated system of economic specialisation (Kuijpers, 2018). As we saw in Chapter Four, contemporary copper and bronze making and working in north Wales do not generally appear to have been carried out by occupational specialists, despite the level and variety of technical skills required. There is no compelling archaeological evidence to indicate that copper mining on the Great Orme was organised differently.

If we distinguish between material and economic specialisation, it is possible to suggest, from the breadth and strength of their mining skill-set, that some of the participants may have identified themselves as knowledgeable 'specialists' in the mining field of activity and been recognised as such by others. This status perhaps also extended to those with both the physical and mental courage to deal with the uncomfortable, constricted, dark, damp and dangerous conditions encountered underground, which researchers surveying or excavating in the mine today vividly describe (e.g., Lewis, 1990d, p. 8; James, 2016, p. 90). It is, however, important to recognise that notions of what constitutes 'skill' are socially mediated, not universal constructs (Bamforth and Finlay, 2008, p. 2; Kuijpers, 2012, p. 140, 2018) – a point that similarly applies

to perceptions of darkness, danger and discomfort, and the related emotions of fear/fearlessness (e.g., Cooney, 2016, p. 161; cf. James, 2016). Although it is to be supposed that Bronze Age communities in north Wales did perceive differences in skill and courage between people in relation to mining (Kuijpers, 2018), perhaps in some of the ways outlined here, they may not have defined, identified and valued these concepts in ways that would be familiar to us in the 21st-century Western world. The notion of 'skill' in prehistory is difficult to define objectively. From the context of our own industrial-world perspectives on making by hand (e.g., Burke and Spencer-Wood, 2019, p. 6), we are perhaps ill-equipped to assess objectively what constituted 'skilled' or 'unskilled' behaviour in relation to mining in Bronze Age north Wales. Similarly, just as the embodied experience of the modern mining geologist surveying the Great Orme mine workings may be significantly different to that of the novice caver on their first trip underground, so might both have little in common with that of the Bronze Age miner. What provoked fear could in each case have been very different. The Bolivian tin miners described in the opening paragraph were, for example, frightened of the influence of the 'evil eye' underground, causing them to modify their behaviour when entering the workings in ways intended to minimise this risk (Nash, 1993, p. 188). How the Great Orme's prehistoric miners may have perceived some aspects of the materiality of the mine and the practice of mining will be examined in more detail in a later section.

Bodily abilities/disabilities

In any given community, we can expect there to be people of different shapes and sizes, and with varying physical and mental abilities. The physical configuration of the Great Orme mine workings, particularly those at considerable distances underground, indicates that while upper body strength would have been an important physical attribute when mining in such spaces, flexibility and a slimmer, more gracile, stature must also have been advantageous. While this may reference age or gender, it also points to the participation of a range of physiques and bodily characteristics. There is likewise no a priori reason to exclude people from the mining field of activity whose practical engagement with the world was affected by what we would now consider to be some form of innate physical or learning disability.

Prehistoric people are also likely to have acquired injuries and infirmities during their lifetime that would have impacted their embodied experience. The individual known as the 'Amesbury Archer', an adult male, had suffered a traumatic knee injury at some point in his life that would have made walking difficult (Fitzpatrick, 2009, p. 176). At least some of the rich assemblage of artefacts accompanying his burial could have been bound up in marking or reacting to this aspect of his identity when alive. Like other components of personhood such as age and gender, understandings of and responses to disabilities are assigned and socio-culturally specific constructs, not inherent matters of fact. We cannot assume that the way people with disabilities were treated in Bronze Age north Wales, nor what it meant to be disabled in that society, was in any way like our own practices or experiences (Southwell-Wright, 2013). Of course, there is no direct evidence for the presence of individuals with any form of disability at the Great Orme mine, but by acknowledging the possibility, we at least begin to move towards recognising the full range of identities that were made, reworked and sustained through the practice of mining there during the Bronze Age. As we shall see, people's identities would have been shaped in fundamental ways by injuries incurred during mining accidents.

Relations of obligation and debt

O'Brien (2015, p. 251) has pointed out that it is important to consider whether prehistoric miners always worked of their own free will, given other researchers' suggestion that slaves and slavery were common – and even normative features – of the European Neolithic and Bronze Ages (e.g., Taylor, 2005, p. 225; Bartelheim, 2013, p. 174; Kristiansen and Earle, 2015, p. 239; Kristiansen and Suchowska-Ducke, 2015, p. 369; Radivojević, et al., 2019). Slavery is difficult to identity archaeologically and archaeologists who use the terms 'slave' and 'slavery' rarely define their meaning (e.g., Kristiansen and Earle, 2015, p. 239). This is problematic as 'slave' as an identity refers to a socio-historically situated relationship between people and does not correlate to one single, defining social form. Ethnographic studies show that, while it can be represented by a range of traits, including, for example, losing control over one's own destiny, being forced to work, losing social standing, being physically mistreated, or regarded as an object for exchange, not all these features may be present in every situation (Taylor, 2005, p. 229; Marshall, 2015, p. 4). When considering the British Bronze Age, we should also avoid inappropriate analogies with the use of slaves in the later Classical world (Stöllner, 2015, p. 76). Rather, in any type of mining operation, whether small-scale and episodic, or large-scale and intensive, the occurrence of slavery and the form it took in that specific archaeological context must be demonstrated, not assumed.

The status and condition of slavery are also not equivalent to and the inevitable outcome of being in captivity, where one is brought against one's will into a new group from one's own community, as a result of warfare or more peaceful transfers of people to, for example, secure a truce or make an alliance. Not all captives are necessarily slaves: slavery is only one of the varied ways that a captive can be absorbed into the new social group (Cameron, 2015, pp. 26, 36; Marshall, 2015, pp. 4–5).

There is no direct evidence for the use or configuration of slavery in Bronze Age north Wales. As discussed in Chapter Four, the exchange of inalienable gifts, such as metalwork, was a means to gain access to the labour of others (Brück, 2015, p. 51). The provision of labour was one way in which both individuals and communities in this region could have fulfilled obligations created through gift giving. The indebtedness arising from such interactions would have been an important means of making and consolidating relationships between different communities and kinship networks across this region and beyond. Over the Bronze Age life of the Great Orme mine, the groups mining there may have included people whose presence and indentured labour were a direct outcome of acts of exchange occurring elsewhere in the taskspace. This may have been a common feature of the organisation of labour not only in mining but in other fields of communal activity in north Wales in the second and early first millennia BC, such as harvesting or house building.

It is, however, questionable to what extent the elements of ownership, coercion, exploitation and control that typify the condition of slavery ethnohistorically and in the present had any relevance in Bronze Age Britain. The notion that is it possible to exploit someone or something for one's own gain is rooted in the development of the ideology of competitive individualism in contemporary Western capitalism, a particular historical construct in which the self is perceived to be separate to the 'other' (e.g., Brück and Fontijn, 2013, pp. 203–4; Brück, 2015, p. 53). In contrast, within the composite, 'dividual' concept of identity characterising the social landscape of Bronze Age Britain, people and things appear to have shared similar qualities. Objects, like human bone, were fragmented for curation as tokens or heirlooms as part of the mortuary rite. The mourners at a cremation burial at Bedd Branwen on Anglesey, for example, broke a bone pommel in half and detached it from its metal blade, before depositing one half in the grave and taking away the other components for curation or dispersal elsewhere (Brück, 2015, pp. 48–9). Burnt fragments of human bone were sometimes mixed with charcoal from the pyre in Earlier Bronze Age cremation burials in north Wales, as in the pits dug into the ring cain at Moel Goedog I, south-western Gwynedd (Lynch, 1984, p. 8; Tellier, 2018, p. 114) (Figure 4.5). By the Later Bronze Age, cremation deposits typically contained only token fragments of human bone, or no bone at all, only charcoal or grain, suggesting that it was ideologically appropriate to replace the human body with these other materials (Davies and Lynch, 2000, p. 212; Tellier, 2018, p. 117). By the end of the second millennium BC, the boundaries between people and things appear to have blurred completely, with no distinction being made between them (Brück, 2015, p. 50, 2019, pp. 40–2, 63). This does not mean only that objects could replace people – or vice versa – in exchange transactions, but rather that things were active social agents, sharing the same attributes as people. Similarly, a person (or their labour) circulated as a gift was not reduced to the status of a passive entity but was a powerful agent instrumental in the making and reworking of relations within and between communities (Brück, 2019, p. 228). The status of those participating in the Great Orme mining field of activity as the outcome of exchange, and their relations with others mining at the source, would have reflected this potency. The categories of identity this created, which were most probably far removed from our own notions of dependency and obligation evoked by the idea of indebtedness, would have been reinforced and sustained as these individuals worked.

Mining and the making of a 'community of labour'

The understandings of the self and society sustained and expressed in the mining field of activity at the Great Orme mine most probably transformed gradually over the course of the second millennium BC and into the first millennium BC due to changes in, for example, mortuary rites and the increased monumentality of settlement in the wider taskscape. Taking part in the mining field of activity at this source would also have actively created, reworked and redefined people's identities and the configuration of their relations with others, in ways that may sometimes have been only momentary or temporary but could also have persisted in some instances long after the closure of a given mining event. The term 'community' has been referred to repeatedly throughout this study, in the sense of people who may not always – or ever – have co-dwelt and who shared familial and kinship ties (however those were constituted and understood in Bronze Age north Wales), as well as a common worldview or set of cosmological beliefs. In this definition, community was both a 'socio-spatial unit' (Yaeger and Canuto, 2000, pp. 5–6) and a fluid and dynamic social construct reproduced as people interacted with each other through the practice of fields of activity like farming, building a roundhouse or burying the dead. It existed at the intersection between "people, place, and premise" (Watanabe, 1992, quoted in Yaeger and Canuto, 2000, p. 5).

The mining field of activity centred on the Great Orme potentially provided a temporal and spatial context for the formation of other social groupings that both encompassed and cut across these core structures and ideas of community. This was the 'mining community': the community of labour that was created and defined as people participated in the mining enterprise (Wager, 2009,

p. 106). It came into being from moment to moment on a variety of scales through the interactions between people working alongside each other. Instances like the bailing out of a flooded stope, or tasks such as moving materials out of and into the workings at the end of a mining event, for example, may have involved all those present at the source. In other activities, such as dragging a container of ore to the surface or foraging for wood for a cooking fire, smaller numbers of people probably worked together. At either scale of interactivity, participation in such acts would have renewed a shared sense of group identity and values between people who did not all necessarily routinely live or labour together.

This ongoing process of creating and recreating a community of labour at the mine was most probably the tacit and unintended outcome of engagements between people as they worked. Its existence could, however, have been thrown into sharp relief at moments of heightened danger, drama or practical difficulty. It may have been while labouring with others in a working on the verge of collapse, or digging aside the debris caused by a cave in, that the relations between people arising specifically from their shared endeavour at the mine were consciously recognised, acknowledged and reinforced. They are likely to have been expressed and sustained in non-discursive ways during routine activities, such as through the coordinated and complimentary form and tempo of bodily gestures as people worked together on a particular task. Some archaeologists have described how, when crushing in a group during ore processing experiments, each person picks up the rhythm and pace of the others, aligning and syncopating each percussive blow (Doonan, 1994, p. 86). This would have helped the task to flow. The noise created would also have signalled how work was progressing across the source, while its intensity and rhythm may have provided an aural indication of who was working with whom and where. The sensory experience of mining would have been even more acute in the underground environment.

Of course, people's affiliations, identities and values within the 'mining community' would have been expressed and reinforced through speech, not just physical action. There must have been much talking around the organisation and progress of the work, such as decisions about where to begin mining or to deposit spoil. For people who did not habitually reside together, a mining event would also have provided an opportunity to pass on news, dissect past events and make plans and agreements relating to broader concerns to be played out in other times and places. Even gossip or phatic conversation ('small-talk') would have affirmed relations between people (Elyachar, 2010). It would have provided further acknowledgement of the place of each individual within broader networks of alliance, affiliation and kinship, as well as their positioning in the community of labour at the mine. In a world in which the social conditions of life were relatively scattered, each mining event may have had considerable significance for the participants, because of the potentials it provided for discourse and group interaction, harmony (and, at times, discord), as well as the acquisition of ore.

As well as the identity of the 'mining community' as a group, participation in the mining field of activity at the Great Orme mine would also have made and remade the identities of individuals in ways that were specific to the *chaînes opératoires* of mining. These conceptions of self could have taken many forms, which may have included identity and status as a 'miner' or at least as a member of this particular community of labour. Such categories of self, from which people with no involvement in any aspect of the mining enterprise would have been excluded, were distinct from and intersected with those arising from interactions elsewhere in the wider taskscape. They would have emerged and re-emerged through an individual's embodied associations with things – particular tools, their form and the materials from which they were made, how they were held and handled; with the mine's spatial layout – where they worked; and with others – choices they made about whom to work alongside. All these distinctions would have been observed and acknowledged by others, in both unspoken and discursive ways. They would also have been literally embodied by the residues of action practitioners in the mining field of activity carried on their bodies. Individuals participating in contributory tasks at or close to the surface, for instance, would have been less smeared in mud than those mining in constricted workings further underground. Nash (1993, p. 174) draws attention to the fact that young, inexperienced Bolivian tin miners have many accidents before they learn the caution and respect for their working environment that is second nature to their more experienced colleagues. At the Great Orme mine, cuts and bruises may have served to distinguish the young, less able or experienced from the older and more skilled. Away from the source, they would have marked out such individuals as members of the community who had taken part in a mining event.

Such minor injuries and grime are, of course, only temporary. Someone's social classification as a 'miner' was probably also carried forward into the wider taskscape in more enduring ways. These could, for example, have included the dexterity with which they were able to apply skills and know-how acquired through mining to the practice of other familiar fields of activity, like agriculture, or the way they were able to talk in an informed way about specific tasks, and draw on in other contexts ideas or metaphors relevant to mining (Edmonds, 1999a, p. 49). Giovanna Fregni (2014, pp. 49–50) has pointed out that a tool used repeatedly by a skilled craftsperson does not just fit in the hand but moulds the hand to itself, indelibly shaping the practitioner's gestures and physical form to express this aspect of their identity even when they are away from their workbench. The heavy use-wear on the skeletons from the Austrian Hallstatt salt mine cemetery illustrates how routine participation in the process of mining could also produce a similar embodied identity (Kern, *et al.*, 2016, pp. 139–141), signalled by, for example, over-developed shoulder muscles on one side of the body. Even though episodic, repeat visits to the Great Orme mine are also likely to have shaped the bodies of the participants in a tangible, if less extreme, manner, as would accidents, such as dislocating a knee or crushing a finger.

Mining and the (re)making of ways of seeing the world

The anthropologist Nigel Barley, in a humorous account of his experiences doing ethnographic fieldwork among the Dowayo of north Cameroon, describes how he gradually realises that the *chaînes opératoires* of many tasks carried out by the Dowayo, from threshing millet to burying the dead, are structured in terms of what happens, and why, during circumcision. Until he develops a more complete understanding of this practice, he finds it difficult to appreciate how these other routine tasks are also constantly caught up in reproducing ideas about what it means to be Dowayo (Barley, 1986, p. 124). Similarly, for recent tin miners in the Bolivian Highlands, the task of mining and the physical removal of ore from the ground were inextricably entangled in an extremely complex belief system of seemingly contradictory elements. This encompassed a ritual cycle grounded in pre-conquest agricultural rites concerned with maintaining both the fertility of the land and harmony with the supernatural, as well as cosmological beliefs derived from post-conquest Catholicism (Nash, 1993, chap. 5). As we have seen, this ontological framework included a belief in apotropaic magic (Nash, 1993, p. 160).

These examples illustrate that the character and content of the Bronze Age mining field of activity at the Great Orme mine would – like all other technologies – have been fundamentally shaped by the contemporary society's worldview: the often tacit set of ideas and values by which people make sense of themselves, their own existence and what is happening in the world around them. These core principles structured and provided a logic to their ways of understanding and inhabiting the world and so guided their technological choices and how they behaved. This system of belief was constantly made and remade through the materiality and practise of the taskscape, including the mining field of activity (e.g., Edmonds, 1990, pp. 56–7, 1999a, pp. 21–31; Childs, 1998, 1999). It also intersected with identity categories, such as gender and age. Ethnographic accounts, for example, show how women in some societies were forbidden from approaching the source, particularly when menstruating (e.g., Childs, 1998, p. 129). In these contexts, it was not only a woman's gendered identity that excluded her from mining, but societal ideas about the nature of female reproductive potency (Reid and MacLean, 1995, p. 149; Wager, 2009, p. 110). The composition of the Bronze Age mining community on the Great Orme would likewise have embodied similar processes of intersection – although the structuring principles drawn upon and their practical enactment may have been profoundly different. Mining at the Great Orme mine is also likely to have referenced and realised many different aspects of contemporary local cosmology and belief (Johnston, 2008, p. 207). Those revolving around the breaking apart, mixing and re-forming of substances will be briefly examined here.

Fragmentation, regeneration and rebirth

The *chaînes opératoires* of many Bronze Age fields of activity in north Wales, as elsewhere in Britain and Ireland, were characterised by processes of fragmentation, burning, amalgamation and remaking (e.g., Brück, 2002, 2006a, 2006b, 2019). Making an object of copper or bronze, for example, variously involved applying heat to solid rock (ore) to transform it into molten metal by smelting; mixing (alloying) copper metal with other substances (tin and lead); solidifying molten metal into new shapes by casting; recycling worn or 'old' metalwork by remelting and recasting after first fragmenting it, a step that itself typically required heating the metal object to be broken up in a fire to around 500–600ºC (Brück, 2019, p. 101; Knight, 2019, p. 267). At Pentrwyn on the Great Orme, as we have seen, the Later Bronze Age metal makers had to crush the smelting slag to release droplets of copper metal.

Elsewhere in the taskscape, the makers of the ceramic urns deposited in Chalcolithic/Earlier Bronze Age graves at Penmaenmawr in the Great Orme hinterland, as well as at the Brenig and on Anglesey, deliberately crushed and added igneous rocks rich in dark ferromagnesian minerals (mafic rock) as a temper to prevent shrinkage and cracking during drying and firing (Williams and Jenkins, 2019). Cremation and its associated mortuary rites, as we have seen, resulted in the disintegration of the individual human body by burning and its blending with fragments from other bodies as well as materials such as charcoal, although there is no evidence for the deliberate crushing of human bone as part of this sequence in north Wales (Tellier, 2018, p. 114). Grain had to be dry roasted (parched) and then ground before it could be cooked (Brück, 2019, p. 230). Heat-mediated cooking and brewing – or perhaps more esoteric activities – at burnt mounds gave rise to distinctive spreads of fire-cracked stones, like that produced during the later second millennium BC at Bryn Cefni, Anglesey (Smith, *et al.*, 2002). The Later Bronze Age ceramic vessels excavated at the Breiddin hill fort contained inclusions of igneous rocks that had first been coarsely crushed (Musson, *et al.*, 1991, p. 119). All these examples, which together span the Bronze Age, indicate how practices of fragmentation endured throughout this period.

Shared across many different technologies, these processes of dissolution by heat-assisted or manual fragmentation, followed by mixing and remaking, formed the foundation of a particular set of worldviews in which life and death were linked in an unending series of transformative cycles of growth, decay and renewal. These beliefs would have structured people's understandings of a whole range of different aspects of self and society, furnishing people with a way to think about both biological and social growth (e.g., Brück, 2002, p. 157, 2019). Brück (2006b, p. 307) gives the example of human reproduction, which, as it involves the mingling and combining of substances and attributes, could have been understood as analogous to the production of pottery and metalwork. The materiality and practice of making things like bronze may also have enabled people to conceptualise significant rites of passage, such as marriage, which involved the breakup and reconfiguration of existing household and kinship groups. They could likewise have provided metaphors for the 'dividual' conception of personhood characterising constructions of social identity in Bronze Age north Wales.

Aspects of this ideational framework would have been tacitly expressed and reproduced as the miners went about the routines of working at the Great Orme copper mine. As we saw in Chapter Three, the *chaîne opératoire* of mining at this source involved a series of practical transformations. Rock was fractured, broken and disaggregated, either by hand using tools or on occasion with the aid of heat by fire-setting. After extraction, the mined rock was crushed and concentrated, in a process that separated the mined extract into its constituent minerals and rock and reconfigured their quantity, type and size range to form the desired concentrate. The miners may also on occasion have explicitly realised ideas about the creation and structure of self and society by carrying out specific rituals of fragmentation and transformation. The two well-used hammerstones at the bottom of a surface pit-working (EVB1) with dated evidence for copper mining activity in the Earlier Bronze Age at Alderley Edge mine, eastern England, are perhaps the material representation of such practices (Figure 3.3). Both tools seem to have been intentionally split lengthwise and then broken into quarters before deposition, rather than fracturing through normal wear and tear (Timberlake and Prag, 2005, p. 47). At the Great Orme mine, an explanation for the high degree of fragmentation of the bone tools, despite their use in soft-rock areas, is that the miners deliberately broke their tools as a means of marking the end of a specific mining episode (James, 2011, p. 431). This event may have been understood as analogous to a death, realised through the destruction of a tool with a strong metaphysical association to a valued, living animal. The two broken rib tools that appear to have been placed on a thin bed of clayey silt just above the surface of a bedrock shelf in a narrow working at West Vein 6 in the Tourist Cliff (David, 2002, p. 99) can perhaps be explained in this way (although not all bone tools identified during excavation as apparently intentional deposits had been broken before deposition, e.g., David, 2003, p. 96). The mine, like contemporary roundhouse settlement, may have been perceived as having a fluid 'lifecycle': as members of the mining community died, others transitioned into adulthood and new ones joined through alliances such as marriage, so new veins were sought and worked, extraction in some areas ceased and mining began again in workings previously explored. These activities were both part of the practical ebb and flow of each mining event and meaningfully entangled in the biography of the living mining community.

Significant moments in the lifecycle of the groups labouring at the source may even have influenced the timing and tempo of each mining episode. Bronze Age communities in north Wales perhaps smelted and cast copper and bronze, both processes of physical transformation, to mark moments in which boundaries (actual and metaphorical) were transgressed (e.g., Brück, 2002, 2006b, p. 306; Webley, Adams and Brück, 2020, p. 186). In such a scenario, smelting at Pentrwyn on the Great Orme could have been carried out not as a trial or assay smelt – an interpretation that views prehistoric metallurgy as the rational application of objective scientific knowledge (Kuijpers, 2012, p. 139) – but as a means of facilitating the reworking of the social fabric on the occasion of a death or burial, a marriage or other rite of passage, or even as part of the preparations for an agricultural festival. This could explain the small-scale of the smelt: the performance was the point, not the amount of copper metal produced. The possible close association between mining and smelting on the headland, at least during the Later Bronze Age, hints that people purposefully mined ore from the source to meet this need. In this way, the *chaîne opératoire* of copper mining would have been integral to how people recognised and responded to moments of transition, reconfiguration and rebirth.

Reproducing the mine and the mining community

For Bronze Age communities in north Wales, the breaking up of an object did not necessarily diminish its value (Brück, 2015, p. 48). As we have seen, the curation and circulation of the fragmented remains of the cremated dead played an important role in the maintenance and reworking of social relationships. Elsewhere in Britain, broken pots and other items of 'refuse' appear to have been intentionally deposited in ditches and building foundations, to mark boundaries and/or significant moments in the lives of a settlement and those who dwelt there (e.g., Brück, 1999a; Nowakowski, 2002). Such foundation practices appear to have continued into the first millennium BC, as may be indicated by the soil accumulations containing a range of objects, such as broken pottery vessels, animal bones and – in one instance, a socketed axehead fragment – identified beneath the Later Bronze Age banks at several early hillforts in north Wales, including Castell Odo and the Breiddin (Waddington, 2013, pp. 94–7) (Figures 4.2, 4.8). The Guilsfield hoard, like other large Later Bronze Age assemblages of deposited broken bronzework, may have been the outcome of the curation, circulation and sharing of metal fragments. As with the manipulation of tokens of cremated human bone, this would have been a means of conveying and reproducing relations between people and of symbolising the relational character of contemporary personhood (Brück, 2019, pp. 95–6).

Collectively, these practices hint that the Great Orme miners did not view the pieces of broken and crushed dolomitic rock they left behind after the *chaînes opératoires* of ore extraction and processing as culturally valueless material or 'waste' (cf. Chapter Three). The careful and in some instances strategic disposal of spoil within the workings may have been planned to facilitate movement around the mine and conceived as fundamental to the growth and regeneration of the lives of both the mine and the community who worked it. This significance may have extended to the discarding of other material residues from the mining field of activity, such as cobblestone and bone tools. In the same way that the intentional deposition of broken metalwork and items of production equipment, like casting moulds, could have been seen as an essential step for a successful smelt or casting (Brück, 2019, p. 95; Webley, Adams and Brück, 2020, p. 120) so the Bronze Age miners may have intentionally left their tools in the Great Orme mine workings (Figure 2.13). Some tools, like the five cobblestones excavated at the base of a narrow cutting in the bedrock floor on the shallow East 7 ore vein, as well as other examples from more difficult to access locations, seem to have been deliberately placed in groups or 'caches' (Lewis, 1996, p. 100; David, 2001, p. 118; James, 2011, p. 370). Did the miners on occasion leave valued tools behind to guarantee the renewal of the vein and the reproduction of the mine? Such deposits could also have acted as practical memory aides, guiding a returning group to the starting point of the next mining season (Jovanović, 1980, p. 34). Better data on the find context, use-wear and other characteristics of the bone and stone assemblages identified as possible caches are needed to explore these suggestions further (e.g., Beecroft, 2005).

Mining as a technology of tradition

However they were valued, increasingly vast quantities of spoil, discarded or deposited tools and other debris, such as the remains of old cooking fires or shelters, gradually accumulated in and around the Pyllau Valley over the life of the mine. The rock outcrops in the area of mineralised ground became scarred with entrances to underground workings. As people returned to the Great Orme mine within living memory of previous visits, they would have recognised their own handiwork or the vestiges of activity by recent, known ancestors. Repeated encounters with these echoes of a past *chaîne opératoire* would, over generations, have served to establish and consolidate customary routines of activity – a technology of tradition – at the mine, giving rise to the apparent uniformity in the tools and techniques used there over the 400–800 years of Bronze Age activity. This may have been reinforced through oral tradition, as the stories told about this place – particular workings, past accidents or rich veins – were passed down from generation to generation.

The mining community would also have come across the material traces of working carried out by other, much earlier, hands. Although we cannot know exactly how these residues were regarded, there is evidence from elsewhere in the north Wales taskscape that the past held significance to communities there during the second and earlier first millennia BC. Earlier Bronze Age mortuary practices stressed personal and collective remembrance of actual or imagined ancestors as existing burial monuments were reopened and modified to accommodate new inhumation and cremation burials (e.g., Tellier, 2018, p. 111; Brück, 2019, pp. 19–24). By the Later Bronze Age, as discussed in Chapter Four, people rarely built new burial mounds, but still occasionally reused existing burial sites. At the flat cremation cemetery of Cae Capel Eithin on Anglesey, for example, later urned cremations were placed among the Earlier Bronze Age cremations burials (Davies and Lynch, 2000, pp. 211–2). Burnt and unburnt human bone was now typically deposited as token fragments in post- and stone-holes, middens, boundary ditches and pits, caves, rivers, lakes and bogs, rather than in formal burial contexts (e.g., Davies and Lynch, 2000, p. 212; Brück, 2019, pp. 51–4). These practices suggest that maintaining and expressing metaphorical links between the living and the dead continued to be important throughout the Bronze Age. By the end of this period, even when the remains were those of family or community members, memorialising the identities of specific, known individuals was much less important than it had been in earlier periods. Rather, during the second and into the early first millennia BC, mortuary *chaînes opératoires* increasingly came to evoke associations between particular places in the landscape and the generalised, mythologized ancestral dead, as part of a relational worldview in which the self was a fluid composite of specific people, places and things (Brück, 2019, p. 56).

In the first half of the first millennium BC, this desire to maintain physical links to the past can be seen in the complex occupation histories of the settlements at Moel y Gerddi and Erw Wen and at the earliest ringwork enclosures of Meillionydd and Castell Odo, where later roundhouses were deliberately built over the foundations of previous buildings. It is also evident at Llandygai/Llandegai near Bangor, where in the 9th century cal BC a settlement was built reoccupying the remains of older earthwork monuments (Gibson, 2018, pp. 106; Waddington, 2013, pp. 49, 94) (Figures 4.2, 4.5). Like earlier mortuary rites, going about the various fields of activity at Llandygai/Llandegai or Moel y Gerddi would have continuously worked to renew, renegotiate or even challenge the histories imbued in these places, gradually creating new concepts of tenure and reshaping the ties that bound people to the land (Barrett, 1994, chap. 6; Edmonds, 1999a, pp. 68–9, 1999b, p. 488).

Each episode of mining on the Great Orme was similarly an occasion that, tacitly or otherwise, brought relations between past and present to the fore, as people engaged practically with the physical remains of previous working – by, for example, reusing cached tools, following an old tunnel underground, or scavenging in a grassed-over spoil tip for ore previously missed – and as they went about their traditional mining practices. Enacting this technology of tradition is likely to have reproduced and consolidated the identities of both the mining community and its individual members within real and imagined lines of descent. The miners perhaps understood much older workings to have been created by distant, maybe mythical, ancestors. Both mining and the mine itself, as a material representation of the actions of earlier generations, may have increasingly figured in origin myths by which the communities working there made sense of their place in the world. The human remains found in the workings can perhaps be understood from this perspective: an intentional inhumation burial that expressed the position of the mine and its community of labour in particular genealogies and reinforced the ties between them.

As Williams and Le Carlier de Veslud (2019, p. 1188) point out, in much of the mineralised zone, great effort was needed to extract perhaps only small quantities of ore. Together with the mine's long history of prehistoric exploitation, this reveals a powerful and prolonged urge for this material. Was the significance of mining on the Great Orme, as a practical embodiment of the bonds between people, the past and this particular place, partly responsible for the intense and sustained drive to extract ore there? Could the act of mining have been as important as its material outcome, an inextricable component of the value of the resulting metal? As the same communities made countless visits to mine on the Great Orme throughout the second millennium BC, these connections would have been strengthened. They were perhaps preserved beyond the mine and an episode of mining to encompass the mined ore. Like the power and resonance retained by long-dead ancestors, the practical and metaphorical associations of community, place and particular traditions of practice given material form by the extract from the Great Orme mine could have been carried forward into the tasks of smelting and casting. This potency may have endured beyond the actual memory or lived experience of mining, or any personal or collective familiarity with the mine: an ingredient in the biography of the cast copper metal that people talked about and acknowledged.

It is common in the archaeological literature to refer to the practice of copper mining as being 'in decline' by the latter part of the Later Bronze Age (e.g., O'Brien, 2015, pp. 298–302; Timberlake and Marshall, 2018, pp. 427–8; Williams and Le Carlier de Veslud, 2019, p. 1185). Alongside the geoenvironmental factors discussed in Chapter Four, it has been suggested that Bronze Age communities ceased mining on the Great Orme as the transition to iron technology reduced the demand for copper metal to a level that could be met by recycling items already in circulation. This is thought to have occurred in combination with improved access to copper metal originating from ore extracted at mines in continental Europe (e.g., O'Brien, 2015, pp. 300–2; Williams and Le Carlier de Veslud, 2019, p. 1194). The idea of 'decline' is borrowed from the modern economic concept of the 'industry lifecycle', where it is characterised as the phase in the evolution of a business when it is no longer able to maintain growth, sales drop and it loses its appeal to its consumers (Investopedia Team, 2020). As we have seen throughout this study, however, the articulation between making and using copper and bronze for Bronze Age communities cannot be characterised as a straightforward economic relationship between producers and consumers, nor in terms of a simple supply and demand equation. Nor, as we saw in Chapter Three, is there any archaeological evidence from the mine itself to support a diminishing or dwindling of skill, knowledge or tool efficacy during the life of the Great Orme mine that could correspond to a period of reduced 'input', whether of time, resources or interest.

The standard period divisions we use to refer to the span of time from the late third to the first millennia BC do not correspond exactly to changes in the currency of particular materials. Consistencies in the evidence for copper and bronze making and working from the Later Bronze Age to the end of the Earliest Iron Age (i.e., from c.1150 to 600 cal BC) indicate that its character did not change markedly during this period (Webley, Adams and Brück, 2020, p. 70). Nor did iron suddenly replace

bronze for the communities in north Wales and the rest of Britain. Iron would not have afforded superior mechanical properties to the bronzes being made in Britain during the early first millennium BC (e.g., Northover, 1984, p. 141; Thomas, 1989, p. 280). There is a marked decrease from c.800 BC onwards, during the first three or four centuries of the Iron Age, in the amount of metalwork of any kind being deposited as single finds, in settlement or hoards, and even in rivers (Davies and Lynch, 2000, p. 214; Brück, 2019, p. 237). Bronze continued to be used for decorated and small personal objects, such as brooches and horse gear throughout the Middle and Late Iron Age, even as iron began to be produced in significant quantities for tools and other everyday objects (Davies and Lynch, 2000, p. 187; Webley, Adams and Brück, 2020, pp. 127–8).

Although the introduction and development of iron-working technology undoubtedly had profound social repercussions, these were felt and expressed gradually over a long period, not as a rapid, widespread event (Webley, Adams and Brück, 2020, p. 127). The introduction of iron as a new material is therefore unlikely to have been directly responsible for the end of Bronze Age mining at the Great Orme mine. Nor was the cessation of mining at this source a direct outcome of copper and bronze recycling during the Later Bronze Age. As discussed in Chapter Four, this practice was taking place from at least 1600 BC, from the beginning of customary prehistoric mine-working at the Great Orme mine. For much of the Bronze Age in Britain, recycling coexisted alongside mining. What may have changed, however, is the significance afforded to making and working bronze. Brück (2019, p. 239) suggests that the practice of this field of activity expressed the relationship between death, regeneration, fertility and the renewal of life for Bronze Age communities. During the early first millennium BC, however, the transformations in land division and use described in Chapter Four may gradually have given rise to a focus on human, animal and agricultural fertility as the means of social reproduction. Bronze and bronze working may have become less relevant to the making of ideas about the self and society. In turn, the social relevance of mining to communities in north Wales could then have started to fade. In this context, the Middle Iron Age to Late Iron Age date recently obtained from spoil deposits excavated at Llanymynech copper mine, in Powys, mid-Wales, is particularly interesting (S. Timberlake 2022, personal communication, 23 May). It shows that there is still much more to know about the end of prehistoric copper mining in Britain. (For discussion of the possible Iron Age date from the Great Orme mine [CAR-1281; 2450±60 BP; 760–410 cal BC; 95% probability], see Chapter Two.) The end of Bronze Age mining on the Great Orme most probably therefore came about through a varied and complex interplay of material and socio-economic issues (Timberlake, 1994, p. 134). It was not driven simply by a single geological fact, economic priority, nor in response to an emerging new material, a point which is reinforced when considering why modern mines cease to be worked: witness, for example, the significant socio-political dimension to the UK government's decision to close most coal mines in Britain from the 1960s onwards.

Like 'decline', using terms like 'closure' and 'abandonment' to describe the end of prehistoric copper mining (e.g., O'Brien, 2015, pp. 298–302; Timberlake and Marshall, 2018, pp. 427–8; Williams and Le Carlier de Veslud, 2019, p. 1185) may also be unhelpful. Both imply a deliberate, conscious decision to cease mine-working. I suggest that, at the Great Orme mine, the end should instead be understood as a gradual reorientation of purpose and meaning as the extraction of copper ore and the making and working of bronze slowly ceased to be active, routine fields of activity for Later Bronze Age and Iron Age communities in north Wales. This transformation is likely to have taken many generations. The significance of the mine, for example, did not stop when the customary Bronze Age phase of mining ended. As Edmonds (1999b, p. 490) remarks, places did not "go away" when they were no longer actively used, their original purpose altered or forgotten. Rather, 'old' sites like mines, stone quarries, monuments and unoccupied settlements continued to orientate the actions of people in the Iron Age present, contributing to their understandings of self and their place in the world (e.g., Gosden and Lock, 1998; Holtorf, 1998; Nowakowski, 2002, pp. 146–7; Frieman, 2020, pp. 159–60). There is evidence that features like former mines and stone quarries featured prominently in the cosmologies of Iron Age communities in north Wales. Waddington (2013, pp. 99–101) makes this point in relation to the unusual cluster of three Late Bronze–Earliest Iron Age double ringwork enclosures (Meillionydd, Conion and Castell Caeron) built on the lower slopes below the Neolithic stone axe quarry of Mynydd Rhiw on the Llŷn Peninsula, to the south-west of the Great Orme (Figures 1.2, 4.2). Although radiocarbon dating indicates that stone extraction ceased there around 3050 cal BC (Burrow, 2011; Barker, 2015), material from this site was still being curated in the first half of the first millennium BC, as finds of stone cores from settlement contexts at Meillionydd indicate. During this period, Mynydd Rhiw and its hill appear to have been part of a way of seeing the world that mythologised old stone quarries, a conception of place that made this location appropriate for the construction of enclosures whose architecture and landscape setting were intended to reproduce and display a group's position and identity within still relatively dispersed social networks.

By the later first millennium BC, there would still have been noticeable traces of Bronze Age mining in the Pyllau Valley on the Great Orme, but these would slowly have been transformed as the mined cliff faces gradually

became overgrown and revegetated, the spoil tips grassed over and carpeted by clusters of pink and white flowers, like Thrift (*Armeria maritima*), which today grows in abundance on the mining spoil tips in the Pyllau Valley (Countryside Council for Wales, 2008, p. 5). The mine would not, however, have 'disappeared' from people's knowledge of the headland. In common with nearby Kendrick's Cave, a site with Upper Palaeolithic deposits that was then used as a workshop and 'museum' display area in the latter decades of the 19th century AD (Rees and Nash, 2017), it is unlikely ever to have been totally abandoned, although the character and tempo of activity there would have shifted with time. It could, for example, have been used as a shelter by people and animals, and to store or hide things. More significantly, like places such as Mynydd Rhiw, the Great Orme mine most probably also figured in the 'sacred geographies' (Brück, 2019, pp. 157–8) of the Iron Age communities along this stretch of the north Wales coast. It is in the context of these processes of veneration and remembrance that we can perhaps understand why Pen y Dinas on the headland was chosen as the site for a hillfort in the 4th/3rd century BC, some 500 years or more after the probable end of Bronze Age extraction at this source (Figure 4.7).

Mining, politics and power

The shaping, remaking and sustaining of the concepts of self and society we have discussed in this chapter would have been entangled with relations of power and political interests (Dobres and Hoffman, 1994; Dobres, 2000, pp. 115–7). Prehistoric copper mining, including at the Great Orme mine, is often linked to the performance of formalised power structures on a regional or wider scale (e.g., O'Brien, 2015, chap. 10; Williams and Le Carlier de Veslud, 2019, p. 1192). This fits into a broader interpretative framework in which the exchange of copper and bronze metalwork is understood as the basis for social differentiation and the development of contemporary political institutions in Bronze Age Britain and Europe, in particular the formation of 'elites'. The political authority of these select individuals is thought to have derived from the manipulation of prestige goods such as metalwork in competitive bouts of display ('conspicuous consumption') and exchange (e.g., Earle, *et al.*, 2015; Kristiansen and Suchowska-Ducke, 2015; Vandkilde, 2019). It is possible to disagree with many different aspects of such models of prehistoric social organisation in Britain and these arguments will not be rehearsed again here (see e.g., Barrett, 1985, 1989a, 1994, 2012; Barrett and Needham, 1988; Bradley and Edmonds, 1993; Edmonds, 1998; Kienlin and Stöllner, 2009; Kienlin, 2013; Brück and Fontijn, 2013; Brück, 2015, 2019). Rather, this section will briefly focus on the nature of power in the small-scale, communally organised society of Bronze Age north Wales. We will then examine potential contexts for the performance of power at the Great Orme mine and while smelting during the Later Bronze Age at nearby Pentrwyn.

In the relational conception of personhood characterising the communities in this region, social status was not fixed, nor power inherent to particular roles, such as that of chief. Rather, the power and authority of an individual or group were mutable and context-dependent, derived from the wider set of relationships comprising the people, places and things of which they were a part. It would therefore have been difficult for one person or community to wield and retain absolute power over others and hence for a static, hierarchically organised ruling elite to develop (Brück, 2019, pp. 5, 235–6). As we have seen, in north Wales as elsewhere in Britain, there is no compelling evidence for this form of social organisation, neither in terms of pre-eminent individuals within communities, nor the enduring primacy of one community over another. Nor would it have been acceptable to treat others in any way one wished, any more than it was possible to discard objects freely (Brück, 2015, p. 53), a further argument against the presence of slaves in the Great Orme mining community.

There would still, however, have been opportunities for the production of social differences and imbalances in power and authority between people. Circular monuments, constructed and used in north Wales during the Chalcolithic and into the first few centuries of the second millennium BC, seem to be socially undifferentiated arenas. They lack specific focal points or spatial divisions, perhaps with the intention of minimising or removing status differences between people (Barnatt and Smith, 2004, pp. 32–3). Even within circular spaces, however, like the large henge monument (Henge B) at Llandegai near Bangor (Gibson, 2018, pp. 100–1) (Figure 4.5), each person could have occupied a particular place relative to others. This positioning could have changed over the course of an individual's life and from generation to generation (Edmonds, 1999a, pp. 145–8). Procession to such monuments and not just the activities carried out within them may have provided a further means of establishing and reproducing social distinctions between people during ceremonies: there would have been those who led and those who followed, and others who were excluded altogether (Barrett, 1994, p. 15). Excavation of roundhouses likewise points to the structuring and differentiation of space in and around these buildings (e.g., Brück, 1999a, 2019, pp. 131–3; Waddington, 2013). The architecture of henges, stone circles and roundhouse settlements could, therefore, have provided spaces empowering some – individuals, groups or whole communities – to gain a selective advantage over others (Edmonds, 1999a, pp. 146–7), sustaining or transforming existing distinctions between people, or enabling new ones.

Relations of power and authority would also have been continuously contested and reconstituted through the interpersonal relations reproduced by gift giving, a practice that generates asymmetric relations of obligation (Brück, 2019, p. 227), and as people went about the 'workaday' technologies of farming, fishing, digging field banks and preparing food making up the taskscape. Like any other field of activity, the enactment of mining on the Great Orme would have provided the participants with an active medium for practical, material, symbolic and ideational manipulation and differentiation, at both inter-community and more localised, even individual scales. This would have occurred in countless routine ways that were overt and intentional, but also subtle, non-discursive and unforeseen (Dobres and Hoffman, 1994; Dobres, 2000, pp. 118–9; Wager, 2009, p. 109).

The purposeful placing of spoil or redirection of water could, for example, have been used to impede others' access to particular areas of the mine to further political interests in a conscious exercise of power. Those who were less able or experienced may on occasion have inadvertently 'got in the way' of the task in hand, leading to tension and conflict. The social impact of accidents may have strengthened community cohesion, as we have seen, but they could also have constituted moments of rupture and conflict, particularly if a specific individual or group was considered at fault. If somebody was injured, others may have taken their place. By causing different people to work together, accidents could have reconfigured the structure of the mining community in unpredictable ways. These effects could have been exacerbated if the participants were from different social groups with relatively distant kinship ties. Social distance may have introduced an element of practical competition – to be the faster worker, haul the heaviest load or mine the most ore – with consequent power imbalances between individuals or groups.

The bounded spaces of the workings would have enabled some people to work together while excluding others, embodying and signalling their relative positioning in the social order to each other and to the other members of the community who were watching (Edmonds, 1999a, p. 45). Mining a constricted ore vein in which there was only sufficient room for one person to extract ore would have assigned others to different roles, such as preliminary ore processing, challenging and contesting categories of identity and socio-political influence. More generally, certain tasks and contexts of action in the mining field of activity at the Great Orme mine were probably not open to every member of that 'community of labour'. Those perceived to possess the required attributes – whether knowledge, skill, age and experience or a particular gendered identity – would have been sanctioned to work in certain areas of the workings, and perhaps also in certain ways, that others were not. Some may also have had access to categories of information that were denied to others, a process of inclusion and exclusion with the potential for conscious or tacit political manipulation as the practical experience of participation opened understandings not granted to those excluded.

While it is difficult to pinpoint specific instances of knowledge exclusivity in action at the mine itself, the location of the nearby Pentrwyn smelting site hints that, during the Later Bronze Age at least, knowledge of how to transform ore into metal was not open to all those labouring at the mine (Figure 1.5). This is despite the relatively wide distribution of contemporary evidence for bronze casting and smithing in Britain indicating that knowledge of the later stages of metalworking was not restricted to specific groups or individuals during the Later Bronze Age (Webley, Adams and Brück, 2020, p. 181). As we saw in Chapter Three, although questions remain about how representative the evidence from Pentrwyn is, it may indicate that smelting was, at least on occasion, embedded in the mining field of activity. Some of the same people may have been involved in both activities. At first glance, the site's topographic setting appears to indicate a desire for physical separation from the rest of the headland and a need for privacy and secrecy during smelting. Located at the foot of an east-facing vertical cliff (Smith, Chapman, et al., 2015, p. 54), it is hidden from the view of anyone on the summit plateau and cannot be seen from most other parts of the headland, nor from the north Wales coastline further east. It is certainly well out of sight of the workings in the Pyllau Valley. The design of the furnace, of a size to match the lung capacity of a single smith and with only one blowpipe slot, certainly points to seclusion and use by a single individual at any one time (Chapman and Chapman, 2013, p. 16).

The impression that the site is perched on a cliff edge is, however, enhanced by the sheer cutting, c.4m high, that was made during road construction in the late 19th century AD. When Pentrwyn was first discovered in 1997, archaeological material was identified on a narrow ledge approximately 4m by 2m. This was all that remained of what was once a wider terrace, an estimated 80 to 90% of which was destroyed when the road was built and the extent of which today continues to be reduced by erosion. The site was positioned along what was formerly a natural routeway around the headland and is only 300m from the natural quay close to Pigeons' Cave. It would also have been visible from the sea (Chapman and Chapman, 2013, pp. 3, 24; Smith, Chapman, et al., 2015, pp. 53–4, 69). The cliffs around the north-eastern tip of the Great Orme headland appear to have been a focus for late prehistoric activity: Chalcolithic/Earlier Bronze Age finds have been recorded in Snail Cave, only 250m south of the smelting site at the foot of the same cliff (Smith, Walker, et al., 2015, p. 126),

while the area around Pigeons' Cave, on the shoreline, was the site for the deposition of a hoard of gold and bronze objects assigned typologically to the Later Bronze Age (HER PRN GAT4577 [accessed online 22-07-2019]). Between Pentrwyn and Pigeons' Cave is the Lloches yr afr rock shelter, which was also truncated by road construction and from where prehistoric artefacts (a flint microlith, animal bone, pottery, hearths and a small fragment of bronze) of indeterminate date were discovered (Davies, 1973, 1974, 1989, pp. 95–7). Prehistoric ore processing is likely to have taken place at the spring of Ffynnon Galchog, only 400m away from the smelting site on the summit plateau (Lewis, 1990b) (Figure 1.5). Taken together, this evidence hints that this stretch of headland along the Pentrwyn cliffs was neither practically nor conceptually as isolated in the second and early first millennia BC as its currently precarious position suggests. Perhaps the Later Bronze Age miners selected this imposing location simply for the shelter it offered from the prevailing wind, as suggested in Chapter Three. Or did they choose this spot for smelting – a potentially dangerous and ritually charged transformative process – not primarily to restrict access nor for strictly practical reasons but because it referenced the powerful cosmological associations of the nearby caves, springs and the sea (Brück, 2019, pp. 182–6)?

'Ownership' and 'control' of the ore source

It is probable that political concerns and power relations would have played out in relation to access rights to the Great Orme mine and its ore. One interpretation is that the mining community on the headland was socio-politically subservient to a different social group, perhaps based in the fertile lowlands of north-east Wales or the Welsh borders and who directly controlled either mining at the source and/or the distribution of its metal (e.g., Needham, 2012; Williams and Le Carlier de Veslud, 2019, p. 1192). Although we know from the settlement and burial evidence that there were differences between regions, this model is difficult to reconcile with the shifting, context-dependent and largely fluid expression and understandings of power and authority that emerge when all the different strands of archaeological evidence from north Wales are pulled together. The communities on and around the Great Orme and its hinterland clearly interacted with those in north-east Wales (and beyond). The striking gold cape found in the rich Earlier Bronze Age Mold burial in that area (Appendix 6, Figure 4.5) is not, however, an unambiguous indicator of the 'wealth' or territorial supremacy of the community where it was found. As we saw in an earlier section, alternative readings of this evidence are possible. Seeing the community of mining on the Great Orme headland as in some way subordinate to or under the control of another group located in the north-east Wales lowlands is also part of a 'core and periphery' model that in this instance places communities in the 'uplands' in relations of dependency to those in the 'lowlands' (e.g., Frankenstein and Rowlands, 1978; Rowlands, 1980, pp. 34–5). This way of thinking assumes that 'upland' and 'lowland' communities were geographically and socially distinct entities. In contrast, we have seen how the evidence points to the existence of permeable boundaries between social groups in Bronze Age north Wales, and to the entanglement of the communities on and around the Great Orme and its hinterland into extensive networks of connections stretching in different directions in ways that were dynamic and fluid.

As discussed in Chapter Four, there is no evidence that, throughout the Bronze Age, the movement, casting and smithing of copper and bronze in Britain, including in north Wales, were under any kind of centralised control. Against this background, the case for the different treatment of contemporary mining is difficult to make convincingly (e.g., Webley, Adams and Brück, 2020, p. 10). It may be that the notion of the Great Orme's ore or its metal as a practical 'resource' that was 'exploited' and 'controlled' for social and economic advantage is fundamentally misplaced (cf. e.g., Lynch, 2000, p. 38; O'Brien, 2015, pp. 290–1; Williams and Le Carlier de Veslud, 2019, p. 1189). Perhaps we need to reassess how the Bronze Age communities mining at that source understood the natural world and the materials they encountered? As we saw earlier in this chapter, the materiality of the mine – its rocks and minerals, the tools used to mine – seems to have been entangled in, not separated from, people's understandings of self and society, in a relationship that was perhaps more subtle, nuanced and reciprocal than terms such as 'exploitation' and 'control' suggest.

It is therefore unlikely that, throughout the Bronze Age, relations of power and authority involving the Great Orme mine centred around 'control' or private 'ownership' of the source by either powerful individuals or groups. Rather than being the outcome of 'competitive individualism' (Brück, 2019, p. 239), the creation and maintenance of tenure at the mine was most probably realised through the making and remaking of community identities and the customary re-enactment of mining traditions, as described earlier. When disagreements, imbalances in power or clashing political interests did, however, arise through mining on the Great Orme, how might they have been resolved? The extent to which armed conflict was endemic during the Bronze Age in Britain and Europe is currently a major topic of research and debate (with the different approaches summarised succinctly by

Dolfini and colleagues in their introduction to the recent volume 'Prehistoric Warfare and Violence') (Dolfini, et al., 2018). Some researchers argue that it was widespread, institutionalised, inter-regional and professionalised by the emergence of a warrior class during this period (e.g., Horn and Kristiansen, 2018; O'Brien, 2018). Others suggest that warfare, as a socially situated and context-dependent practice, was not endemic in the past but varied in scale and frequency across time and space (e.g., Armit, et al., 2006, p. 10; Vandkilde, 2014, p. 13). Booth, et al. (2021, p. 7), for example, argue that the osteological evidence from Britain as a whole indicates less interpersonal violence by the Earlier Bronze Age than in the preceding Neolithic. Conflict may not have been endemic even in the Late Bronze and Earliest Iron Age when the deposition of swords, such as the two leaf-shaped blade fragments found in the Cors Bodwrog hoard on Anglesey (HER PRN GAT5056 [Accessed online 13-03-2020] became far more common than in earlier periods (e.g., Osgood, 1998; Schulting, 2011, p. 32) (Figure 4.6). Neither does the appearance of hillforts point conclusively to a more martial society by the first half of the first millennium BC (Armit, 2007; Waddington, 2013, p. 18).

Examples from ethnography are also reminders that armed conflict is only one possible strategy in response to the tensions and social contradictions arising as people go about the daily business of their lives. Among the Toro of Uganda, for example, iron ore was an important means of attaining considerable wealth and status and so its discovery was the focus of intense social competition. To forestall the potential this held for division in the community, a public festival took place at the ore source before mining began (Childs, 1998, pp. 128–9). This served to announce the discoverer's ownership of the vein and provided an arena in which members of the discoverer's family, clan and neighbours could ask permission from him to mine other, spatially separated, parts of the source. In this way, the Toro both anticipated and thwarted the high likelihood for conflict when ore was discovered (Pfaffenberger, 1998, p. 295). Like any other field of activity, armed conflict in Bronze Age north Wales would have been embedded in wider understandings about the self and society (Armit, 2007, p. 35). For communities who perhaps viewed ore not as a physical resource to be controlled but as an active component in a relational cosmology legitimised through customary routines of practice, the mine may not have been considered an object that one group could attempt to take by force from another.

6

A fundamentally social archaeology of Bronze Age mining

This book has aimed to explore the relationship between technology and society in Bronze Age Britain, from the fresh perspective of copper mining on the Great Orme during the second and early first millennia BC as a fundamentally social phenomenon. It has situated the mining field of activity at the Great Orme mine within the broader routines and social dynamics of the contemporary taskscape in north Wales. By adopting an expanded time frame, extending from the Chalcolithic to the end of the Earliest Iron Age, it offers an interpretation of activity at this source that considers the emergence, enactment and cessation of the *chaînes opératoires* of mining, processing and smelting there as part of an historical trajectory of action. Overall, it repositions the understanding of mines and mining during the Bronze Age more centrally within mainstream accounts of Bronze Age social life.

The opening chapter posed a number of research questions, namely: what were the material conditions of Bronze Age copper ore extraction at the Great Orme mine? What was its practical and social context? How did mine-working there relate to coeval routines of dwelling and subsistence? And how did it articulate with other stages in the local copper production sequence, and with the circulation of copper and bronze, tin and lead within and between communities? This chapter begins by reviewing the conclusions reached in answer to each question, beginning with the evidence for the material conditions of mining at the Great Orme mine. It then considers whether the interpretation presented here, of mining at this source as a predominately small-scale activity, downplays the importance of this site to contemporary social life. The value of the *chaîne opératoire* methodology applied in this study to the interpretation of mining more generally is then assessed through a brief consideration of the evidence for Bronze Age mine-working at the Mynydd Parys mine, another copper ore source close to the Great Orme headland. Finally, the discussion closes with some proposals for fruitful directions for future research at the Great Orme mine and for the study of prehistoric mines and mining in general.

Material conditions

Analysis of the radiocarbon dates associated with prehistoric activity at the Great Orme mine shows that the 'customary' mining phase – when people regularly and repeatedly mined there during the Bronze Age – most probably started in the early 18th century cal BC and continued until around the mid-10th century cal BC (at 68% probability) (Chapter Two). We do not know exactly when it began and ended, although it probably lasted for around 400–800 years. There are strong contextual reasons for suggesting that extraction of ore from this source probably began earlier, from at least the latter part of the third millennium BC. These include, for example, the presence of Great Orme copper in some objects from c.2150 BC, at least three centuries prior to the earliest dated activity at the mine. The removal of minerals for pigments may have been a precursor to metal mining at this site, a proposition that is worth assessing further in view of the hints of such activity at other copper mines in England and Wales with dated Bronze Age mine-working (e.g., Timberlake and Marshall, 2018). It may have continued later, into the Earliest Iron Age, although secure dated evidence for this is currently lacking.

Customary Bronze Age mining-working at the Great Orme mine was most likely characterised by extraction unfolding piecemeal, with broadly contemporaneous extraction in many different areas and later mining in locations that had been a focus for previous activity (Chapter Two). The myriad prehistoric workings represent the culmination of countless such essentially small-scale episodes of extraction. Although there would have been peaks and troughs in the quantities of copper metal produced in different periods throughout the Bronze Age life of the mine, only limited amounts of ore may have been extracted at any one time. Trench working may not have been a major feature of the mine's prehistoric development. The 'Great Opencast' (Figure 2.4) is most probably the remains of a collapsed underground cavern, the Lost Cavern. Detailed geoengineering study is, however, needed to verify these hypotheses. There is no secure dated evidence for the systematic and progressive exhaustion of one part of the source after another, nor to link working in different locales, such as the Lost Cavern or Loc.18, with specific periods in the mine's development. The main mining implements used were ones of bone with, to a lesser extent cobblestone mining tools (Chapter Three). The latter were employed either on their own or with bone tools. It is probable that wooden tools and equipment were also prevalent although none have survived due to the alkaline groundwater conditions. The use of bronze tools has been suggested but research into how they were used and how widespread this practice was is still ongoing. Fire-setting was used on occasion in some parts of the mine, but it does not appear to have been the main mining technique.

While the scale of production cannot be described as 'industrial', the available evidence strongly suggests Bronze Age mining on the Great Orme was not a disorganised activity but was, in general, carefully choreographed: planned and organised across a range of spatial and temporal scales (Chapter Three). For example, animal butchery appears to have been carried out with the eventual making and use of bone mining tools in mind. The Bronze Age miners made active technological choices, preferring, for example, specific species and skeletal elements for use as bone tools, e.g., cattle long bones over ribs. While some selections have an obvious practical explanation, like the suitability of a cattle radius or tibia for prising mineralised ore from soft, 'rotted' dolomitic limestone, culturally situated understandings, such as the social significance of cattle to contemporary communities, would also have played an important role in toolmaking and use. The selection, making, use and deposition of bone and cobblestone mining tools appear to have been highly standardised and carefully curated – assembled, organised and maintained so as to be perpetuated with no or limited change over time – throughout the customary Bronze Age mining period, for at least 400–800 years (Chapter Three). No evidence for any decline in the efficacy of the tools used, nor in the technical proficiency of those wielding them, can be identified over the Bronze Age life of the mine, such as might have been expected to characterise later periods of dwindling production. Overall, while there is likely to have been some diachronic variation in the tempo, scale, character and practical organisation of mining at this source, there is no compelling evidence for any dramatic or radical changes during the c.800 years of known prehistoric activity there.

Practical and social context

What has this study shown about how Bronze Age mining at the Great Orme mine related to coeval routines of dwelling and subsistence? The surviving archaeology of the Great Orme is probably reasonably representative of the character, scale and patterning of prehistoric activity on the headland, with no evidence for dramatic or abrupt changes during the Bronze Age life of the mine (Chapter Four). Patterns of dwelling on this promontory and in its immediate hinterland complement those seen throughout the rest of north Wales. There is no evidence that the mine itself was a focus for unusual or unusually intense activity, at least during the customary Bronze Age mine-working period. The initiation of this phase does not correlate convincingly to any kind of watershed in ways of being elsewhere in the taskscape of north Wales. The settlement, burial and environmental evidence from the Great Orme headland, its hinterland and north Wales more broadly indicates that the *chaînes opératoires* of mining were tied into fields of activity occurring at other times and places, particularly those concerned with animals and the land, such as raising and sustaining herds, and selecting wood for fuel and to make tools (Chapter Three). Like these other fields of activity, copper mining at the Great Orme mine was carried out by essentially small-scale communities, whose social and material conditions were fairly scattered, although becoming less dispersed in some areas by the end of the Earliest Iron Age (Chapter Four). The timing of mining was probably embedded in the agricultural cycle and so seasonal, with the customary phase of Bronze Age activity at the mine marked by numerous, repeated, episodic mining events, rather than continuous activity. The generally careful and consistent planning and organisation of mining at the source suggests the retention and perpetuation of knowledge and working practices from one mining event to another. This points to each mining episode taking place in living memory of previous visits, although the intervals between each occurrence may have varied, as might the pace of mine-working.

An important context to activity at the Great Orme mine during the 13–12th centuries BC was the deteriorating climate (Chapter Three). While detailed consideration of the complex relationship between climate deterioration and the anthropogenic response during this period was beyond the scope of this study, the changing climate most probably had practical consequences for the organisation of the mining field of activity, probably affecting not only how it was carried out but also perhaps its tempo and pace. These effects would not have been straightforward adaptive responses but part of a socio-historically situated and dynamic dialogue between people, places and the material conditions they encountered as they went about the routine business of their lives. The end of the customary Bronze Age phase of mining, for example, cannot be linked directly to climate change and its effects. Nor is it likely to have been due solely to geoenvironmental factors, such as rising water tables, but to how people understood and responded to these over time in the context of changing practices elsewhere in the wider taskscape, particularly bronze recycling.

Routine life for Bronze Age communities in north Wales is likely to have involved a high degree of mobility over a variety of terrain, although residential mobility appears to have become less common from the Chalcolithic onwards (Chapter Four). The Great Orme's coastal and riverine settings would have made it fully accessible to communities from elsewhere in Wales and from across north-western and central England, south-western Scotland and the Isle of Man (Chapter Five). Journeys were probably typically only over distances that could be travelled in a couple of days, whether overland or via inland waterways and coastal sea routes, although they could in some cases be longer, albeit for relatively few people. In the context of these patterns of mobility, it is probable that the Great Orme mineralised source was first encountered by people involved in routine fields of activity, such as herding or clearing woodland, rather than during active prospection for suitable ores for metal smelting (Chapter Four).

Articulation with the local copper production sequence

The tasks involved in the mining field of activity may have been dispersed across the landscape of the Great Orme (Chapter Three). There is no firm evidence that ore processing, smelting and smithing were ever centralised

in large-scale, permanent work camps at or close to the mine, although such evidence may not have survived the impact of early/late modern mining on the landscape of the Pyllau Valley. Although the significance of the smelting evidence from nearby Pentrwyn is ambiguous, this site's probable original extent and spatial proximity to the Great Orme mine hint that, at least during the very latter part of the customary Bronze Age mining phase, smelting was undertaken as mining progressed, and that members of the mining community also smelted (Chapter Three). However, (semi-)processed ore may also have been transported by sea for smelting elsewhere in the Great Orme's hinterland or even further afield. The presence of a natural harbour close to the mine means that this option was potentially available to the Bronze Age miners, even though contemporary evidence from elsewhere in Western Europe points to smelting generally occurring close to the ore source. In the absence of conclusive evidence for large-scale, permanent smelting furnaces on the Great Orme, any smelting on the headland may always have been carried out non-intensively and repeatedly on a very small-scale, using the technologically simple process identified at Pentrwyn. This practice could perhaps have involved no more than a handful of people. The Great Orme mine does not seem to have been linked into a centrally coordinated and strictly controlled system for the dispersal of ore or metal away from this source to either metalworkers or consumers. Instead, the mine's products most probably participated in complex, varied and locally specific ways in networks of production and exchange, even where these appear to have stretched over long distances.

Copper and bronze metalworking in north Wales and the rest of Britain, like the scale and character of contemporary mining on the Great Orme, appears to have been customarily small-scale, community organised and dispersed across the taskscape (Chapter Four). It may gradually have become somewhat less dispersed and more centralised by the end of the Bronze Age. The same people who mined and perhaps smelted the Great Orme ore may not have been involved in its casting and working. There is no strong evidence that bronze smiths in north Wales were full-time economic specialists, nor that the smithing of metal from the Great Orme mine was formally organised or centrally controlled. The 'Acton Park' metalworking tradition, in which metal from ore extracted at the Great Orme mine played a major role, can perhaps be better understood as the outcome of a series of technological choices linked to socially situated material preferences, rather than as a direct representation of the way metalworking practices were organised across time and space. Although the intensity of mine-working on the Great Orme is highly unlikely to have been constant throughout the Bronze Age life of the mine, the idea that there was a 'peak' in productivity (represented by increased ore output) around the middle of the second millennium BC, linked to the intensification of agriculture as a major social change, is not supported by the archaeological evidence from the mine nor from elsewhere in north Wales (Chapter Four). Nor is there production evidence for any abrupt or significant intensification in the scale and organisation of metalworking in Britain during the Bronze Age nor for routine smithing on a large scale in north Wales, even into the Late Iron Age.

Articulation with the movement of copper, bronze, tin and lead

There is no direct evidence to show whether the mined extract typically left the Great Orme mine and its hinterland as (semi-)processed ore or as ingots of unalloyed copper metal, like those found in the Salcombe assemblage from southern Britain (Chapter Four). Variations from one mining event to the next in the social composition of the community labouring at the source could have influenced the form of the mined output, with different products being produced for different individuals or communities, or for differing purposes (Chapter Five). Copper/bronze artefact distribution patterns are not, however, straightforward representations of the routes along which the output from the Great Orme mine travelled, nor the mechanisms by which it circulated (Chapter Four). Instead, they are a culmination of the entire complex of interactions in which a thing participated, both before being placed in the ground and any subsequent post-depositional processes. From finds of objects made of non-local materials, we can nonetheless suggest that communities along the north Wales coast were situated within geographically extensive networks of social relations, stretching in different directions, into the North of England, Scotland, Ireland and continental Europe as far as Sweden. The output from the mine, whatever its form, would have moved through these connections, as would the other raw materials needed to make copper and bronze. The finding of an unalloyed tin object in north Wales, in the Llangwyllog hoard, is evidence that this metal moved over long distances from its source, which was most probably in south-west Britain or Brittany (Figure 4.6). The lead added during bronze casting could present more local scales of movement, perhaps even within north Wales.

These connections most probably comprised links that were only rarely direct (Chapter Four). They were instead most probably customarily the outcome of socially and politically meaningful but fundamentally local interactions of exchange, obligation and reciprocity: not only of copper, bronze and its material constituents (tin and lead) but also of other things and ideas. In this way, communities were able to obtain materials and objects, like tin, from even distant sources. From the Great Orme's topographic situation, it is possible to suggest that members of communities from across western and central England, south-western Scotland, the Isle of Man and of course Wales could have participated in mining on the headland. The output from this activity could therefore have been dispersed in many different directions (Chapter Five). It is therefore highly unlikely that the Great Orme mine was at the epicentre of a coordinated, centrally controlled and unidirectional long-distance flow of metal (Chapter Four).

The reproduction of ideas about the self and society

In what ways was copper mining at the Great Orme mine during the Bronze Age not simply a technological process, but also a political activity, active in creating, reinforcing or expressing relations and distinctions between individuals and groups? Burial and settlement evidence indicates that society throughout the Bronze Age in north Wales was typical of that elsewhere in Britain during this period (Chapter Five). It was small in scale, communally organised, and socially and geographically dispersed. In its horizontal social structure, ties of actual, socially constituted or mythologised lineage and genealogical ancestry were emphasised. The extent and physical organisation of the mine point to the accumulation over long periods of the knowledge and expertise needed to extract ore from this source (Chapter Five). Together with the generally consistent choreography of the mining field of activity, this hints that the mine was visited and worked by the same communities for generations. I propose that mining provided an opportunity for people who did not habitually live or labour alongside one another and/or who were not close or biological kin to engage with one another, thereby creating a dynamic spatially and socio-historically situated 'community of labour'. This could have included adult men and women, and children, as well as individuals with differing physical and cognitive abilities. We cannot, however, identify how the tasks in the mining field of activity were habitually assigned based on age, gender or actual/perceived skills and experience or the intersections between these different constructs.

Certain tasks and contexts of action in the mining field of activity at the Great Orme mine were probably not open to every member of that 'community of labour' (Chapter Five). The piecemeal pattern of mine-working at this source can be seen as an expression of the differences between people, with groups with different identities, affiliations and relations working at the same time in different areas. It could also represent activity by miners with differing levels of technical aptitude and experience. Participation in or exclusion from the mining field of activity and its various *chaînes opératoires* would have reproduced what it meant to be a person, including how people understood concepts of self-identity like age and gender. Following

Brück (e.g., 2019), I argue that, in common with the other members of contemporary society, the miners would have had a relational conception of personhood, in which part of the self transcended the confines of the physical body and was constantly reproduced through a person's relationships with other people, places and things. Within this ideological framework, participation in mining would have drawn on ideas about the self – specifically those relating to age, skill and experience, bodily (dis)abilities, and one's place in networks of obligation and debt – arising through engagement in other routine activities elsewhere in the taskscape. Mining and all its contributory tasks would have provided an arena for these conceptualisations to be restated, modified and reworked, in ways that on occasion would have extended beyond the duration of a single mining event.

Political concerns and power relations would also have played out in relation to access rights to the Great Orme mine and its ore. How this occurred can best be understood by situating the mine-working in its contemporary socio-political context. Although the archaeological evidence for this is typically ambiguous and challenging to interpret, it is possible to argue for the community-based character of contemporary copper and bronze production; for the structure of Bronze Age society in north Wales as context-dependent and largely fluid; and that the material aspects of the mine were entangled in people's understandings of self and society in complex, dynamic ways. From these perspectives, rather than 'ownership' or 'control' of the ore source by a powerful community or an individual in a position of authority over others, we can instead propose that the creation and maintenance of tenure at the mine arose through the reproduction of community identities and the customary re-enactment of mining traditions.

Understanding the structure of contemporary social relations

The character and content of mining at the Great Orme mine would have been fundamentally shaped by the worldviews of the participants (Chapter Five). It would have referenced and realised many different aspects of contemporary cosmology and belief, such as those making up an ideational framework concerned with the breaking apart, mixing and reforming of substances. Aspects of this belief would in turn have been tacitly expressed, reproduced and reworked as the mining field of activity unfolded. The timing and tempo of each mining episode was most probably influenced not only by 'natural' phenomena and the seasonal round of tending to animals and the land (Chapter Three), but also by significant moments in the lifecycle of the communities coming together to mine on the headland (Chapter Five). These may have included a death or burial, a marriage or other rite of passage. In this way, copper mining at the

Great Orme mine would have been integral to how people recognised and responded to liminal events.

Customary Bronze Age mining-working at the Great Orme mine can be characterised as a 'technology of tradition', with the same tools and techniques being used for the Bronze Age life of the mine, over 400–800 years (Chapter Five). During this time, each mining event provided a means to maintain and express the metaphorical and actual ties linking the past to the present, consolidating and reproducing the identities of the mining community and its individual members within real and imagined lines of descent. I suggest that the significance of mining at the Great Orme mine, as the practical embodiment of the relations between people, the past and this particular place, could have been partly responsible for the intense and sustained drive to extract ore there. The echoes of this significance may have continued to resonate in the copper or bronze objects into which the ore from this source was fashioned. The physical residues of this technology of tradition, experienced by communities dwelling on and around the Great Orme and its hinterland long after extraction finally ceased, may have been the focus for ideologies of veneration and remembrance stretching into the Iron Age. If the mine did indeed hold a place in Iron Age cosmologies, this perhaps contributed to the choice of Pen y Dinas on the headland as the site for construction of a hillfort towards the end of the first millennium BC (Chapter Five).

Repositioning the significance of small-scale mining

The Great Orme copper mine is a site of international importance, due to its size and the distance over which at least some of the objects made from its ore are known to have travelled. The model of small-scale, repeated and episodic mining for this site developed in this study complements that proposed for the other much smaller Bronze Age copper mines in mid-Wales as well as in central England, at Ecton and Alderley Edge (e.g., Timberlake, 2017) (Figure 3.3). While the oft-highlighted difference of the Great Orme mine is its size, this may reflect the longer duration of small-scale mining there compared to the activity at these other mines. But does the notion that Bronze Age mine-working at the Great Orme mine was not industrial in scale, marked by gradual transformations in intensity rather than abrupt 'boom and bust' (O'Brien, 2015, pp. 298–300), downplay the importance of this site to contemporary social life?

Prehistoric copper mining on a small-scale has been interpreted as a marginal activity (Budd, Gale, et al., 1992, p. 37). Describing any act of mining as marginal implies that it occurred on the periphery of routine life, both in terms of the spatial location of the sites where it took place and its social significance. Through analogy with present-day artisanal mining within a capitalist economic system, small-scale may be taken to signify a peripheral and disorganised mode of operation, carried out part-time by generally unskilled practitioners who delve and scrabble for ore as a way of ameliorating poverty. In fact, the term 'small-scale' in this context simply describes the parameters of the mining field of activity. It does not assign to it any particular set of work relations nor a specific social value nor degree of importance. Nor is it a marker of a lack of complexity, skill or technological sophistication. The findings from a study of prehistoric copper smelting at Colchis in Western Georgia, for example, demonstrate that any supposed correlation between production scale and organizational complexity must be demonstrated; it cannot be assumed (Erb-Satullo, Gilmour and Khakhutaishvili, 2017, pp. 109–10). The Colchis study also shows how even production sequences characterised by spatial dispersal and without any evidence for exclusive control can still be highly productive. Top-down centralised control is not a prerequisite for productivity (Erb-Satullo, Gilmour and Khakhutaishvili, 2017, p. 122). A model of customarily small-scale production at the Great Orme mine during the Bronze Age is therefore not inevitably inconsistent with the standardisation of bronze metalwork made in Britain from the mid-second millennium BC onwards. A distinctive and notable feature of the mining field of activity at the Great Orme mine, which may have set it apart from the character of working at other, smaller sources, was that it most probably drew people from geographically scattered communities together. In this way, it created a time and place in the taskscape for the making, reworking, negotiating, contesting, resisting and sustaining of individual and group identities over extended social scales.

Contemporary Bronze Age mines and mining: the value of a *chaîne opératoire* research methodology

This study has considered Bronze Age mining on the Great Orme in the context of several different contemporary fields of activity: dwelling and settlement; tending to animals and the land; the circulation and exchange of things; the smelting and casting of copper and bronze. At the same time, however, as customary Bronze Age mine-working on this headland was beginning, it is likely that people were also mining copper ore only 20km along the coast to the east of the Great Orme, on Mynydd Parys, north-east Anglesey (Timberlake, 1988, 2017; Jenkins, 1995; Jenkins, *et al.*, 2021) (Figure 4.8). Mining on this low hill (147m AOD) is estimated to have taken place for around *175–440 years (68% probability)*. Starting in the Earlier Bronze Age, sometime in the earlier 20th to early 19th century BC (*probably 2085–1890 cal BC, 68% probability*), and continuing until at least the mid-17th to early 15th century cal BC (*1735–1485 cal BC, 68% probability*) (Timberlake, 2010; Jenkins, *et al.*, 2021), miners used cobblestone mining tools and fire-setting to follow the copper mineral veins where they outcropped close to the summit. This created opencast inclined drift workings to depths of 10m or more. Much of the evidence for this activity was destroyed by both stone quarrying and very extensive opencast copper mining on the hillside in the late 18th and early 19th centuries AD. It can now best be identified underground, where surviving Bronze Age workings are intersected by historic-era passages. Many kilometres of the latter remain to be explored for traces of prehistoric mining. No secure evidence for associated settlements or work camps has been identified. It is therefore difficult to assess accurately the scale of extraction on Mynydd Parys during the Bronze Age. The size of the source hints that the amount of mine-working there in prehistory could have been very extensive (Timberlake, 1988, 2009; Jenkins, 1995; Jenkins, *et al.*, 2021).

The beginning of ore extraction on Mynydd Parys predates the modelled start of working on the Great Orme by at least a century or more. At the same time as copper mining was happening at the former, it was also occurring further away, at various locations in a 25km radius around the Plynlimon massif in mid-Wales – most notably at Copa Hill mine, Cwmystwyth – and along Alderley Edge and on Ecton Hill in north-west central England (Timberlake, 1998, 2003b; Timberlake and Prag, 2005; Timberlake, *et al.*, 2014, 2017–2018, p. 99, 2019) (Figure 3.3). Finds of cobblestone mining tools from Bradda Head and Langness on the Isle of Man also tentatively point to prehistoric copper mining there, although the mines have not yet been located (Pickin and Worthington, 1989; Doonan and Eley, 2000). The geographic sequence of development of these mines over time was complex and non-linear (Timberlake, 2017) (Figure 6.1). Broadly speaking, the latest research re-evaluating the radiocarbon chronologies from all these sites indicates that copper mining in Britain commenced in mid-Wales during the Chalcolithic; it then occurred along the north Wales coast at Mynydd Parys; before beginning in England around 1900 cal BC, first at Alderley Edge and then at Ecton (The Lumb) around a century later (Timberlake and Marshall, 2013, 2018, 2019). It is probable that the beginning of the customary Bronze Age mining period at the Great Orme mine also overlapped with extraction at the Alderley Edge mines (Jenkins, *et al.*, 2021). Only on the Great Orme, however, did working continue after c.1600–1500 cal BC in Britain, although there was contemporary activity in south-west Ireland, at Derrycarhoon copper mines, until the 10th century cal BC (O'Brien and Hogan, 2012; Timberlake and Marshall, 2018, p. 428; O'Brien, 2019).

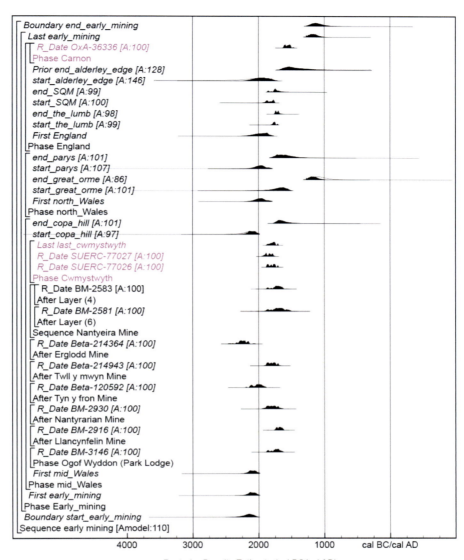

Figure 6.1: Bayesian-modelled dating sequence for early mines in England and Wales (version 7) with the relative dates of the Penparc working, Cwmystwyth, the most recently identified Bronze Age mine in mid-Wales, highlighted [SUERC-77026 and 77027]. Also included for comparison is the relative date for an antler pick found during tin streaming in the Carnon Valley, Cornwall [OxA-36336] (illustration by P. Marshall, HE, reproduced from Timberlake, et al., 2017–2018, fig. 7).

There are obvious technological and material parallels between the mining fields of activity on Mynydd Parys and the Great Orme. These include, for example, evidence for repeated instances of working; extraction following the ore vein; ore concentration by crushing and washing; and the use of fire-setting and hard fine-grained beach cobbles as mining tools (Jenkins, 1995; Timberlake, 2017; Jenkins, et al., 2021, p. 272). We can also envisage that mining in both places had a similar tempo and temporality, and that it was tied into the same networks of people and things that stretched across and beyond the north Wales coast during this period. As at the Great Orme – and at Copa Hill (Timberlake, 2009, p. 288) – contemporary mining on Mynydd Parys is likely to have been episodic and repeated. Layered deposits of clayey silt identified during excavation are possible evidence of periods of hiatus between different episodes of backfilling (Jenkins, et al., 2021, p. 269). Each visit to the source most probably involved only small groups of people. Although we do not know whether the same communities mined at each site, it is probable, given how close the two mines are to each other and the probability of overlapping activity, that extraction on this hill was part of the local practical and historical conditions out of which copper mining on the Great Orme emerged during the Earlier Bronze Age. Existing traditions of mining on Mynydd Parys are likely to have furnished the mining community on the nearby headland with some of the practical and conceptual understandings needed to extract copper ore from that source.

There are, however, key differences between the two sites. They have, for example, very different geologies. On Mynydd Parys, the miners most likely selected oxidised

A FUNDAMENTALLY SOCIAL ARCHAEOLOGY OF BRONZE AGE MINING   155 and supergene minerals present in small quantities along faults at the junction with veins of copper sulphide, which occurred mainly in hard quartz host rock, with some areas of softer shales and slates. They may also have been seeking the sizeable pieces of native copper – up to 15kg – that formerly occurred close to the surface at this source. There is no evidence that they used bone mining tools for this task (although any bone is unlikely to have survived in the strongly acidic rock environment other than as localised phosphate anomalies in the deposited spoil). The cobblestone mining tools are generally smaller in weight than those found at the Great Orme mine, with approximately 10% of the assemblage displaying notching or pecking for hafting. There seems to have been more frequent use of fire-setting than at the Great Orme mine (Timberlake, 1988, p. 13, 2009, pp. 267–8, 2010, p. 290; Jenkins, 1995; Jenkins, *et al.*, 2021, pp. 269–72). Topographically, both mines are situated on prominent landmarks in areas of predominately low relief, but unlike the Great Orme mine, the source at Mynydd Parys is located inland, a few kilometres from the sea (Jenkins, *et al.*, 2021, p. 263).

These points of divergence between the two sites are important. The *chaîne opératoire* of mining on Mynydd Parys does not seem to have been exactly the same as the technical sequence of working on the Great Orme. The mining community on the headland did not simply copy the tools, techniques and patterns of working at the neighbouring site. Rather, they adapted their know-how to make different technological choices about the best way to proceed in response to the specific physical characteristics of the source they encountered. This hints that, while mining on the Great Orme and Mynydd Parys were both closely tied into the other fields of activity in the broader taskscape, such as woodland clearance and grazing (Jenkins, *et al.*, 2021, p. 276), there may have been subtle or perhaps even more profound differences between the two sites in the character, scale and patterning of these links.

Crucially, prehistoric mining on Mynydd Parys, as on the Great Orme, was a "total social fact" (Mauss, 1979 [1935]), a field of activity structured according to an ontological framework characterised by a relational understanding of personhood enmeshed with a worldview shaped around the fragmentation, mixing and recombining of substances. The ways in which these ideas about the self and society were expressed, sustained, reworked, negotiated and contested at the Mynydd Parys mine would have arisen from people's embodied practical engagement with the materiality of that place: its distinctive and unusual rocks and minerals, distance from the sea, the types of tools used. As extraction progressed, for example, the spatial layout of the workings would have generated opportunities for people to work together that potentially reshaped individual and group identities and affiliations in ways that may have been new

or distinct to those I have outlined for the Great Orme mine. While the socially informed perspectives developed here can be brought into consideration of any mine, the specific conclusions about mining on the Great Orme drawn in this study do not represent a universal model that can be applied to all British Bronze Age mines, even though there may be points of similarity between activities at different sites. The form and layout of the Mynydd Parys mine and its historically situated and materially distinct landscape setting would all have led to new ways of being in the world, transformations whose impact would have extended into the wider taskscape and which may have persisted beyond the end of mining there. This same argument also applies to the other known early mines in Wales and England. The detailed findings of a long-term programme of investigation at these sites by the Early Mines Research Group (EMRG), led by Simon Timberlake, mean that the mining field of activity at each site and its specific landscape context are now well understood (e.g., Timberlake, 2010; Timberlake and Marshall, 2019). Similarities in the mining tools and techniques used indicate that, for perhaps 200 years at the beginning of the Earlier Bronze Age, the miners in these different areas were connected through a shared knowledge and experience of mining (Timberlake and Marshall, 2018, p. 426). These links could have involved the movement of both people and/or ideas. Either would have been made possible through the patterns of mobility discussed in Chapter Four.

As at the Great Orme and Mynydd Parys, there are, however, crucial differences between the other Earlier Bronze Age mines. These include, for example, the use of antler as the primary mining tool at Ecton mine, or the mountainous position – at 400–500m OD – of the Copa Hill and Nantyreira mines (both in mid-Wales) (Figure 3.3). I suggest that the study of prehistoric mining in Wales and England should now be expanded to explore in detail the variability, or elements of continuity and discontinuity, between the structure and logic of the *chaînes opératoires* of extraction at these different mines. Each *chaîne* should be assessed within the historically constituted taskscape at different scales of analysis, from discrete instances of extraction in a single working to the role played by the entire mining field of activity in sustaining and reworking regional (and even wider) networks of kinship and affiliation. For the reasons outlined throughout this volume, this approach should take as its starting point an understanding of prehistoric mining as a fundamentally socially embedded technology, in which the materials, gestures and routines of people mining for copper ore were inextricably entwined in a "seamless web" (Dobres and Hoffman, 1994, p. 247) with contemporary social relations, practices, structures, beliefs and ideologies. In my view, such an approach offers the potential to develop a better understanding of the character of the relationship (if any) between the different prehistoric copper mines in Wales, England and Ireland, as well as of regional variations in how the mining field of activity articulated with the tasks of making, casting and circulating copper and bronze. As I hope I have shown, a *chaîne opératoire* research methodology combined with Ingold's construct of the 'taskscape' fully integrates mines and mining into contemporary society, thereby offering richer insights into the reproduction of social life in Britain during the late third to early first millennia BC: for, as Ingold (1993b, p. 158) points out, every single task in a society takes its meaning from its position relative to every other.

Future directions

There are numerous possible avenues for future research into prehistoric activity at the Great Orme mine. The following align most closely to the approach offered throughout this book.

Bayesian chronological modelling

This study shows the potential of Bayesian modelling as a technique for producing more precise chronologies of prehistoric activity at the Great Orme mine, even with a legacy dataset. This technique needs to be firmly embedded in future archaeological research designs for the site. Future dating programmes for the prehistoric mine should be planned around recommended best practice in context recording, radiocarbon dating sample selection and Bayesian modelling (e.g., Bayliss, 2015; Hamilton and Krus, 2018). It can be challenging to micro sequence the mine's complex, multi-period deposits, even after careful excavation, but this approach nonetheless offers ways to significantly improve the robustness of the dated information and the usefulness of the chronological models produced. The key recommendations for future date sampling programmes are:

1. Samples for dating should be selected from locations that will enable specific, clearly defined research questions to be answered, such as defining the comparative chronology between workings accessed via the Tourist Cliff and those reached through the other mineralised rocky scarps in the Pyllau Valley. Samples for dating should also be chosen to target specific events of importance to the mine's development, such as the dated location of the earliest and latest prehistoric mining at the source. This will help to establish with greater confidence when Bronze Age mining probably began and ended. A comprehensive, systematic radiocarbon dating programme combined with selective excavation at the mine and its nearby production sites could eventually extend the duration of Bronze Age extraction on the Great Orme. It would also enable the temporal relationship between the source

and activity at Pentrwyn and the ore washing sites of Ffynnon Rhufeinig and Ffynnon Galchog to be better understood.

2. Sampling best practice should be followed, by selecting only short-lived, single entity samples for dating. At the mine, this would be animal bone finds identified as definite or probable mining tools (Timberlake, *et al.*, 2014, p. 187), rather than the ubiquitous charcoal. This will help to avoid the inclusion of unknown age offsets in the chronological model. For sample quality assurance, at least 10% of the dates should be replicated by, for example, submitting two samples of either the same or different types of material from a single context (Hamilton and Krus, 2018, p. 9).

3. Bayesian chronological models should be continually updated, in line with improvements in statistical methods and radiocarbon calibration data, the acquisition of more 'quality' radiocarbon dates (as defined in Chapter Two), and as we ask new archaeological questions (Bayliss, 2015, p. 680). Hence, in this study the original model produced by Peter Marshall (Model 1) was reproduced using the latest version of the OxCal computer programme and the IntCal calibration curve (Model 1_v2) (Chapter Two). It is anticipated that future researchers will in turn review and revise the new model (Model 2) on which many of the conclusions in this book are based.

Settlement landscape and environment

It is clear from the discussion in Chapters Three and Five that the contemporary settlement landscape and vegetational setting of the Great Orme mine are still poorly understood. As Watkins and colleagues (2007, pp. 179–80) point out, in a region as ecologically diverse as north Wales there are likely to have been significant local differences in the variety and type of plant species represented. A useful starting point would be local pollen studies combined with a coordinated programme of remote sensing on the Great Orme headland to locate potential buried settlement and 'farmscape' features, as well as possible smelting locations. The methodology and epistemological framework adopted by La Trobe-Bateman (2020) in her recent PhD study on the use of laser scanning (lidar) data for the analysis and interpretation of early fieldscapes in Snowdonia could provide a useful model for the latter.

Socio-historic context: links to Neolithic stone quarrying

As we explored in Chapter Five, copper mining at the Great Orme mine during the Bronze Age was a consequence of the reworking of existing understandings of the significance of this headland, its topographic setting, vegetation, residues of earlier activity, rocks and minerals. An important aspect of these specific local historic conditions for the Bronze Age mining community is likely to have been earlier stone axe quarrying and production at Graig Lwyd (e.g., O'Brien, 2010, p. 6), an igneous rock source around 10km from the headland, just over two hours' walk away or a shorter boat journey directly across Conwy Bay (Figure 1.2).

The Graig Lwyd source was a relatively important Neolithic axe-producing location during the fourth and third millennia BC (Williams, *et al.*, 1998; Williams and Kenney, 2011). As we saw in Chapter Five, there is another significant Neolithic stone-axe source in north-west Wales, at Mynydd Rhiw on the Llŷn Peninsula, and such sites most probably featured strongly in Iron Age cosmologies. Further study of the intersections between the mining field of activity on the Great Orme and the *chaînes opératoires* of production at these earlier sites – from the perspective that each one was a "total social fact" (Mauss, 1979 [1935]) – would potentially lead to a better understanding of how communities in north Wales came about the practical and conceptual knowledge needed to begin extracting the copper minerals they encountered on the Great Orme headland.

Object life histories and the technological choice to mine Great Orme ore

As discussed in Chapter Two, research by Williams (e.g., Williams and Le Carlier de Veslud, 2019) indicates that, during the mid-second millennium BC, there was marked usage of Great Orme ore to make Acton Park-type metalwork. Detailed study of the life histories of those objects for which contextual information is available, both those made from copper originating at the Great Orme mine, as well as those that seem to have been made using metal from other sources, could reveal valuable insights into this technological choice. This would in turn enhance our understanding of the significance of the Great Orme mine and mining to Bronze Age social life during this period, and the ways in which this changed over the course of the second millennium BC.

The Great Orme mine and new ways of thinking about people and things

While updating this book from the original research completed at the beginning of the new millennium, I have become increasingly intrigued by newly developing strands of archaeological thought dealing with materiality and relations between people, plants, animals and objects (e.g., Harris and Cipolla, 2017). Aspects of this theoretical framework have already been used to explore, for example, the relationship between human activity and geology at Neolithic stone sources (e.g., Dickinson, 2019), with exciting results. The Bronze Age miners at the Great Orme mine perhaps shared a worldview in which no ontological distinctions existed between themselves and the rocks and minerals they extracted; an idea touched on briefly when discussing 'ownership' of the source in Chapter Five. Approaching the archaeology of the Great Orme mine from this perspective has the potential to greatly transform our understanding of the significance of this place to Bronze Age communities and to yield rich new insights about the mining field of activity as a social construct.

Bibliography

Adams, S., Brück, J. and Webley, L. (2017) *The social context of technology: non-ferrous metalworking in later prehistoric Britain and Ireland [data-set]*. York: Archaeology Data Service [online]. Available at: doi:10.5284/1046749 (Accessed: 3 August 2022).

Adkins, L., Adkins, R. and Leitch, V. (2008) *The handbook of British archaeology*. 2nd edn. London: Constable and Robinson.

Allen, D., *et al.* (1979) 'Excavations at Hafod y Nant Criafolen, Brenig Valley, Clwyd, 1973–74', *Post-Medieval Archaeology*, 13(1), pp. 1–59.

Allnatt, V. (2017) *WMID-EFB938: a Bronze Age unidentified object* [online]. Available at: https://finds.org.uk/database/artefacts/record/id/875357 (Accessed: 4 August 2022).

Ambers, J. and Bowman, S. (1994) 'British Museum natural radiocarbon measurements XXIII', *Radiocarbon*, 36(1), pp. 95–111.

Ambers, J. and Bowman, S. (1998) 'Radiocarbon measurements from the British Museum: datelist XXIV', *Archaeometry*, 40, pp. 413–435.

Ambers, J. and Bowman, S. (1999) 'Radiocarbon measurements from the British Museum: datelist XXV', *Archaeometry*, 41, pp. 185–195.

Ambert, P., *et al.* (2009) 'The copper mines of Cabrières (Hérault) in southern France and the Chalcolithic metallurgy', in Kienlin, T. L. and Roberts, B. W. (eds) *Metals and societies. Studies in honour of Barbara S. Ottaway (Universitätsforschungen zur Prähistorischen Archäologie, Band 169)*. Bonn: Verlag Dr Rudolf Habelt GMBH, pp. 285–295.

Anon (1998) *Landscapes of historic interest in Wales. Part 2 of the register of landscapes, parks and gardens of special historic interest in Wales. Part 2.1: Landscapes of outstanding historic interest in Wales*. Cardiff: Cadw.

Anon (2008) *Wet woodland – carr* [blog post]. Available at: https://www.woodlands.co.uk/blog/flora-and-fauna/wet-woodland-carr/ (Accessed: 21 June 2019).

Anon (2013) *Cape* [online]. Available at: https://www.britishmuseum.org/collection/object/H_1836-0902-1 (Accessed: 12 May 2020).

Aris, M. (1996) *Historic landscapes of the Great Orme. Early agriculture and copper-mining*. Llanrwst, Wales: Gwasg Carreg Gwalch.

Armit, I., *et al.* (2006) 'Warfare and violence in prehistoric Europe: an introduction', in Pollard, T. and Banks, I. (eds) *War and sacrifice: studies in the archaeology of conflict*. Leiden: Brill, pp. 1–11.

Armit, I. (2007) 'Hillforts at war: from Maiden Castle to Taniwaha Pā', *Proceedings of the Prehistoric Society*, 73, pp. 25–37.

Bamforth, D. B. and Finlay, N. (2008) 'Introduction: archaeological approaches to lithic production skill and craft learning', *Journal of Archaeological Method and Theory*, 15(1), pp. 1–27.

Barber, M. (2003) *Bronze and the Bronze Age. Metalwork and society in Britain c2500–800 BC*. Stroud, Gloucestershire: Tempus.

Barker, L. (2015) *Neolithic axe factory, Mynydd Rhiw. Site record* [online]. Available at: https://coflein.gov.uk/en/site/302263/ (Accessed: 11 May 2021).

Barley, N. (1986) *The innocent anthropologist. Notes from a mud hut*. London: Penguin.

Barnatt, J. and Smith, K. (2004) *The Peak District. Landscapes through time*. Macclesfield: Windgather Press.

Barrett, J. C. (1985) 'Hoards and related metalwork', in Clarke, D. V., *et al.* (eds) *Symbols of power at the time of Stonehenge*. Edinburgh: National Museum of Antiquities of Scotland, pp. 95–106.

Barrett, J. C. (1988) 'The living, the dead and the ancestors: Neolithic and Early Bronze Age mortuary practices', in Barrett, J. C. and Kinnes, I. A. (eds) *The archaeology of context in the Neolithic and Bronze Age: recent trends*. Sheffield: Department of Archaeology and Prehistory, University of Sheffield, pp. 30–41.

Barrett, J. C. (1989a) 'Food, gender and metal: questions of social reproduction', in Stig Sørensen, M. L. and Thomas, R. (eds) *The Bronze Age–Iron Age transition in Europe. Aspects of continuity and change in European Societies c.1200 to 500 BC (BAR International Series 483ii)*. Oxford: BAR Publishing, pp. 304–320.

Barrett, J. C. (1989b) 'Time and tradition: the rituals of everyday life', in Nordström, H.-Å. and Knape, A. (eds) *Bronze Age studies. Transactions of the British-Scandinavian Colloquium in Stockholm, May 10–11, 1985*. Stockholm: Statens Historiska Museum, pp. 113–126.

Barrett, J. C. (1994) *Fragments from antiquity. An archaeology of social life in Britain, 2900–1200 BC*. Oxford: Blackwell Publishing Ltd.

Barrett, J. C. (1999) 'Rethinking the Bronze Age environment', *Quaternary Proceedings*, 7(6), pp. 493–500.

Barrett, J. C. (2012) 'Are models of prestige goods economies and conspicuous consumption applicable to the archaeology of the Bronze to Iron Age transition in Britain?', in Jones, A. M., *et al.* (eds) *Image, memory and monumentality. Archaeological engagements with the material world: a celebration of the academic achievements of Professor Richard Bradley (Prehistoric Society Research Paper No. 5)*. Oxford: Oxbow Books & Prehistoric Society, pp. 12–14.

Barrett, J. C. and Fewster, K. J. (1998) 'Stonehenge: is the medium the message?', *Antiquity*, 72, pp. 847–852.

Barrett, J. C. and Needham, S. P. (1988) 'Production, circulation and exchange: problems in the interpretation of Bronze Age metalwork', in Barrett, J. C. and Kinnes, I. A. (eds) *The archaeology of context in the Neolithic and Bronze Age: recent trends*. Sheffield: Department of Archaeology and Prehistory, University of Sheffield, pp. 127–140.

Barrett, M.-T., Brown, D. and Plunkett, G. (2019) 'Refining the statistical parameters for constructing tree-ring chronologies using short-lived species: Alder (*Alnus glutinosa Gaertn*)', *Dendrochronologia*, 55, pp. 16–24.

Bartelheim, M. (2013) 'Innovation and tradition. The structure of the early metal production in the North Alpine region', in Burmeister, S., *et al.* (eds) *Metal matters. Innovative technologies and social change in prehistory and antiquity (Forschungscluster 2)*. Rahden/Westf.: Verlag Marie Leidorf, pp. 169–180.

Bayliss, A. (2015) 'Quality in Bayesian chronological models in archaeology', *World Archaeology*, 47(4), pp. 677–700.

Baynes, E. N. (1909) 'The excavation of two barrows at Ty'n-y-pwll, Llanddyfnan, Anglesey', *Archaeologia Cambrensis*, 9(Sixth Series), pp. 312–332.

Bedlington, D. J. (1994) *Holocene sea-level changes and crustal movements in north Wales and Wirral*. Unpublished PhD thesis, Durham University.

Beecroft, S. (2005) *Evidence of ritual activity at the Great Orme, Llandudno*. Unpublished MA thesis, University of Liverpool.

Beeton, H. (2019) *LVPL-F7963B: a Bronze Age casting waste* [online]. Available at: https://finds.org.uk/database/artefacts/record/id/984603 (Accessed: 4 August 2022).

Bell, M., *et al.* (2007) 'Report on pollen analysis from Melyd Avenue, Prestatyn', in Bell, M. (ed.) *Prehistoric coastal communities: the Mesolithic in western Britain (CBA Research Report 149)*. York: Council for British Archaeology, pp. 305–306.

Bender, B. (1991) 'Women in prehistory. By Margaret Ehrenberg', *Proceedings of the Prehistoric Society*, 57(2), pp. 213–214.

Bender, B., Hamilton, S. and Tilley, C. (1997) 'Leskernick: stone worlds; alternative narratives; nested landscapes', *Proceedings of the Prehistoric Society*, 63, pp. 147–178.

Berger, D., *et al.* (2022) 'The Salcombe metal cargoes: new light on the provenance and circulation of tin and copper in Later Bronze Age Europe provided by trace elements and isotopes', *Journal of Archaeological Science*, 138, pp. 1–28.

Bibby, D. I. (1984) 'Round huts on the Great Orme's Head, Llandudno', *Bulletin of the Board of Celtic Studies*, 31, pp. 293–303.

Blanchard, I. (1978) 'Labour productivity and work psychology in the English mining industry', *The Economic History Review*, 31(1), pp. 1–24.

Booth, T. J., *et al.* (2021) 'Tales from the supplementary information: ancestry change in Chalcolithic–Early Bronze Age Britain was gradual with varied kinship organization', *Cambridge Archaeological Journal*, First View [online]. Available at: doi:10.1017/S0959774321000019 (Accessed: 24 March 2021).

Booth, T. J., Chamberlain, A. and Pearson, M. P. (2015) 'Mummification in Bronze Age Britain', *Antiquity*, 89, pp. 1155–1173.

Bradley, R. (1998) *The passage of arms. An archaeological analysis of prehistoric hoard and votive deposits.* 2nd edn. Oxford and Oakville: Oxbow Books.

Bradley, R. and Edmonds, M. R. (1993) *Interpreting the axe trade. Production and exchange in Neolithic Britain.* Cambridge: Cambridge University Press.

Bray, P. J. (2012) 'Before 29Cu became Copper: tracing the recognition and invention of metalleity in Britain and Ireland during the 3rd millennium BC', in Allen, M. J., Gardiner, J., and Sheridan, A. (eds) *Is there a British Chalcolithic? People, place and polity in the later 3rd millennium.* Oxford: Oxbow, pp. 56–70.

Bray, P. J., *et al.* (2015) 'Form and flow: the "karmic cycle" of copper', *Journal of Archaeological Science*, 56, pp. 202–209.

Bray, P. J. (2019) 'Biography, prosopography and the density of scientific data: some arguments from the metallurgy of Early Bronze Age Britain and Ireland', in Armada, X.-L., Murillo-Barroso, M., and Charlton, M. (eds) *Metals, minds and mobility. Integrating scientific data with archaeological theory.* Oxford and Philadelphia: Oxbow Books, pp. 123–133.

Bray, P. J. and Pollard, A. M. (2012) 'A new interpretative approach to the chemistry of copper-alloy objects: source, recycling and technology', *Antiquity*, 86, pp. 853–867.

Briggs, C. S. (1985) 'Problems of the early agricultural landscape in upland Wales, as illustrated by an example from the Brecon Beacons', in Spratt, D. and Burgess, C. (eds) *Upland settlement in Britain. The second millennium BC and after (BAR British Series 143).* Oxford: BAR Publishing, pp. 285–316.

Britnell, W., *et al.* (1982) 'The excavation of two round barrows at Trelystan, Powys', *Proceedings of the Prehistoric Society*, 48(1), pp. 133–201.

Britton, D. (1963) 'Traditions of metal-working in the Later Neolithic and Early Bronze Age of Britain: part I', *Proceedings of the Prehistoric Society*, 29, pp. 258–325.

Brock, F., *et al.* (2010) 'Current pretreatment methods for AMS radiocarbon dating at the Oxford Radiocarbon Accelerator Unit (ORAU)', *Radiocarbon*, 52(1), pp. 103–112.

Brocklehurst, S. (2001) *An ore processing replication experiment investigating Bronze Age ore processing at the Great Orme's Head, north Wales.* Unpublished BSc dissertation, University of Sheffield.

Bronk, C. R. and Hedges, R. E. M. (1989) 'Use of the CO_2 source in radiocarbon dating by AMS', *Radiocarbon*, 31, pp. 298–304.

Bronk Ramsey, C. (2009) 'Bayesian analysis of radiocarbon dates', *Radiocarbon*, 51(1), pp. 337–360.

Brown, T. (2008) 'The Bronze Age climate and environment of Britain', *Bronze Age Review*, 1, pp. 7–22.

Brück, J. (1999a) 'Houses, lifecycles and deposition on Middle Bronze Age settlements in Southern England', *Proceedings of the Prehistoric Society*, 65, pp. 145–166.

Brück, J. (1999b) 'Ritual and rationality: some problems of interpretation in European archaeology', *European Journal of Archaeology*, 2(3), pp. 313–344.

Brück, J. (2002) 'Body metaphors and technologies of transformation in the English Middle and Late Bronze Age', in Brück, J. (ed.) *Bronze Age landscapes. Tradition and transformation.* Oxford: Oxbow Books, pp. 149–160.

Brück, J. (2006a) 'Death, exchange and reproduction in the British Bronze Age', *European Journal of Archaeology*, 9(1), pp. 73–101.

Brück, J. (2006b) 'Fragmentation, personhood and the social construction of technology in Middle and Late Bronze Age Britain', *Cambridge Archaeological Journal*, 16(3), pp. 297–315.

Brück, J. (2007) 'The character of Late Bronze Age settlement in southern Britain', in Haselgrove, C. and Pope, R. (eds) *The Earlier Iron Age in Britain and the near continent.* Oxford: Oxbow Books, pp. 24–38.

Brück, J. (2009) 'Women, death and social change in the British Bronze Age', *Norwegian Archaeological Review*, 42(1), pp. 1–23.

Brück, J. (2015) 'Gifts or commodities? Reconfiguring Bronze Age exchange in northwest Europe', in Suchowska-Ducke, P., Scott Reiter, S., and Vandkilde, H. (eds) *Forging identities. The mobility of culture in Bronze Age Europe (BAR International Series 2771).* Oxford: British Archaeological Reports, pp. 47–56.

Brück, J. (2016) 'Hoards, fragmentation and exchange in the European Bronze Age', in Hansen, S., Neumann, D., and Tilmann, V. (eds) *Raum, Gabe und Erinnerung: Weihgaben und Heiligtümer in prähistorischen und antiken Gesellschaften.* Berlin: Edition Topoi, pp. 75–92.

Brück, J. (2019) *Personifying prehistory. Relational ontologies in Bronze Age Britain and Ireland.* Oxford: Oxford University Press.

Brück, J. and Fontijn, D. (2013) 'The myth of the chief: prestige goods, power, and personhood in the European Bronze Age', in Fokkens, H. and Harding, A. (eds) *The Oxford handbook of the European Bronze Age.* Oxford: Oxford University Press, pp. 197–215.

Brück, J. and Jones, A. M. (2018) 'Finding objects, making persons: fossils in British Early Bronze Age burials', in Harrison-Buck, E. and Hendon, J. A. (eds) *Relational identities and other-than-human agency in archaeology.* Louisville: University of Colorado Press, pp. 237–262.

Buck, C. E. and Meson, B. (2015) 'On being a good Bayesian', *World Archaeology*, 47(4), pp. 567–584.

Budd, P., Gale, D., *et al.* (1992) 'Early mines in Wales: a reconsideration', *Archaeology in Wales*, 32, pp. 36–38.

Budd, P., Pollard, M., *et al.* (1992) 'The early development of metallurgy in the British Isles', *Antiquity*, 66, pp. 677–686.

Burke, C. and Spencer-Wood, S. M. (2019) 'Introduction', in Burke, C. and Spencer-Wood, S. M. (eds) *Crafting in the world. Materiality in the making*. Cham: Springer International Publishing, pp. 1–16.

Burrow, S. (2011) 'The Mynydd Rhiw quarry site: recent work and its implications', in Davis, V. and Edmonds, M. (eds) *Stone axe studies III*. Oxford and Oakville: Oxbow Books, pp. 247–260.

Cameron, C. M. (2015) 'Commodities or gifts? Captive/slaves in small-scale societies', in Marshall, L. W. (ed.) *The archaeology of slavery. A comparative approach to captivity and coercion*. Carbondale: SIU Press, pp. 24–40.

Caseldine, A. E. (1990) *Environmental archaeology in Wales*. Lampeter: Department of Archaeology, Saint David's University College.

Caswell, E. and Roberts, B. W. (2018) 'Reassessing community cemeteries: cremation burials in Britain during the Middle Bronze Age (c.1600–1150 BC)', *Proceedings of the Prehistoric Society*, 84, pp. 329–357.

Chamberlain, A. (2014) *Gazetteer of caves, fissures and rockshelters in Wales containing human remains* [online]. Available at: http://caveburial.ubss.org.uk/wales/wales.htm (Accessed: 2 March 2021).

Chambers, F. M., Kelly, R. S. and Price, S. M. (1988) 'Development of the late-prehistoric cultural landscape in upland Ardudwy, north-west Wales', in Birks, H. H., *et al.* (eds) *The cultural landscape – past, present and future*. Cambridge: Cambridge University Press, pp. 333–348.

Chambers, F. M. and Price, S. M. (1988) 'The environmental setting of Erw-wen and Moel y Gerddi: prehistoric enclosures in upland Ardudwy, north Wales', *Proceedings of the Prehistoric Society*, 54, pp. 93–100.

Chapman, D. A. and Chapman, S. G. (2013) *Reconstructing and testing the Pentrwyn pit furnaces. Late Bronze Age copper smelting on the Great Orme* [online]. Available at: https://www.academia.edu/4835945/Pentrwyn_Late_Bronze_Age_furnaces_1_Reconstructin g_and_testing_the_Pentrwyn_pit_furnaces._Late_Bronze_Age_copper_smelting_on_the_Gre at_Orme (Accessed: 8 June 2018).

Childs, S. T. (1998) '"Find the ekijunjumira". Iron mine discovery, ownership and power among the Toro of Uganda', in Knapp, A. B., Pigott, V. C., and Herbert, E. W. (eds) *Social Approaches to an industrial past. The archaeology and anthropology of mining*. London and New York: Routledge, pp. 123–137.

Childs, S. T. (1999) '"After all, a hoe bought a wife": the social dimensions of ironworking among the Toro of East Africa', in Dobres, M.-A. and Hoffman, C. R. (eds) *The social dynamics of technology. Practice, politics, and world views*. Washington and London: Smithsonian Institution Press, pp. 23–45.

Clark, P. (2004) *The Dover Bronze Age boat*. Swindon: English Heritage.

Conwy County Borough Council (2011) *Great Orme country park and local nature reserve management plan, 2011–2016*. Unpublished report.

Cooney, G. (2016) 'Coming in and out of the dark', in Dowd, M. and Hensey, R. (eds) *The archaeology of darkness*. Oxford and Philadelphia: Oxbow Books, pp. 161–165.

Countryside Council for Wales (2008) *Core management plan including conservation objectives for Gogarth Special Conservation Area (SCA)*. Unpublished report.

Cowie, T. and O'Connor, B. (2009) 'Some Early Bronze Age stone moulds from Scotland', in Kienlin, T. L. and Roberts, B. W. (eds) *Metals and societies. Studies in honour of Barbara S. Ottaway (Universitätsforschungen zur prähistorischen Archäologie, Band 169)*. Bonn: Verlag Dr. Rudolf Habelt GMBH, pp. 313–327.

Craddock, B. (1990) 'The experimental hafting of stone mining hammers', in Crew, P. and Crew, S. (eds) *Early mining in the British Isles. Proceedings of the Early Mining Workshop at Plas Tan y Bwlch Snowdonia National Park Study Centre, 17–19 November, 1989*. Blaenau Ffestiniog: Plas Tan y Bwlch, Snowdonia National Park Study Centre, p. 58.

Craddock, P. (1992) 'A short history of firesetting', *Endeavour*, 16(3), pp. 145–150.

Craddock, P. (1995) *Early metal mining and production*. Edinburgh: Edinburgh University Press.

Crew, P. (1990) 'Firesetting experiment at Rhiw Goch, 1989', in Crew, P. and Crew, S. (eds) *Early mining in the British Isles. Proceedings of the Early Mining Workshop at Plas Tan y Bwlch Snowdonia National Park Study Centre, 17–19 November, 1989*. Blaenau Ffestiniog: Plas Tan y Bwlch, Snowdonia National Park Study Centre, p. 57.

David, G. C. (1992) 'Great Orme Bronze Age mine, Llandudno (SH 771 831)', *Archaeology in Wales*, 32, p. 58.

David, G. C. (1993) 'Great Orme Bronze Age mine, Llandudno (SH 7710 8310)', *Archaeology in Wales*, 33, p. 48.

David, G. C. (1994) 'Great Orme Bronze Age mine, Llandudno (SH 7710 8310)', *Archaeology in Wales*, 34, p. 46.

David, G. C. (1995a) 'Great Orme Bronze Age mine, Llandudno (SH 7710 8310)', *Archaeology in Wales*, 35, pp. 38–39.

David, G. C. (1995b) *Great Orme excavations since 1991.* Llandudno. Unpublished report.

David, G. C. (1996) 'Great Orme Bronze Age mine, Llandudno (SH 7710 8310) Conwy', *Archaeology in Wales*, 36, pp. 59–60.

David, G. C. (1997) 'Great Orme, Bronze Age mine, Llandudno (SH 771 831)', *Archaeology in Wales*, 37, p. 56.

David, G. C. (1998) 'Great Orme, Bronze-Age mine, Llandudno (SH 771 831)', *Archaeology in Wales*, 38, pp. 94–96.

David, G. C. (1999) 'Great Orme, Bronze-Age mine, Llandudno (SH 771 831)', *Archaeology in Wales*, 39, pp. 78–79.

David, G. C. (2000) 'Great Orme, Bronze Age mine, Llandudno (SH 771 831)', *Archaeology in Wales*, 40, pp. 73–75.

David, G. C. (2001) 'Great Orme, Bronze-Age mine, Llandudno (SH 771 831)', *Archaeology in Wales*, 41, pp. 118–119.

David, G. C. (2002) 'Great Orme, Bronze-Age mine, Llandudno (SH771831)', *Archaeology in Wales*, 42, pp. 99–100.

David, G. C. (2003) 'Great Orme, Bronze Age mine, Llandudno (SH 771 831)', *Archaeology in Wales*, 43, pp. 94–96.

David, G. C. (2004) 'Great Orme Bronze Age mine, Llandudno (SH 771 831)', *Archaeology in Wales*, 44, pp. 144–145.

David, G. C. (2005) 'Great Orme, Bronze Age mine, Llandudno (SH771 831)', *Archaeology in Wales*, 45, pp. 144–146.

Davidson, A., Smith, G. and Roberts, J. (2007) *Archaeological evaluation, excavation and watching brief during the A497 road improvement scheme, Gwynedd (GAT project no. 1692, report no. 625)* [online]. Available at: https://walesher1974.org/her/groups/GAT/media/GAT_Reports/GATreport_625_compressed .pdf (Accessed: 5 September 2022).

Davies, J. L. and Lynch, F. (2000) 'The Late Bronze Age and Iron Age', in Lynch, F., Aldhouse-Green, S., and Davies, J. L. *Prehistoric Wales*. Stroud, Gloucestershire: Sutton Publishing, pp. 139–219.

Davies, M. (1973) 'Lloches-yr-Afr, Great Orme (SH 7792 8383)', *Archaeology in Wales*, 13, pp. 14–15.

Davies, M. (1974) 'Lloches-yr-afr (SH 7792 8383)', *Archaeology in Wales*, 14, p. 9.

Davies, M. (1989) 'Cave archaeology in North Wales', in Ford, T. D. (ed.) *Limestones and caves of Wales*. Cambridge: Cambridge University Press, pp. 92–101.

Davies, M. and Stone, T. (1977) 'Upper Kendrick's Cave – a new archaeological site in Llandudno', *Cambrian Caving Council Annual Journal*, 5, pp. 10–11.

Dickinson, S. (2019) 'Moving mountains: reciprocating with rock in the Neolithic', in Teather, A., Topping, P., and Baczkowski, J. (eds) *Mining and quarrying in Neolithic Europe: a social perspective (Neolithic Studies Group Seminar Papers 16)*. Oxford and Philadelphia: Oxbow Books, pp. 163–178.

Dinnis, R. and Ebbs, C. (2013) 'Cave deposits of North Wales: some comments on their archaeological importance and an inventory of sites of potential interest', *Cave and Karst Science*, 40(1), pp. 28–34.

Dobres, M.-A. (2000) *Technology and social agency. Outlining a practice framework for archaeology.* Oxford: Blackwell Publishers.

Dobres, M.-A. and Hoffman, C. (1994) 'Social agency and the dynamics of prehistoric technology', *Journal of Archaeological Method and Theory*, 1(3), pp. 211–258.

Dolfini, A., *et al.* (2018) 'Interdisciplinary approaches to prehistoric warfare and violence: past, present, and future', in Dolfini, A., *et al.* (eds) *Prehistoric warfare and violence.* Cham: Springer International Publishing, pp. 1–18.

Doonan, R. (1994) 'Sweat, fire and brimstone: pre-treatment of copper ore and the effects on smelting techniques', *The Journal of the Historical Metallurgy Society*, 28(2), pp. 84–97.

Doonan, R. and Eley, T. (2000) 'The Langness ancient mining survey', in Darvill, T. (ed.) *Billown Neolithic landscape project, Isle of Man (5th Report: 1999).* Bournemouth and Douglas: Bournemouth University and Manx National Heritage, pp. 45–53.

Dresser, Q. (1985) 'University College Cardiff radiocarbon dates I', *Radiocarbon*, 27(2B), pp. 338–385.

Dunbar, E., *et al.* (2016) 'AMS 14C dating at the Scottish Universities Environmental Research Centre (SUERC) Radiocarbon Dating Laboratory', *Radiocarbon*, 58(1), pp. 9–23.

Dutton, A., *et al.* (1994) 'Prehistoric copper mining on the Great Orme, Llandudno, Gwynedd', *Proceedings of the Prehistoric Society*, 60, pp. 245–286.

Dutton, L. A. (1988) 'Great Orme copper mine, Llandudno (SH766 832)', *Archaeology in Wales*, 28, p. 46.

Dutton, L. A. (1990) 'Surface remains of early mining on the Great Orme', in Crew, P. and Crew, S. (eds) *Early mining in the British Isles. Proceedings of the Early Mining Workshop at Plas Tan y Bwlch Snowdonia National Park Study Centre, 17–19 November, 1989.* Blaenau Ffestiniog: Plas Tan y Bwlch, Snowdonia National Park Study Centre, pp. 11–14.

Earle, T., *et al.* (2015) 'The political economy and metal trade in Bronze Age Europe: understanding regional variability in terms of comparative advantages and articulations', *European Journal of Archaeology*, 18(4), 633–657.

Edmonds, M. R. (1990) 'Description, understanding and the *chaîne opératoire*', *Archaeological Review from Cambridge*, 9(1), pp. 55–70.

Edmonds, M. R. (1994) 'Review of Pierre Lemonnier (ed.) "Technological choices: transformation in material culture"', *Antiquity*, 68, pp. 472–475.

Edmonds, M. R. (1995) *Stone tools and society. Working stone in Neolithic and Bronze Age Britain*. London and New York: Routledge.

Edmonds, M. R. (1998) 'Sermons in stone: identity, value and stone tools in Later Neolithic Britain', in Edmonds, M. and Richards, C. (eds) *Understanding the Neolithic of north-western Europe*. Glasgow: Cruithne Press, pp. 248–276.

Edmonds, M. R. (1999a) *Ancestral geographies of the Neolithic. Landscapes, monuments and memory*. London and New York: Routledge.

Edmonds, M. R. (1999b) 'Inhabiting Neolithic landscapes', *Journal of Quaternary Science*, 14(6), pp. 485–492.

Elyachar, J. (2010) 'Phatic labor, infrastructure, and the question of empowerment in Cairo', *American Ethnologist*, 37(3), pp. 452–464.

Erb-Satullo, N. L., Gilmour, B. J. J. and Khakhutaishvili, N. (2017) 'Copper production landscapes of the South Caucasus', *Journal of Anthropological Archaeology*, 47, pp. 109–126.

Finley, J. B., *et al.* (2020) 'Multidecadal climate variability and the florescence of Fremont societies in Eastern Utah', *American Antiquity*, 85(1), pp. 93–112.

Fitzpatrick, A. P. (2009) 'In his hands and in his head: the Amesbury Archer as a metalworker', in Clark, P. (ed.) *Bronze Age connections. Cultural contact in prehistoric Europe*. Oxford and Oakville: Oxbow Books, pp. 176–188.

Fitzpatrick, A. P., Hamilton, D. and Haselgrove, C. (2017) 'Radiocarbon dating and Bayesian modelling of the Late Iron Age cremation burial cemetery at Westhampnett (West Sussex/GB)', *Archäologisches Korrespondenzblatt*, 47(3), pp. 359–381.

Fowler, C. (2001) 'Personhood and social relations in the British Neolithic with a study from the Isle of Man', *Journal of Material Culture*, 6(2), pp. 137–163.

Frankenstein, S. and Rowlands, M. J. (1978) 'The internal structure and regional context of Early Iron Age society in south-western Germany', *Bulletin of Institute of Archaeology University of London*, 16, pp. 73–112.

Fregni, E. G. (2014) *The compleat metalsmith: craft and technology in the British Bronze Age*. Unpublished PhD thesis, University of Sheffield.

Fregni, E. G. (2019) 'Looking over the shoulder of the Bronze Age metalsmith: recognising the crafter in archaeological artefacts', in Burke, C. and Spencer-Wood, S. M. (eds) *Crafting in the world. Materiality in the making*. Cham: Springer International Publishing, pp. 37–49.

Frieman, C. J. (2020) 'Editorial', *European Journal of Archaeology*, 23(2), pp. 159–161.

Gale, D. (1995) *Stone tools employed in prehistoric metal mining. A functional study of cobblestone tools from prehistoric metalliferous mines in England and Wales in relation to mining strategies by use-wear analysis and cobble morphometry*. Unpublished PhD thesis, University of Bradford.

Ghey, E., *et al.* (2007) 'Characterising the Welsh roundhouse: chronology, inhabitation and landscape', *Internet Archaeology*, 23 [online]. Available at: https://intarch.ac.uk/journal/issue23/johnston_index.html (Accessed: 16 March 2021).

Gibson, A. (2018) 'Llandegai A – sanctuary or settlement?', *Archaeologia Cambrensis*, 167, pp. 95–108.

Goldenberg, G. (2015) 'Prähistorische Kupfergewinnung aus Fahlerzen der Lagerstätte Schwaz-Brixlegg im Unterinntal, Nordtirol', in Stöllner, T. and Oeggl, K. (eds) *Bergauf Bergab, 10.000 Jahre Bergbau in den Ostalpen*. Bochum: VML Verlag Marie Leidorf, pp. 151–163.

Gosden, C. and Lock, G. (1998) 'Prehistoric histories', *World Archaeology*, 30(1), pp. 2–12.

Green, H. S. (1985) 'A Bronze Age stone mould from New Mills, Newtown, Powys', *Bulletin of the Board of Celtic Studies*, 32, pp. 273–274.

Gregory, R. A., *et al.* (2000) 'A retrospective assessment of 19th-century finds from a Little Orme quarry', *Archaeology in Wales*, 40, pp. 3–8.

Guilbert, G. (2018) 'Historical excavation and survey of hillforts in Wales: some critical issues', *Internet Archaeology*, 48 [online]. Available at: doi: 10.11141/ia.48.3 (Accessed: 17 March 2021).

Gwilt, A. (2015) *NMGW-04216C: a Bronze Age hoard* [online]. Available at: https://finds.org.uk/database/artefacts/record/id/727558 (Accessed: 4 August 2022).

Gwilt, A., *et al.* (2019) *A Late Bronze Age hoard from Abergele Community, Conwy (Treasure Case 17.07)*. Unpublished report.

Gwilt, A. (2021) *A probable Middle Bronze Age hoard from Conwy Community, Conwy (Treasure Case 17.12)*. Unpublished report.

Gwyn, D. and Thompson, D. (1999) *Historic landscape characterisation – Creuddyn and Arllechwedd. Historic landscape character areas (GAT project no. G1527, report no. 318)* [online]. Available at: http://www.walesher1974.org/her/groups/GAT/media/GAT_Reports/GATreport_318_compressed.pdf (Accessed: 3 August 2022).

Hamilton, W. D. and Krus, A. M. (2018) 'The myths and realities of Bayesian chronological modeling revealed', *American Antiquity*, 83(2), pp. 187–203.

Harris, O. (2020) 'Joanna Brück personifying prehistory: relational ontologies in Bronze Age Britain and Ireland (Oxford: Oxford University Press, 2019, 308 pp., 52 figs, hbk, ISBN 978-0-19-876801-2)', *European Journal of Archaeology*, 11(3), pp. 520–522.

Harris, O. and Cipolla, C. (2017) *Archaeological theory in the new millennium. Introducing current perspectives.* London: Routledge.

Harush, O., *et al.* (2020) 'Social signatures in standardized ceramic production – a 3-D approach to ethnographic data', *Journal of Anthropological Archaeology*, 60.

Healy, F., *et al.* (2018) 'When and why? The chronology and context of flint mining at Grime's Graves, Norfolk, England', *Proceedings of the Prehistoric Society*, 84, pp. 277–301.

Hedges, R. E. M., *et al.* (1996) 'Radiocarbon dates from the Oxford AMS system: Archaeometry datelist 21', *Archaeometry*, 38, pp. 181–207.

Herbert, E. W. (1984) *Red gold of Africa. Copper in precolonial history and culture.* Madison and London: University of Wisconsin Press.

Herbert, E. W. (1993) *Iron, gender and power. Rituals of transformation in African societies.* Bloomington and Indianapolis: Indiana University Press.

Herbert, E. W. (1998) 'Mining as microcosm in precolonial sub-Saharan Africa. An overview', in Knapp, A. B., Pigott, V. C., and Herbert, E. W. (eds) *Social Approaches to an industrial past. The archaeology and anthropology of mining.* London and New York: Routledge, pp. 138–154.

Hicklin, J. (1858) *The hand-book to Llandudno and its vicinity.* London: Whittaker and Co.

Hind, D. (2000) *Landscape and technology in the Peak District of Derbyshire: the fifth and fourth millennia.* Unpublished PhD thesis, University of Sheffield.

Holtorf, C. (1998) 'The life-histories of megaliths in Mecklenburg-Vorpommern (Germany)', *World Archaeology*, 30(1), pp. 23–38.

Hopewell, D. (2013a) *Evaluation of scheduling proposals 2012–13 (GAT project no. G2246, report no. 1125)* [online] Available at: http://www.walesher1974.org/her/groups/GAT/media/GAT_Reports/GATreport_1225_PartI_compressed.pdf (Accessed: 5 September 2022).

Hopewell, D. (2013b) 'Great Orme. Hwylfa'r Ceirw field system SH 7653 8399C', *Archaeology in Wales*, 52, pp. 187–188.

Horn, C. and Kristiansen, K. (eds) (2018) *Warfare in Bronze Age society.* Cambridge: Cambridge University Press.

Hughes, S. (2003) 'The achievements and future of the uplands initiative', in Browne, D. M. and Hughes, S. (eds) *The archaeology of the Welsh uplands.* Aberystwyth: Royal Commission on the Ancient and Historical Monuments of Wales, pp. 117–124.

Hughes, T. P. (1979) 'The electrification of America: the system builders', *Technology and Culture*, 20(1), pp. 124–161.

Hunt, A. (1993) *An analysis of animal bone assemblages from the Bronze Age copper mines of the Great Orme, Wales.* Unpublished BSc dissertation, University of Sheffield.

Iles, L. (2018) 'Forging networks and mixing ores: rethinking the social landscapes of iron metallurgy', *Journal of Anthropological Archaeology*, 49, pp. 88–99.

Ingold, T. (1993a) 'The reindeerman's lasso', in Lemonnier, P. (ed.) *Technological choices. Transformation in material cultures since the Neolithic.* London and New York: Routledge, pp. 108–125.

Ingold, T. (1993b) 'The temporality of landscape', *World Archaeology*, 2, pp. 152–174.

Ingold, T. (1995) 'Work, time and industry', *Time and Society*, 4(1), pp. 5–28.

Ingold, T. (1997) 'Eight themes in an anthropology of technology', *Social Analysis: the International Journal of Anthropology*, 41(1), pp. 106–138.

Investopedia Team (2020) *Industry life cycle* [online]. Available at: https://www.investopedia.com/terms/i/industrylifecycle.asp (Accessed: 4 December 2021).

Jackson, J. S. (1979) 'Metallic ores in Irish prehistory: copper and tin', in Ryan, M. (ed.) *The origins of metallurgy in Atlantic Europe. Proceedings of the Fifth Atlantic Colloquium, Dublin, 30th March to 4th April 1978.* Dublin: Stationery Office, pp. 107–125.

Jackson, J. S. (1980) 'Bronze Age copper mining in counties Cork and Kerry', in Craddock, P. (ed.) *Scientific studies in early mining and extractive metallurgy (British Museum Occasional Paper 20).* London: British Museum Press, pp. 9–29.

James, D. (1990) 'Prehistoric copper mining on the Great Orme's Head', in Crew, P. and Crew, S. (eds) *Early mining in the British Isles. Proceedings of the Early Mining Workshop at Plas Tan y Bwlch Snowdonia National Park Study Centre, 17–19 November, 1989.* Blaenau Ffestiniog: Plas Tan y Bwlch, Snowdonia National Park Study Centre, pp. 1–4.

James, S. E. (2011) *The economic and environmental implications of faunal remains from the Bronze Age copper mines at Great Orme, North Wales.* Unpublished PhD thesis, University of Liverpool.

James, S. E. (2016) 'Digging into the darkness: the experience of copper mining in the Great Orme, north Wales', in Dowd, M. and Hensey, R. (eds) *The Archaeology of darkness.* Oxford and Philadelphia: Oxbow Books, pp. 88–97.

Jenkins, D. A. (1994) 'Geological context of the Great Orme mines. In: Dutton, A., *et al.* Prehistoric copper mining on the Great Orme, Llandudno, Gwynedd', *Proceedings of the Prehistoric Society*, 60, pp. 252–255.

Jenkins, D. A. (1995) 'Mynydd Parys copper mines (SH 441 904)', *Archaeology in Wales*, 35, pp. 35–37.

Jenkins, D. A., Owen, A. and Lewis, C. A. (2001) 'A rapid geochemical survey of the Bronze Age copper mines on the Great Orme, Llandudno', in Millard, A. (ed.) *Archaeological Science '97. Proceedings of the conference held at the University of Durham 2nd–4th September 1997 (BAR International Series 939)*. Oxford: BAR Publishing, pp. 164–169.

Jenkins, D.A., *et al.* (2021) 'Copper mining in the Bronze Age at Mynydd Parys, Anglesey, Wales', *Proceedings of the Prehistoric Society*, 87, pp. 261–291.

Jenkins, D. A. and Lewis, C. A. (1991) 'Prehistoric mining for copper in the Great Orme, Llandudno', in Budd, P., *et al.* (eds) *Archaeological Sciences 1989. Proceedings of a conference on the application of scientific techniques to archaeology, Bradford, September 1989 (Oxbow Monograph 9)*. Oxford: Oxbow, pp. 151–161.

Johnston, R. (2008) 'Copper mining and the transformation of environmental knowledge in Bronze Age Britain', *Journal of Social Archaeology*, 8(2), pp. 190–213.

Jones, S. G. (1994) *Deciphering the 'metallic arts' of the Bronze Age: a proposed criteria for the identification of ore-washing sites associated with early copper mining in Wales*. Unpublished MSc dissertation, University of York.

Jovanović, B. (1980) 'Primary copper mining and the production of copper', in Craddock, P. (ed.) *Scientific studies in early mining and extractive metallurgy (British Museum Occasional Paper 20)*. London: British Museum Press, pp. 31–40.

Jowett, N. (2017) 'Evidence for the use of Bronze Age mining tools in the Bronze Age copper mines on the Great Orme, Llandudno', *Archaeology in Wales*, 56, pp. 63–69.

Jowett, N. (2019) *Great Orme Bronze Age copper mines*. Llandudno: Great Orme Mines Ltd.

Kamp, K. A. (2001) 'Where have all the children gone?: The archaeology of childhood', *Journal of Archaeological Method and Theory*, 8, pp. 1–34.

Kelly, R. S. (1982) 'The excavation of a Medieval farmstead at Cefn Graeneg, Clynnog, Gwynedd', *Bulletin of the Board of Celtic Studies*, 29, pp. 859–908.

Kelly, R. S. (1988) 'Two late prehistoric circular enclosures near Harlech, Gwynedd', *Proceedings of the Prehistoric Society*, 54, pp. 101–172.

Kern, A., *et al.* (eds) (2016) *Kingdom of salt. 7000 years of Hallstatt*. 2nd edn. Vienna: Natural History Museum.

Kienlin, T. L. (2013) 'Copper and bronze: Bronze Age metalworking in context', in Fokkens, H. and Harding, A. (eds) *The Oxford Handbook of the European Bronze Age*. Oxford: Oxford University Press, pp. 414–436.

Kienlin, T. L. and Stöllner, T. (2009) 'Singen copper, alpine settlement and Early Bronze Age mining: is there a need for elites and strongholds?', in Kienlin, T. L. and Roberts, B. W. (eds) *Metals and societies. Studies in honour of Barbara S. Ottaway (Universitätsforsch. Prähist. Arch. 169)*. Bonn: Verlag Dr. Rudolf Habelt GMBH, pp. 67–104.

Knapp, A. B., Pigott, V. C. and Herbert, E. W. (eds) (1998) *Social approaches to an industrial past*. London and New York: Routledge.

Knight, M., *et al.* (2019) 'The Must Farm pile-dwelling settlement', *Antiquity*, 93, pp. 645–663.

Knight, M. G. (2019) 'Going to pieces: investigating the deliberate destruction of Late Bronze Age swords and spearheads', *Proceedings of the Prehistoric Society*, 85, pp. 251–272.

Knight, M. G. (2021) 'There's method in the fragments: a damage ranking system for Bronze Age metalwork', *European Journal of Archaeology*, 24(1), pp. 48–67.

Kristiansen, K. and Earle, T. (2015) 'Neolithic versus Bronze Age social formations: a political economy approach', in Kristiansen, K., Šmejda, L., and Turek, J. (eds) *Paradigm found. Archaeological theory present, past and future (essays in honour of Evžen Neustupný)*. Oxford and Philadelphia: Oxbow Books, pp. 234–247.

Kristiansen, K. and Suchowska-Ducke, P. (2015) 'Connected histories: the dynamics of Bronze Age interaction and trade 1500–1100 bc', *Proceedings of the Prehistoric Society*, 81, pp. 361–392.

Kuijpers, M. H. G. (2012) 'The sound of fire, taste of copper, feel of bronze, and colours of the cast: sensory aspects of metalworking technology', in Stig Sørensen, M. L. and Rebay-Salisbury, K. (eds) *Embodied knowledge. Perspectives on belief and technology*. Oxford: Oxbow Books, pp. 137–150.

Kuijpers, M. H. G. (2018) 'The Bronze Age, a world of specialists? Metalworking from the perspective of skill and material specialization', *European Journal of Archaeology*, 21(4), pp. 550–571.

La Trobe-Bateman, E. (2020) *Snowdonia's early fieldscapes*. Unpublished PhD thesis, University of Sheffield.

Latour, B. (1988) 'How to write "The Prince" for machines as well as for machinations', in Elliott, B. (ed.) *Technology and social change*. Edinburgh: Edinburgh University Press, pp. 20–43.

Lemonnier, P. (1992) *Elements for an anthropology of technology (Anthropological Papers, Museum of Anthropology, University of Michigan No.88)*. Ann Arbor, Michigan: the Museum of Anthropology, University of Michigan.

Lemonnier, P. (ed.) (1993) *Technological choices. Transformation in material cultures since the Neolithic*. London and New York: Routledge.

Leroi-Gourhan, A. (1971 [1943]) *Evolution et techniques: L'homme et la matière*. Paris: Albin Michel.

Lewis, C. A. (1988) 'Great Orme copper mines, Llandudno (SH 771 831)', *Archaeology in Wales*, 28, pp. 45–46.

Lewis, C. A. (1989) 'Great Orme copper mines, Llandudno (SH 7710 8315)', *Archaeology in Wales*, 29, pp. 42–43.

Lewis, C. A. (1990a) 'Ffynnon Galchog, Great Orme, Llandudno (SH 7753 8365)', *Archaeology in Wales*, 30, pp. 43–44.

Lewis, C. A. (1990b) 'Firesetting experiments on the Great Orme, 1989', in Crew, P. and Crew, S. (eds) *Early mining in the British Isles. Proceedings of the Early Mining Workshop at Plas Tan y Bwlch Snowdonia National Park Study Centre, 17–19 November, 1989*. Blaenau Ffestiniog: Plas Tan y Bwlch, Snowdonia National Park Study Centre, pp. 55–56.

Lewis, C. A. (1990c) 'Great Orme copper mine, Llandudno (SH 771 831)', *Archaeology in Wales*, 30, p. 43.

Lewis, C. A. (1990d) 'Underground exploration of the Great Orme copper mines', in Crew, P. and Crew, S. (eds) *Early mining in the British Isles. Proceedings of the Early Mining Workshop at Plas Tan y Bwlch Snowdonia National Park Study Centre, 17–19 November, 1989*. Blaenau Ffestiniog: Plas Tan y Bwlch, Snowdonia National Park Study Centre, pp. 5–10.

Lewis, C. A. (1993) *Underground survey of mine workings at Great Orme's Head, Llandudno*. Unpublished report for Cadw.

Lewis, C. A. (1994) 'Bronze Age mines of the Great Orme: interim report', *Bulletin of the Peak District Mines Historical Society (Special Publication: Mining Before Powder)*, 12(3), pp. 31–36.

Lewis, C. A. (1996) *Prehistoric mining at the Great Orme. Criteria for the identification of early mining*. Unpublished MPhil thesis, University of Wales, Bangor.

Lynch, F. (1983) 'Report of the excavation of a Bronze Age barrow at Llong near Mold', *Flintshire Historical Society Journal*, 31, pp. 13–28.

Lynch, F. (1984) 'Moel Goedog Circle I. A complex ring cairn near Harlech', *Archaeologia Cambrensis*, CXXXIII, pp. 8–50.

Lynch, F. (1986) 'Excavation of a kerb circle and ring cairn on Cefn Caer Euni, Merioneth', *Archaeologia Cambrensis*, CXXXV, pp. 81–120.

Lynch, F. (1995) *A Guide to ancient and historic Wales: Gwynedd*. London: HMSO.

Lynch, F. (2000) 'The Later Neolithic and Earlier Bronze Age', in Lynch, F., Aldhouse-Green, S., and Davies, J. L. *Prehistoric Wales*. Stroud, Gloucestershire: Sutton Publishing, pp. 79–138.

Lynch, F., Aldhouse-Green, S., and Davies, J. L. (2000) *Prehistoric Wales*. Stroud, Gloucestershire: Sutton Publishing.

MacGregor, N. (2012) *A history of the world in 100 objects*. London: Penguin.

Manley, J. (1985) 'Fields, cairns and enclosures on Ffridd Brynhelen, Clwyd', in Spratt, D. and Burgess, C. (eds) *Upland settlement in Britain. The second millennium BC and after (BAR British Series 143)*. Oxford: BAR Publishing, pp. 317–349.

Manley, J. (1990) 'A Late Bronze Age landscape on the Denbigh Moors, northeast Wales', *Antiquity*, 64, pp. 514–526.

Marshall, L. W. (2015) 'Introduction: the comparative archaeology of slavery', in Marshall, L. W. (ed.) *The archaeology of slavery. A comparative approach to captivity and coercion*. Carbondale: SIU Press, pp. 1–23.

Mauss, M. (1979 [1935]) *Sociology and psychology: essays [trans. B. Brewster]*. London: Routledge and Kegan Paul.

McInstosh, F. (2008) *LVPL-936AE0: a unknown waste* [online]. Available at: https://finds.org.uk/database/artefacts/record/id/240421 (Accessed: 4 August 2022).

McKeown, S. A. (1994) 'Appendix 2. The analysis of wood remains from Mine 3, Mount Gabriel', in O'Brien, W. (ed.) *Mount Gabriel. Bronze Age mining in Ireland*. Cork: Galway University Press, pp. 265–280.

Mighall, T. M., *et al.* (2000) 'Prehistoric copper mining and its impact on vegetation: palaeological evidence from Mount Gabriel, Co. Cork, south-west Ireland', in Nicholson, R. A. and O'Connor, T. P. (eds) *People as an agent of environmental change*. Oxford: Oxbow, pp. 19–29.

Mighall, T. M., *et al.* (2002) 'A palaeoenvironmental investigation of sediments from the prehistoric mine of Copa Hill, Cwmystwyth, mid-Wales', *Journal of Archaeological Science*, 29, pp. 1161–1188.

Mighall, T. M. and Chambers, F. M. (1993) 'The environmental impact of prehistoric mining at Copa Hill, Cwmystwyth, Wales', *The Holocene*, 3(3), pp. 260–264.

Mighall, T. M. and Chambers, F. M. (1995) 'Holocene vegetation history and human impact at Bryn y Castell, Snowdonia, north Wales', *The New Phytologist*, 130(2), pp. 299–321.

Moore, P. D. (1993) 'The origin of blanket mire, revisited', in Chambers, F. M. (ed.) *Climate change and human impact on the landscape*. London: Chapman and Hall, pp. 217–224.

Morgan, D. E. (1990) *Bronze Age metalwork from Flintshire (North West Archaeological Trust Report No. 4)*. Preston: North West Archaeological Trust.

Musson, C., *et al.* (1991) *The Breiddin hillfort. A later prehistoric settlement in the Welsh Marches (CBA Research Report 76)*. London: Council for British Archaeology.

Must Farm (2021) *Logboats 1 to 9, 2011: Bronze Age river discoveries* [online]. Available at: http://

www.mustfarm.com/bronze-age-river/discoveries/ (Accessed: 6 May 2020).

Nash, J. (1993) *We eat the mines and the mines eat us. Dependency and exploitation in Bolivian tin mines.* New York: Colombia University Press.

Needham, S. (1989) 'Selective deposition in the British Early Bronze Age', in Nordström, H.-Å. and Knappe, A. (eds) *Bronze Age studies. Transactions of the British-Scandinavian Colloquium in Stockholm, May 10–11, 1985.* Stockholm: Statens Historiska Museum, pp. 45–61.

Needham, S. (1996) 'Chronology and periodisation in the British Bronze Age', *Acta Archaeologica*, 67, pp. 121–40.

Needham, S. (2012) 'Putting capes into context: Mold at the heart of a domain', in Britnell, W. J. and Silvester, R. J. (eds) *Reflections on the past: essays in honour of Frances Lynch.* Welshpool: Cambrian Archaeological Association.

Needham, S. and Bridgford, S. (2013) 'Deposits of clay refractories for casting bronze swords', in Brown, N. and Medlycott, M. (eds) *The Neolithic and Late Bronze Age enclosures at Springfield Lyons, Essex (East Anglian Archaeology Report 149).* Chelmsford: Essex County Council, pp. 47–74.

Needham, S., Parham, D. and Frieman, C. J. (2013) *Claimed by the sea: Salcombe, Langdon Bay, and other marine finds of the Bronze Age (CBA Research Report 173).* York: Council for British Archaeology.

Northover, J. P. (1980a) 'Appendix. The analysis of Welsh Bronze Age metalwork', in Savory, H. N. (ed.) *Guide catalogue of the Bronze Age collections.* Cardiff: National Museum of Wales, pp. 229–243.

Northover, J. P. (1980b) 'Bronze in the British Bronze Age', in Oddy, W. A. (ed.) *Aspects of early metallurgy (British Museum Occasional Paper No. 17).* London: British Museum Press, pp. 63–70.

Northover, J. P. (1982) 'The metallurgy of the Wilburton hoards', *Oxford Journal of Archaeology*, 1, pp. 69–109.

Northover, P. (1984) 'Iron Age bronze metallurgy in central southern England', in Cunliffe, B. and Miles, D. (eds) *Aspects of the Iron Age in Central Southern Britain (University of Oxford Committee for Archaeology, Monograph No. 2).* Oxford: Oxford University Committee for Archaeology, pp. 126–145.

Nowakowski, J. A. (2002) 'Leaving home in the Cornish Bronze Age: insights into planned abandonment processes', in Brück, J. (ed.) *Bronze Age landscapes. Tradition and transformation.* Oxford: Oxbow Books, pp. 139–148.

O'Brien, W. (1994) *Mount Gabriel. Bronze Age mining in Ireland.* Cork: Galway University Press.

O'Brien, W. (2004) *Ross Island. Mining, metal and society in early Ireland (Bronze Age Studies 6).* Galway:

Department of Archaeology, National University of Ireland.

O'Brien, W. (2007) 'Miners and farmers: local settlement contexts for Bronze Age mining', in Burgess, C., Topping, P., and Lynch, F. (eds) *Beyond Stonehenge: essays on the Bronze Age in honour of Colin Burgess.* Oxford: Oxbow Books, pp. 20–30.

O'Brien, W. (2010) 'Copper axes, stone axes: production and exchange systems in the Chalcolithic of Britain and Ireland', in Anreiter, P., *et al.* (eds) *Mining in European history and its impact on environment and human societies. Proceedings for the 1st Mining in European History Conference of the SFB-HIMAT, 12.–15. November 2009, Innsbruck.* Innsbruck: Innsbruck University Press, pp. 3–8.

O'Brien, W. (2013) 'Copper mining in Ireland during the Later Bronze Age', in Anreiter, P., *et al.* (eds) *Mining in European history and its impact on environment and human societies. Proceedings for the 2nd Mining in European History Conference of the FZ HiMAT, 7–10. November 2012, Innsbruck.* Innsbruck: Innsbruck University Press, pp. 191–197.

O'Brien, W. (2015) *Prehistoric copper mining in Europe 5500–500 BC.* Oxford: Oxford University Press.

O'Brien, W. (2018) 'Hillforts and warfare in Bronze Age Ireland', in Hansen, S. and Krause, R. (eds) *Bronze Age hillforts between Taunus and Carpathian Mountains.* Bonn: Verlag Dr. Rudolf Habelt GMBH, pp. 1–22.

O'Brien, W. (2019) 'Derrycarhoon mine and the supply of copper in later Bronze Age Ireland', in Brandherm, D. (ed.) *Aspects of the Bronze Age in the Atlantic Archipelago and beyond. Proceedings from the Belfast Bronze Age Forum, 9–10 November 2013.* Hagen/Westf.: Curach Bhán Publications, pp. 121–143.

O'Brien, W. (2022) *Derrycarhoon. A Later Bronze Age copper mine in south-west Ireland (BAR International Series 3069).* Oxford: BAR Publishing.

O'Brien, W. and Hogan, N. (2012) 'Derrycarhoon: a Bronze Age copper mine in West Cork', *Archaeology Ireland*, 26(1), pp. 12–15.

Oakden, V. (2012a) *LVPL-3996B1: a Bronze Age metal working debris* [online]. Available at: https://finds.org. uk/database/artefacts/record/id/530012 (Accessed: 4 August 2022).

Oakden, V. (2012b) *LVPL-943AA3: a Bronze Age socketed axehead* [online]. Available at: https://finds.org. uk/database/artefacts/record/id/500215 (Accessed: 4 August 2022).

Oakden, V. (2012c) *LVPL-A47FE3: a Bronze Age axe* [online]. Available at: https://finds.org.uk/database/ artefacts/record/id/504062 (Accessed: 4 August 2022).

Osgood, R. (1998) *Warfare in the Late Bronze Age of North Europe (BAR International Series No. S694).* Oxford: British Archaeological Reports.

Ottaway, B. S. (2001) 'Innovation, production and specialization in early prehistoric copper metallurgy', *European Journal of Archaeology*, 4(1), pp. 87–112.

Ottaway, B. S. and Seibel, S. (1998) 'Dust in the wind: experimental casting of bronze in sand moulds', in Frère-Sautot, M.-C. (ed.) *Paléométallurgie des cuivres. Actes du colloque de Bourg-en-Bresse et Beaune, 17–18 oct. 1997 (Monograph instrumentum 5)*. Montagnac: Éditions monique mergoil, pp. 59–63.

Pany-Kucera, D., Kern, A. and Reschreiter, H. (2019) 'Children in the mines? Tracing potential childhood labour in salt mines from the Early Iron Age in Hallstatt, Austria', *Childhood in the past*, 12(2), pp. 67–80.

Pernicka, E., Lutz, J. and Stöllner, T. (2016) 'Bronze Age copper produced at Mitterberg, Austria, and its distribution', *Archaeologia Austriaca*, 1, pp. 19–56.

Pfaffenberger, B. (1998) 'Mining communities, *chaînes opératoires* and sociotechnical systems', in Knapp, A. B., Pigott, V. C., and Herbert, E. W. (eds) *Social Approaches to an industrial past. The archaeology and anthropology of mining*. London and New York: Routledge, pp. 291–300.

Pickin, J. and Timberlake, S. (1988) 'Stone hammers and fire-setting: a preliminary experiment at Cwmystwyth Mine, Dyfed', *Bulletin of the Peak District Mines Historical Society*, 10(3), pp. 165–167.

Pickin, J. and Worthington, T. (1989) 'Prehistoric mining hammers from Bradda Head, Isle of Man', *Bulletin of the Peak District Mines Historical Society*, 10(5), pp. 274–275.

Preuschen, E. and Pittioni, R. (1954) 'Untersuchungen im Bergbaugebiet Kelchalm bei Kitzbühel, Tirol. Dritter Bericht über die Arbeiten 1946–1953 zur Urgeschichte des Kupferbergwesens in Tirol', *Archaeologia Austriaca*, 15, pp. 7–97.

Radivojević, M., *et al.* (2019) 'The provenance, use, and circulation of metals in the European Bronze Age: the state of debate', *Journal of Archaeological Research*, 27, pp. 131–185.

Randall, M. (1995) *Estimates of Bronze Age copper production at Great Orme*. Unpublished MSc thesis, University of London.

RCAHMW (1956) *An inventory of the ancient monuments in Caernarvonshire, volume I: East. The Cantref of Arllechwedd and the Commote of Creuddyn*. London: HMSO.

Rees, C. and Nash, G. (2017) 'Recent archaeological investigations at Kendrick's Upper Cave, Great Orme, Llandudno', *Proceedings University of Bristol Spelaeological Society*, 27(2), pp. 185–196.

Reid, A. and MacLean, R. (1995) 'Symbolism and the social contexts of iron production in Karagwe', *World Archaeology*, 27, pp. 144–161.

Reimer, P. J., *et al.* (2020) 'The IntCal20 Northern Hemisphere radiocarbon age calibration curve (0–55 cal kBP)', *Radiocarbon*, 62(4), pp. 725–757.

Reynolds, P. J. (2000) 'Butser ancient farm', *Current Archaeology*, 171(3), pp. 92–97.

Roberts, B., Uckelmann, M. and Brandherm, B. (2013) 'Old Father Time: the Bronze Age chronology of Western Europe', in Fokkens, H. and Harding, A. (eds) *The Oxford Handbook of the European Bronze Age*. Oxford: Oxford University Press, pp. 17–46.

Roberts, E. (1998) 'Thoughts on Bronze Age copper ore processing, at the Great Orme', *Journal of the Great Orme Exploration Society*, (1), pp. 7–8.

Roberts, J. (1909) 'Llandudno as it was', *Llandudno Advertiser*, 29 May.

Roberts, J. (2007) 'Short journeys, long distance thinking', in Cummings, V. and Johnston, R. (eds) *Prehistoric journeys*. Oxford: Oxbow Books, pp. 102–109.

Rohl, B. and Needham, S. (1998) *The circulation of metal in the British Bronze Age: the application of lead isotope analysis (British Museum Occasional Paper No. 102)*. London: the British Museum.

Rowlands, M. J. (1971) 'The archaeological interpretation of prehistoric metalworking', *World Archaeology*, 3(2), pp. 210–224.

Rowlands, M. J. (1980) 'Kinship, alliance and exchange in the European Bronze Age', in Barrett, J. C. and Bradley, R. (eds) *Settlement and society in the British Later Bronze Age (BAR British Series 83(i)*. Oxford: BAR Publishing, pp. 15–55.

Rule, J. (1998) 'A risky business. Death, injury and religion in Cornish mining c.1780–1870', in Knapp, A. B., Pigott, V. C., and Herbert, E. W. (eds) *Social approaches to an industrial past. The archaeology and anthropology of mining*. London and New York: Routledge, pp. 155–173.

Sagona, A. (ed.) (1994) 'The quest for red gold', in *Bruising the red earth. Ochre mining and ritual in Aboriginal Tasmania*. Melbourne: Melbourne University Press, pp. 8–38.

Savory, H. N. (1958) 'The Late Bronze Age in Wales. Some discoveries and new interpretations', *Archaeologia Cambrensis*, CVII, pp. 3–63.

Schibler, J., *et al.* (2011) 'Miners and mining in the Late Bronze Age: a multidisciplinary study from Austria', *Antiquity*, 85, pp. 1259–1278.

Schulting, R. J. (2011) 'War without warriors?', in Ralph, S. (ed.) *Archaeology of violence: interdisciplinary approaches*. New York: State University of New York Press, pp. 19–36.

Schulting, R. J. (2020) 'Claddedigaethau mewn ogofâu: prehistoric human remains (mainly) from the caves of Wales', *Proceedings of the University of Bristol Spelaeological Society*, 28(2), pp. 185–219.

Silvester, R. (2003) 'The archaeology of the Welsh uplands: an introduction', in Browne, D. M. and Hughes, S. (eds) *The archaeology of the Welsh uplands*. Aberystwyth: Royal Commission on the Ancient and Historical Monuments of Wales, pp. 3–14.

Smith, D. (1988) *The Great Orme copper mines*. Llandudno: Creuddyn Publications.

Smith, G. (1999) 'Survey of prehistoric and Romano-British settlement in north-west Wales', *Archaeologia Cambrensis*, 148, pp. 22–53.

Smith, G., et al. (2002) 'Excavation of a Middle Bronze-Age burnt mound at Bryn Cefni, Llangefni, Anglesey', *Archaeology in Wales*, 42, pp. 29–36.

Smith, G. (2003) *Prehistoric funerary and ritual monument survey: West Gwynedd and Anglesey, 2002–2003 (GAT project no. G1629, part 1: survey report, report no. 478)*. Unpublished report.

Smith, G., et al. (2012) *Reassessment of two hillforts in north Wales: Pen-y-Dinas, Llandudno and Caer Seion, Conwy. A report for Archaeology in Wales (GAT project no. G1770; report no. 1087)* [online]. Available at: http://www.walesher1974.org/her/groups/GAT/media/GAT_Reports/GATreport_1087_compressed.pdf (Accessed: 15 November 2017).

Smith, G., Chapman, S. G., et al. (2015) 'Rescue excavation at the Bronze Age copper smelting site at Pentrwyn, Great Orme, Llandudno, Conwy, 2011', *Archaeology in Wales*, 54, pp. 53–71.

Smith, G., Walker, E., et al. (2015) 'Snail Cave rock shelter, north Wales: a new prehistoric site', *Archaeologia Cambrensis*, 163(2014), pp. 99–131.

Smith, G., et al. (2017) 'An Early Bronze Age burnt mound trough and boat fragment with accompanying palaeobotanical and pollen analysis at Nant Farm, Porth Neigwl, Llŷn Peninsula, Gwynedd', *Studia Celtica*, 51(1), pp. 1–63.

Smith, G. and Jacques, D. (2009) *Re-assessment of the archive of the excavation of a roundhouse in Pen-y-Dinas hillfort, Great Orme, Llandudno in 1960 (GAT project no. G1770; report no. 823)*. Unpublished report.

Smith, G. and Williams, R. A. (2012) *Pentrwyn copper smelting site excavation, 2011, Great Orme, Llandudno, Conwy. (GAT project no. G2178, report no. 2178)* [online]. Available at: https://walesher1974.org/her/groups/GAT/media/GAT_Reports/GATreport_1029_compressed.pdf (Accessed: 23 February 2022).

Sofaer Derevenski, J. (1994) 'Where are the children? Accessing children in the past', *Archaeological Review from Cambridge*, 13, pp. 7–20.

Sofaer Derevenski, J. (1997) 'Engendering children, engendering archaeology', in Moore, J. and Scott, E. (eds) *Invisible people and processes. Writing gender and childhood into European archaeology*. London and New York: Leicester University Press, pp. 192–202.

Southwell-Wright, W. (2013) 'Past perspectives: what can archaeology offer disability studies?', in Wappett, M. and Arndt, K. (eds) *Emerging perspectives on disability studies*. New York: Palgrave Macmillan US, pp. 67–95.

Stanley, W. O. (1850) 'Proceedings at meetings of the Royal Archaeological Institute. January 4, 1850', *Archaeological Journal*, 7, pp. 68–69.

Stig Sørensen, M. L. (1996) 'Women as/and metalworkers', in Devonshire, A. and Wood, B. (eds) *Women in industry and technology from prehistory to the present day*. London: Museum of London, pp. 45–51.

Stöllner, T. (2003) 'Mining and economy – a discussion of spatial organisations and structures of early raw material exploitation', in Stöllner, T., et al. (eds) *Man and mining. Studies in honour of Gerd Weisgerber on occasion of his 65th birthday (Der Anschnitt Beiheft 16)*. Bochum: VML Verlag Marie Leidorf, pp. 415–446.

Stöllner, T. (2014) 'Methods of mining archaeology (Montanarchäologie)', in Roberts, B. W. and Thornton, C. P. (eds) *Archaeometallurgy in global perspective: methods and syntheses*. New York: Springer, pp. 133–159.

Stöllner, T. (2015) 'Humans approach to resources: Old World mining between technological innovations, social change and economical structures. A key-note lecture', in Hauptmann, A. and Modarressi-Tehrani, D. (eds) *Archaeometallurgy in Europe III. Proceedings of the 3rd International Conference, Deutsches Bergbau-Museum Bochum, June 29–July 1, 2011 (Der Anschnitt Beiheft 26)*. Bochum: Deutsches Bergbau-Museum, pp. 63–82.

Stöllner, T. (2018) 'Mining as a profession in prehistoric Europe', in Alexandrov, S., et al. (eds) *Gold and bronze. Metals, technologies and interregional contacts in the Eastern Balkans during the Bronze Age*. Sofia: National Archaeological Institute with Museum Bulgarian Academy of Sciences, pp. 71–85.

Stöllner, T. (2019) 'Between mining and smelting in the Bronze Age – beneficiation processes in an Alpine copper-producing district. Results of 2008 to 2017 excavations at the "Sulzbach-Moos"-bog at the Mitterberg (Salzburg, Austria)', in Turck, R., Stöllner, T., and Goldenberg, G. (eds) *Alpine copper II. New results and perspectives on prehistoric copper production (Der Anschnitt Beiheft 42)*. Bochum: VML Verlag Marie Leidorf GmbH, pp. 165–190.

Strachan, D. (2010) *Carpow in context. A Late Bronze Age logboat from the Tay*. Edinburgh: Society of Antiquaries of Scotland.

Stuiver, M. and Polach, H. A. (1977) 'Discussion reporting of 14C data', *Radiocarbon*, 19(3), pp. 355–363.

Taylor, R. E. and Bar-Yosef, O. (2014) *Radiocarbon dating. An archaeological perspective*. 2nd edn. New York: Routledge.

Taylor, T. (2005) 'Ambushed by a grotesque: archaeology, slavery and the third paradigm', in Parker Pearson, M. and Thorpe, I. J. N. (eds) *Warfare, violence and slavery in prehistory. Proceedings of a Prehistoric Society Conference at Sheffield University (BAR International Series 1374)*. Oxford: BAR Publishing, pp. 225–233.

Tellier, G. (2018) *Neolithic and Bronze Age funerary and ritual practices in Wales, 3600–1200 BC (BAR British Series 642)*. Oxford: BAR Publishing.

Thomas, P. (2015) 'Holz im bronzezeitlichen Bergbau der Ostalpen', in Stöllner, T. and Oeggl, K. (eds) *Bergauf Bergab, 10.000 Jahre Bergbau in den Ostalpen*. Bochum: VML Verlag Marie Leidorf, pp. 247–253.

Thomas, R. (1989) 'The bronze-iron transition in southern England', in Stig Sørensen, M. L. and Thomas, R. (eds) *The Bronze Age – Iron Age transition in Europe. Aspects of continuity and change in Bronze Age societies c.1200 to 500 BC (BAR International Series 483ii)*. Oxford: BAR Publishing, pp. 263–286.

Timberlake, S. (1988) 'Excavations at Parys Mountain and Nantyreira', *Archaeology in Wales*, 28, pp. 11–17.

Timberlake, S. (1990a) 'Firesetting and primitive mining experiment, Cwmystwyth, 1989', in Crew, P. and Crew, S. (eds) *Early mining in the British Isles. Proceedings of the Early Mining Workshop at Plas Tan y Bwlch Snowdonia National Park Study Centre, 17–19 November, 1989*. Blaenau Ffestiniog: Plas Tan y Bwlch, Snowdonia National Park Study Centre, pp. 53–54.

Timberlake, S. (1990b) 'Review of the historical evidence for the use of firesetting', in Crew, P. and Crew, S. (eds) *Early mining in the British Isles. Proceedings of the Early Mining Workshop at Plas Tan y Bwlch Snowdonia National Park Study Centre, 17–19 November, 1989*. Blaenau Ffestiniog: Plas Tan y Bwlch, Snowdonia National Park Study Centre, pp. 49–52.

Timberlake, S. (1994) 'Archaeological and circumstantial evidence for early mining in Wales', *Bulletin of the Peak District Mines Historical Society*, 12(3), pp. 133–143.

Timberlake, S. (1996) 'Copa Hill, Cwmystwyth (SN 811 751) Ceredigion', *Archaeology in Wales*, 36, pp. 60–61.

Timberlake, S. (1998) 'Survey of early metal mines within the Welsh uplands', *Archaeology in Wales*, 38, pp. 79–81.

Timberlake, S. (2003a) 'Early mining research in Britain: the developments of the last ten years', in Craddock, P. T. and Lang, J. (eds) *Mining and metal production through the ages*. London: British Museum Press, pp. 21–42.

Timberlake, S. (2003b) *Excavations on Copa Hill, Cwmystwyth (1986–1999). An Early Bronze Age copper mine within the uplands of central Wales (BAR British Series 348)*. Oxford: BAR Publishing.

Timberlake, S. (2009) 'Copper mining and metal production at the beginning of the British Bronze Age', in Clark, P. (ed.) *Bronze Age connections. Cultural contact in prehistoric Europe*. Oxford: Oxbow, pp. 239–321.

Timberlake, S. (2010) 'Geological, mineralogical and environmental controls on the extraction of copper ores in the British Bronze Age', in Anreiter, P., *et al.* (eds) *Mining in European history and its impact on environment and human societies. Proceedings for the 1st Mining in European History Conference of the SFB-HIMAT, 12.–15. November 2009, Innsbruck*. Innsbruck: Innsbruck University Press, pp. 289–295.

Timberlake, S., *et al.* (2014) 'Prehistoric copper extraction in Britain: Ecton Hill, Staffordshire', *Proceedings of the Prehistoric Society*, 80, pp. 159–206.

Timberlake, S. (2015a) 'Lloches y Lladron, Nantyricket Mine, River Severn, Hafren Forest (SN 8650 8670)', *Archaeology in Wales*, 54.

Timberlake, S. (2015b) 'Predictive experimental archaeology as a tool in the study of ancient mining and metallurgy', *Experimentelle Archäologie in Europa*, 14, pp. 145–164.

Timberlake, S. (2017) 'New ideas on the exploitation of copper, tin, gold, and lead ores in Bronze Age Britain: the mining, smelting, and movement of metal', *Materials and Manufacturing Processes*, 32(7–8) [online], pp. 709–727. Available at: https://www.tandfonline.com/doi/full/10.1080/10426914.2016.1221113 (Accessed: 25 March 2021).

Timberlake, S., *et al.* (2017–2018) 'A new Early Bronze Age mine site at Cwmystwyth, Ceredigion: archaeological excavations at Penparc (Comet Lode (W)) in 2017', *Archaeology in Wales*, 57–58, pp. 91–99.

Timberlake, S. (2019) 'Some provisional results of experiments undertaken using a reconstructed sluice box: an attempt to try and reproduce the methods of washing and concentrating chalcopyrite at the Middle Bronze Age ore processing site of Troiboden, Mitterberg, Austria', in Turck, R., Stöllner, T., and Goldenberg, G. (eds) *Alpine copper II. New results and perspectives on prehistoric copper production (Der Anschnitt Beiheft 42)*. Bochum: VML Verlag Marie Leidorf, pp. 191–208.

Timberlake, S., *et al.* (2019) 'Aberdovey, Panteidal copper mine (SN 6637 9726)', *Archaeology in Wales*, 59, pp. 117–119.

Timberlake, S. and Craddock, B. (2013) 'Prehistoric metal mining in Britain: the study of cobble stone mining tools based on artefact study, ethnography and experiment', *Chungara Revista de Antropología Chilena*, 45(1), pp. 33–59.

Timberlake, S. and Marshall, P. (2013) 'Understanding the chronology of British Bronze Age mines – Bayesian modelling and theories of exploitation', in Anreiter, P., *et al.* (eds) *Mining in European history and its impact on environment and human societies. Proceedings for the 2nd Mining in European History Conference of the FZ HiMAT, 7–10. November 2012, Innsbruck.* Innsbruck: Innsbruck University Press, pp. 57–64.

Timberlake, S. and Marshall, P. (2014) 'The beginnings of metal production in Britain: A new light on the exploitation of ores and the dates of Bronze Age mines', *Journal of Historical Metallurgy*, 47(Part 1 for 2013), pp. 93–109.

Timberlake, S. and Marshall, P. (2018) 'Copper mining and smelting in the British Bronze Age: new evidence of mine sites including some re-analysis of dates and ore sources', in Ben-Yosef, E. (ed.) *Mining for ancient copper: essays in memory of Professor Beno Rothenberg (Monograph Series 37)*. Indiana & Tel Aviv: Eisensbrauns & Tel Aviv University, pp. 418–431.

Timberlake, S. and Marshall, P. (2019) 'The Bronze Age mines dating project and some new ideas on ore extraction and smelting', in Brandherm, D. (ed.) *Aspects of the Bronze Age in the Atlantic archipelago and beyond. Proceedings from the Belfast Bronze Age Forum, 9–10 November 2013*. Hagen/Westf.: Curach Bhán Publications, pp. 13–31.

Timberlake, S. and Prag, A. J. N. W. (2005) *The archaeology of Alderley Edge. Survey, excavation and experiment in an ancient mining landscape (BAR British Series 396)*. Oxford: John & Erica Hedges.

Ucko, P. J. (1969) 'Ethnography and the archaeological interpretation of funerary remains', *World Archaeology*, 1(2), pp. 262–280.

University of Durham (2021) *Project ancient tin* [online]. Available at: https://projectancienttin.wordpress.com/ (Accessed: 16 May 2022).

Van de Noort, R. (2006) 'Argonauts of the North Sea – a social maritime archaeology for the 2nd millennium BC', *Proceedings of the Prehistoric Society*, 72, pp. 267–287.

Vandkilde, H. (2014) 'Archaeology, theory, and war-related violence: theoretical perspectives on the archaeology of warfare and warriorhood', in Gardner, A., Lake, M., and Sommer, U. (eds) *The Oxford handbook of archaeological theory*. Oxford: Oxford University Press, pp. 1–20.

Vandkilde, H. (2019) 'Bronze Age beginnings – a scalar view from the global outskirts', *Proceedings of the Prehistoric Society*, 85, pp. 1–27.

Waddington, K. (2013) *The settlements of northwest Wales, from the Late Bronze Age to the Early Medieval period*. Cardiff: University of Wales Press.

Wager, E. C. (2002) *The character and context of Bronze Age mining on the Great Orme, north Wales, UK*. Unpublished PhD thesis, University of Sheffield.

Wager, E. C. (2009) 'Mining ore and making people: re-thinking notions of gender and age in Bronze Age mining communities', in Kienlin, T. L. and Roberts, B. W. (eds) *Metals and societies. Studies in honour of Barbara S. Ottaway (Universitätsforsch. Prähist. Arch. 169)*. Bonn: Verlag Dr. Rudolf Habelt GMBH, pp. 105–115.

Wager, E. C. and Ottaway, B. S. (2019) 'Optimal versus minimal preservation: two case studies of Bronze Age ore processing sites', *Journal of Historical Metallurgy*, 52(Part 1 for 2018), pp. 22–32.

Walker, R. (1978) 'Diatom and pollen studies of a sediment profile from Melynllyn, a mountain tarn in Snowdonia, north Wales', *New Phytologist*, 81(3), pp. 791–804.

Wang, Q., Strekopytov, S. and Roberts, B. W. (2018) 'Copper ingots from a probable Bronze Age shipwreck off the coast of Salcombe, Devon: composition and microstructure', *Journal of Archaeological Science*, 97, pp. 102–117.

Ward, M. and Smith, G. (2001) 'The Llŷn cropmarks project. Aerial survey and ground evaluation of Bronze Age, Iron Age and Romano-British settlement and funerary sites in the Llŷn Peninsula of north-west Wales: excavations by Richard Kelly and Michael Ward', *Studia Celtica*, XXXV, pp. 1–87.

Warren, P. T., *et al.* (1984) *Geology of the country around Rhyl and Denbigh (memoir for 1:50 000 geological sheets 95 and 107 and parts of sheets 94 and 106)*. London: HMSO.

Warrilow, W., *et al.* (1986) 'Eight ring-ditches at Four Crosses, Llandysilio, Powys, 1981–85', *Proceedings of the Prehistoric Society*, 52, pp. 53–87.

Watanabe, J. M. (1992) *Maya saints and souls in a changing world*. Austin: University of Texas Press.

Watkins, R. (1991) *Postglacial vegetational dynamics in lowland north Wales*. Unpublished PhD thesis, University College of North Wales, Anglesey.

Watkins, R., Scourse, J. D. and Allen, J. R. M. (2007) 'The Holocene vegetation history of the Arfon Platform, north Wales, UK', *Boreas*, 36(2), pp. 170–181.

Webley, L. and Adams, S. (2016) 'Material genealogies: bronze moulds and their castings in later Bronze Age Britain', *Proceedings of the Prehistoric Society*, 82, pp. 323–340.

Webley, L., Adams, S. and Brück, J. (2020) *The social context of technology. Non-ferrous metalworking in later prehistoric Britain and Ireland (Prehistoric Society Research Paper No.11)*. Oxford: Oxbow Books.

Weisgerber, G. (1989a) 'Montanarchäologie. Grundzüge einer systematioschen Bergbaukunde für Vor- und Frühgeschichte und Antike. Teil I', *Der Anschnitt*, 41, pp. 190–204.

Weisgerber, G. (1989b) 'Montanarchäologie. Grundzüge einer systematischen Bergbaukunde für Vor- und Frühgeschichte und Antike. Teil I', in Hauptmann, A., Pernicka, E., and Wagner, G. A. (eds) *Old World archaeometallurgy (Der Anschnitt, Beiheft 7)*. Bochum: Selbstverlag des Deutsches Bergbau-Museum, pp. 79–98.

Wiles, J. (2007) *Site record: Porth Dafarch barrow I* [online]. Available at: https://coflein.gov.uk/en/site/308082/ (Accessed: 12 March 2020).

Williams, C. J. (1995) *Great Orme mines (British Mining No.52)*. Keighley: The Northern Mine Research Society.

Williams, H. (1924) 'A flat celt mould from the Lledr valley', *Archaeologia Cambrensis*, 7th ser. 4, pp. 212–213.

Williams, J. L., *et al.* (1998) 'Survey and excavation at the Graiglwyd Neolithic axe-factory, Penmaenmawr', *Archaeology in Wales*, 38, pp. 3–21.

Williams, J. L. and Jenkins, D. (1999) 'A petrographic investigation of a corpus of Bronze Age cinerary urns from the Isle of Anglesey', *Proceedings of the Prehistoric Society*, 65, pp. 189–230.

Williams, J. L. and Kenney, J. (2011) 'A chronological framework for the Graig Lwyd Neolithic axe factory, an interim scheme', *Archaeology in Wales*, 50, pp. 13–19.

Williams, J. L. W. and Jenkins, D. (2019) 'The prehistoric pottery of Gwynedd: a petrographic review', *Archaeology in Wales*, 59, pp. 25–34.

Williams, R. A. (2014) 'Linking Bronze Age copper smelting slags from Pentrwyn on the Great Orme to ore and metal', *Journal of Historical Metallurgy*, 47(Part 1 for 2013), pp. 93–110.

Williams, R. A. (2017) 'The Great Orme Bronze Age copper mine: linking ores to metals by developing a geochemically and isotopically defined mine-based metal group methodology', in Montero-Ruiz, I. and Perea, A. (eds) *Archaeometallurgy in Europe IV (2015) (Biblio-theca Praehistorica Hispana XXXIII)*. Madrid: Editorial CSIC, pp. 29–47.

Williams, R. A. (2019) 'Linking ore to metal: characterizing the ores and tracing the metal from the Great Orme Bronze Age copper mine in north Wales', in Brandherm, D. (ed.) *Aspects of the Bronze Age in the Atlantic archipelago and beyond. Proceedings from the Belfast Bronze Age Forum, 9–10 November 2013*. Hagen/Westf.: Curach Bhán Publications, pp. 145–175.

Williams, R. A. and Le Carlier de Veslud, C. (2019) 'Boom and bust in Bronze Age Britain: major copper production from the Great Orme mine and European trade, c. 1600–1400 BC', *Antiquity*, 93, pp. 1178–1196.

Willis, C., *et al.* (2016) 'The dead of Stonehenge', *Antiquity*, 90, pp. 337–356.

Wiseman, R. (2018) 'Random accumulation and breaking: the formation of Bronze Age scrap hoards in England and Wales', *Journal of Archaeological Science*, 90, pp. 39–49.

Wood, M. (1992) 'A mine of information – the Great Orme Mines project', *Earth Science Conservation*, 31, pp. 13–15.

Wood, M. and Campbell, S. (1995) 'Flood for thought on the Great Orme', *Earth Heritage*, 3, pp. 15–19.

Woodbridge, J., *et al.* (2012) 'A spatial approach to upland vegetation change and human impact: the Aber Valley, Snowdonia', *Environmental Archaeology*, 17, pp. 80–94.

Yaeger, J. and Canuto, M. A. (2000) 'Introducing an archaeology of communities', in Canuto, M. A. and Yaeger, J. (eds) *Archaeology of communities. A New World perspective*. London and New York: Routledge, pp. 1–15.

Zahorodnia, O. M. (2014) 'ПРО ПРИЗНАЧЕННЯ ОДНІЄЇ З КАТЕГОРІЙ КІСТЯНИХ ЗНАРЯДЬ ІЗ КАРТАМИШУ [About the purpose of one of bone tools categories from Kartamysh]', Археологія, 1, pp. 15–27.

APPENDICES

Appendix 1: Physical relationships between different areas of the prehistoric mine. The AOD data are reproduced from Lewis (1996, App. A). Level data are from the nearest adjacent working. * indicates no level data available. Areas without definite evidence for prehistoric activity (Lewis, 1996, App. C) have been omitted. Locations with radiocarbon dated sample(s) are shown in italics.

Location	Height Above Sea Level (m AOD)	Apparent Vertical Distance from Present Ground Surface (m)	Approximate Horizontal Distance from Present Ground Surface (m)	Links to Other Locations of Working	Likelihood of Direct Connections to Surface	Location of Original Entrances at Surface	Figure Numbers
1	147	c.5–10	<15	Possible links between Loc.1 and Loc.2	Likely direct connections to surface	Likely in area to south of Water Board building, underlying or immediately south of modern road to summit	2.20, 2.24
2	137.5	c.15	<25	Possible links with Locs. 1 and 3	Likely direct connections to surface	Likely in area on slope to south of modern road to summit and to north of present Great Orme Mines perimeter fence	2.20, 2.24
3	135.5	c.30	100	Possible links with Loc.5, as chert is present in spoil at both locations. Possible links with Loc.2	Possible direct connections southwards to the surface. In addition, or alternatively, upper workings at Loc.3 possibly accessed via workings at Loc.2. Possible direct connections northwards to surface, through Summit Limestones: uncertain as vertical extent of these strata and whether they originally outcropped at surface are unknown. Possible access either northwards or southwards through cherty beds at Loc.5	If accessed from the south, likely in area on slope to south of modern road to summit and to north of present Great Orme Mines perimeter fence. If accessed from the north, possible in lower exposures of surface outcrop of Summit Limestones	2.20, 2.24
5	131	c.27	<90	Possible links with Loc.3	Likely direct connections to surface southwards. Possible direct connections to surface northwards, through Summit Limestones	Likely to south, in area to south of modern road to summit and to north of present Great Orme Mines perimeter fence, but to west of entrances to Locs. 1, 2 and perhaps 3 in same general location. Possibly to north, through surface outcrop of Summit Limestones to north of Roman Shaft	2.20, 2.24, 2.27
7a	133.3	c.4	<45	Likely links to other, as yet unexplored, areas of working	Likely links to surface through Tourist Cliff, and/or possible links to surface through Tourist Cliff via Lost Cavern	To south, in south face of Tourist Cliff	2.20, 2.24, 2.27
7c (Underground)	c.133	>1	<55	Known links with Tourist Cliff and Lost Cavern	Known access to surface via Tourist Cliff, and/or likely access via Lost Cavern	To south, in south face of Tourist Cliff	2.20, 2.27
8	134.2	c.20	<90	Likely links with Loc.10. Presence of prehistoric spoil in upper parts of Loc.8 indicates that other workings originally existed above. Possible links between Loc.8 and these other workings	Possible direct connections to surface from lower areas of working, but likely that these areas also/instead accessed through Loc.10. Possible direct connections to surface from upper areas at Loc.8 and any areas of working above this location. Also/instead, upper parts of Loc.8 possibly accessed via any higher workings	If direct links between surface and lower workings existed, entrances likely to south, in hollow midway between Owen's Shaft and Pyllau farmhouse. If direct links between surface and upper areas of working/higher workings existed, entrances likely to south, in area on slope to north of entrances to Loc.10 and to south of Owen's Shaft	2.20, 2.25
9	133.9*	c.18	<75	Likely links with Loc.10. Possible links with Loc.8	Possible direct connections to surface. Also/instead, likely accessed via Loc.10	If direct connections to surface existed, entrances likely to south, in hollow midway between Owen's Shaft and Pyllau farmhouse	2.20
10	135.8	c.15	<60	Likely links with Locs. 8, 9 and 11	Likely direct connections to surface. Also/instead, likely accessed via Loc.11	Likely to south, in hollow midway between Owen's Shaft and Pyllau farmhouse	2.20, 2.25
11	135.8*	c.10	<25	Likely links with Loc.10	Likely direct connections to surface	Likely to south, in hollow midway between Owen's Shaft and Pyllau farmhouse	2.20, 2.25, 2.27
13	125.3	c.15	<100	Likely that eastern workings at least linked to Loc.14. Likely links with Lost Cavern	Unlikely direct connections to surface. Instead, likely accessed through Tourist Cliff, via lower part of northern wall/base of Lost Cavern	To south, in south face of Tourist Cliff	2.21

Location	Height Above Sea Level (m AOD)	Apparent Vertical Distance from Present Ground Surface (m)	Approximate Horizontal Distance from Present Ground Surface (m)	Links to Other Locations of Working	Likelihood of Direct Connections to Surface	Location of Original Entrances at Surface	Figure Numbers
14	122.3	c.30	<10	Likely links with workings at Loc.13. Likely links with Lost Cavern	Unlikely direct connections to surface. Instead, likely accessed through Tourist Cliff, via lower part of northern wall/base of Lost Cavern	To south, in south face of Tourist Cliff	2.21
17	123	c.10	0	Likely links with Loc.21 as prehistoric workings between Locs. 17 and 21 are known. Likely links with workings to south of and leading into Loc.18, and with Loc.18 itself	Known direct connections to surface, now used as the entrance and exit to tourist route through the mine	To south, in south face of Tourist Cliff	2.21, 2.24
18	117	c.15	<25	Likely links with Locs. 17, 20 and 21. Possible links with Loc.30, through restricted workings northwards. Possible links with Loc.29, through workings in base. Known links with base of Lost Cavern above, and with restricted workings southwards	Likely direct connections to surface. Known access through Tourist Cliff, via Lost Cavern	To south, in south face of Tourist Cliff	2.21, 2.24, 2.28
19	117.0*	c.18	<30	Possible links to workings below, particularly Locs. 31 and 32. Known links with north end of Lost Cavern. Likely links to Loc.7b, Loc.7c (Surface) and other workings in cliff on west side of Lost Cavern. Likely links with Tourist Cliff	Likely direct connections to surface, and/or likely access through Tourist Cliff via workings in cliff on west side of Lost Cavern. Known access via Lost Cavern	To south, in south face of Tourist Cliff	2.21, 2.24
20	115	c.14	<55	Likely links with Loc.18 and Lost Cavern. Possible links with Locs. 30 and 33	No direct connections to surface can be identified. Instead, possibly accessed via Loc.18 and/or through base of Lost Cavern, above and to north of Loc.18	To south, in south face of Tourist Cliff	2.21, 2.24
21	115.5	c.20	<15	Likely links to Loc.17 as prehistoric workings between Locs. 17 and 21 are known. Also likely links to Loc 18	Likely direct connections to surface. Possibly accessed via workings at Loc.17	To south, in south face of Tourist Cliff. Direct entrances may have been lower down face and to the east of those into Loc.17	2.21, 2.24, 2.28
22	108.5	c.40	<80	Likely links to Loc.23. Possible links to Loc.26	Likely direct connections to surface	To south, in vicinity of modern road to the mine to the west of the Tourist Cliff, in the area between Great Orme Mine car park and Pyllau Farm sheepfold	2.22, 2.25
23	109.2*	c.55	<180	Likely links to Loc.22	Possible direct connections to surface. Likely also accessed via Loc.22	To south, in vicinity of modern road to the mine to the west of the Tourist Cliff, in the area between Great Orme Mine car park and Pyllau Farm sheepfold	2.22, 2.25
26	91.5*	c.55	<100	Form of workings suggests mining at this location continued upwards, so possible links with workings along route of access to Loc.22. Possible links with early workings southwards, like those that perhaps existed above Loc.27, a recent working	Possible direct connections to surface. Possible access via higher workings, like those on route of access to Loc.22. Possible access via workings southwards, like those that may have existed above Loc.27	To south, in vicinity of modern road to the mine to the west of the Tourist Cliff, in the area between Great Orme Mine car park and Pyllau Farm sheepfold	2.23, 2.25
29	93.3	c.40	<25	Possible links with stoping above to base of Loc.18	If no natural cave passages, unlikely direct connections to surface, given position of Loc.29, dip of rock strata and likely original profile of ground surface above to the north and south. Likely access to surface via workings above	To south, in south face of Tourist Cliff	2.23, 2.24

Location	Height Above Sea Level (m AOD)	Apparent Vertical Distance from Present Ground Surface (m)	Approximate Horizontal Distance from Present Ground Surface (m)	Links to Other Locations of Working	Likelihood of Direct Connections to Surface	Location of Original Entrances at Surface	Figure Numbers
30	102.5	c.30	<55	Known links with Loc.31. Likely links with Loc.33. Possible links with stoping above, to north end of Loc.18 and south end of Loc.20	If no natural cave passages, unlikely direct connections to surface, given position of Loc.30, dip of rock strata and likely original profile of ground surface above to the north and south. Likely access to surface via workings above, including Locs. 18 and 20. Possible access via Loc.31	To south, in south face of Tourist Cliff	2.23, 2.24
31	93.4/102.5*	c.30–45	<55	Known links with Loc.30. Likely links with Loc.32. Possible links to workings above, including Loc.19. Possible links with workings northwards (yet to be explored because route blocked by roof and wall collapse)	Possible surface-derived spoil at this location (Lewis, 1996, App. C, 15). But, if no natural karst system, direct connections to surface unlikely, given position of Loc.31, dip of rock strata and likely original profile of ground surface above to north and south. Likely access to surface via other workings, including Loc.19, and/or Loc.30 and/or perhaps Loc.32	To south, in south face of Tourist Cliff	2.23, 2.24
32	93.4*	c.45	<30	Likely links with Loc.31. Possible links with workings above, including Loc.19	If no natural karst system, unlikely direct connections to surface, given position of Loc.32, dip of rock strata and likely original profile of ground surface above to north and south. Likely access to surface via other workings, including Loc.19 and/or Loc.31	To south, in south face of Tourist Cliff	2.23, 2.24
33	106.5	c.35	<85	Known links to Loc.34. Likely links to Loc.30. Possible links southwards to workings like Loc.20, through backfilled passage to south	If no natural karst system, unlikely direct connections to surface, given position of Loc.33, dip of rock strata and likely original profile of ground surface above to the north and south. Likely access to surface via other workings, including Loc.30 and perhaps Loc.20	To south, in south face of Tourist Cliff	2.23, 2.24
34	96.0*	c.50	<130	Known links with Loc.33. Likely links with Loc.35. Possible links with workings above, like Loc.14 and perhaps the eastern workings at Loc.13	If no natural karst system, unlikely direct connections to surface, given position of Loc.34, dip of rock strata and likely original profile of ground surface above to the north and south. Likely access to surface via Loc.33. Possible access via Loc.14 and/or 13	To south, in south face of Tourist Cliff	2.23, 2.24
35	107.8*	c.48	<140	Likely links with Loc.34. Possible links with Loc.14 above	If natural karst system, unlikely direct connections to surface, given position of Loc.35, dip of rock strata and likely original profile of ground surface above to north and south. Likely access to surface via other workings, like Loc.34 and perhaps Loc.14	To south, in south face of Tourist Cliff	2.23, 2.24
36	139.16	>1	<12	Definite links into other unexplored workings. Possible links to North-West Corner	Known direct connections to surface through rock outcrop at this location would formerly have been visible at surface	To south, in south face of rock outcrop at this location; possibly also in south face of outcrop at North-West Corner to east	2.20, 2.27
Lost Cavern	c.133	c.1	12+	Known links with Locs. 18 and 19, and workings in Tourist Cliff. Likely links with Locs. 13, 14 and 20	Likely connections to surface via Tourist Cliff	To south, in south face of Tourist Cliff	2.20, 2.24
Loc.7b, Loc.7c (Surface), 7c (West), plus other workings on west side of Lost Cavern	c.133	>1	c.12–45	Known links with Lost Cavern. Likely links with Loc.19	Likely connections to surface via Lost Cavern	To south, in south face of Tourist Cliff	2.20, 2.27
North-West Corner	c.139*	>1	<12	Possible links into Loc.36. Definite links into other unexplored workings	Known direct connections to surface through rock outcrop at this location that would formerly have been visible at surface	To south, in face of rock outcrop at this location	2.20, 2.27
Tourist Cliff	c.135	>1	<12	Known links with Lost Cavern, Loc.7c (Underground) and Loc.18. Likely links with Locs. 7a, 19 and 21	Known direct connections to surface	To south, in south face of cliff	2.20, 2.27

Appendix 2: Radiocarbon dates from the Great Orme mine and its associated production sites, with details of the radiocarbon measurements, sampled material and site of recovery. '–' indicates data unavailable; 'n.d.' not determined; 'n.m.' not measured, estimated using values typical of the material type. BM-2753 – calibrated date range is AD.

Laboratory Number	Radiocarbon Age (BP)	Calculation Details	Laboratory Methods	δ13C(‰)	Sample ID	Sample Location	Location Detail	Calibrated Date Range cal BC, 95% probability
WET ORE PROCESSING SITES								
BETA-148793	3450 ± 70	Stuiver & Polach (1977), corrected for fractionation using reported value (Accessed online; see BETA-127076)	Collagen extraction (acid only); radiometric standard delivery (collagen analysis) (for reference, see BETA-127076)	-19.0 (n.m.)	Animal bone fragments (2) with blue-green copper staining, species n.d.	Ffynnon Rhufeinig	Spoil tip near water source	1945–1545
BM-2753	*1200 ± 60*	*Stuiver & Polach (1977), corrected for fractionation using reported value*	*Pre-treatment, synthesis & measurement in Ambers & Bowman (1998)*	*-22.4*	*Animal bone fragment with blue-green copper staining, species n.d.*	*Ffynnon Galchog*	*Spoil tip near water source*	*680–980*
COPPER SMELTING SITE								
BETA-127076	3310 ± 80	Stuiver & Polach (1977), corrected for fractionation using reported value (https://www.radiocarbon. com/isotopic-fractionation.htm [Accessed online 21-06-21])	Collagen extraction (acid only); radiometric standard delivery (collagen analysis) (https://www. radiocarbon. com/beta-lab.htm [Accessed online 21-06-21])	-25.0 (n.m.)	Wood charcoal, species/form n.d.	Pentrwyn	Small conical pit cutting occupation layer	1870–1420
SUERC-39896	2730 ± 30	Dunbar, *et al.* (2016)	Pre-treatment, synthesis & measurement in Dunbar, *et al.* (2016)	-25.6	Single piece nutshell charcoal, *Corylus* (hazel)	Pentrwyn	Occupation layer	930–810
SUERC-39897	2780 ± 30	Dunbar, *et al.* (2016)	Pre-treatment, synthesis & measurement in Dunbar, *et al.* (2016)	-27.6	Single piece twig wood charcoal, *Corylus* (hazel)	Pentrwyn	Occupation layer	1005–835
SUERC-44867	2727 ± 33	Dunbar, *et al.* (2016)	Pre-treatment, synthesis & measurement in Dunbar, *et al.* (2016)	-25.5	Twig wood charcoal, *Ilex* (holly)	Pentrwyn	Base of small pit feature	965–805
GREAT ORME MINE								
BETA-65894	3010 ± 60	Stuiver & Polach (1977), corrected for fractionation using reported value (Accessed online; see BETA-127076)	Collagen extraction (acid only); radiometric standard delivery (collagen analysis) (for reference, see BETA-127076)	–	Wood charcoal, *Quercus* sp. (oak)	South of Loc.21	Restricted underground working, south of Loc.21	1415–1055
BETA-65895	3220 ± 70	Stuiver & Polach (1977), corrected for fractionation using reported value (Accessed online; see BETA-127076)	Collagen extraction (acid only); radiometric standard delivery (collagen analysis) (for reference, see BETA-127076)	–	Wood charcoal, species/form n.d.	Loc.7c (Underground)	Small underground gallery, west side of Lost Cavern	1675–1300
BETA-65896	2970 ± 70	Stuiver & Polach (1977), corrected for fractionation using reported value (Accessed online; see BETA-127076)	Collagen extraction (acid only); radiometric standard delivery (collagen analysis) (for reference, see BETA-127076)	–	Wood charcoal, species/form n.d.	Loc.7c (Underground)	Small underground gallery, west side of Lost Cavern	1405–1005
BM-2641	3000 ± 50	Stuiver & Polach (1977), corrected for fractionation using reported value	Pre-treatment, synthesis & measurement in Ambers & Bowman (1994)	-25.6	Charcoal from branchwood & twigs, *Quercus* sp. (oak)	Loc.21	Restricted underground working	1400–1055
BM-2645	3290 ± 60	Stuiver & Polach (1977), corrected for fractionation using reported value	Pre-treatment, synthesis & measurement in Ambers & Bowman (1994)	-22.4	Animal bone fragments (3), species n.d.	Tourist Cliff, Loc.007	Small working at surface, east of Vivian's Shaft, junction of East 1 & North 1 ore veins	1735–1440

Laboratory Number	Radiocarbon Age (BP)	Calculation Details	Laboratory Methods	δ13C(‰)	Sample ID	Sample Location	Location Detail	Calibrated Date Range cal BC, 95% probability
BM-2751	3230 ± 50	Stuiver & Polach (1977), corrected for fractionation using reported value	Pre-treatment, synthesis & measurement in Ambers & Bowman (1998)	-21.4	Animal bone, species/ number n.d.	Loc.18	Large underground mined chamber (stope) beneath Lost Cavern, depth of 12m below surface off Vivian's Shaft	1615–1415
BM-2752	3070 ± 50	Stuiver & Polach (1977), corrected for fractionation using estimated value (-21.0‰ for bone)	Pre-treatment, synthesis & measurement in Ambers & Bowman (1998)	Not given	Animal bone, species/ number n.d.	Loc.18	Large underground mined chamber (stope) beneath Lost Cavern, depth of 12m below surface off Vivian's Shaft	1445–1135
BM-2802	3180 ± 80	Stuiver & Polach (1977), corrected for fractionation using reported value	Pre-treatment, synthesis & measurement in Ambers & Bowman (1998)	-26.0	Branchwood charcoal, *Alnus* sp. (alder) (Ambers & Bowman, 1998). Note: Dutton, *et al.* (1994, table 1) suggest *Quercus/Alnus* (alder, oak)	Loc.7c (Surface)	Working/pocket at surface, west side of Lost Cavern, West 4 vein	1625–1230
BM-3117	3000 ± 45	Stuiver & Polach (1977), corrected for fractionation using reported value	Pre-treatment, synthesis & measurement in Ambers & Bowman (1999)	-22.9	Wood charcoal, species/form n.d.	Loc.35	Confined unstable underground working, c.50m depth (not c.30m as stated in Ambers & Bowman (1999)	1395–1060
BM-3118	3000 ± 50	Stuiver & Polach (1977), corrected for fractionation using reported value	Pre-treatment, synthesis & measurement in Ambers & Bowman (1999)	-24.0	Wood charcoal, species/form n.d.	Loc.11	Small underground working accessed from 18m deep level off Owen's Shaft, near survey point 270	1400–1055
CAR-1184	3370 ± 80	Dresser (1985)	Pre-treatment, synthesis & measurement used by the now defunct Radiocarbon Dating Laboratory, University College Cardiff, in Dresser (1985)	–	Wood charcoal, *Alnus glutinosa* (alder)	Tourist Cliff, Loc.007	Small working at surface, east of Vivian's Shaft, junction of East 1 & North 1 ore veins	1885–1500
CAR-1280	2970 ± 70	Dresser (1985)	Pre-treatment, synthesis & measurement used by the now defunct Radiocarbon Dating Laboratory, University College Cardiff, in Dresser (1985)	–	Large pieces of wood charcoal, *Quercus* sp. & *Corylus* (oak, hazel)	Tourist Cliff, N1 ore vein	Working at surface, junction of North 1 & East 1 veins, area around Vivian's Shaft	1405–1005
CAR-1281	2450 ± 60	Dresser (1985)	Pre-treatment, synthesis & measurement used by the now defunct Radiocarbon Dating Laboratory, University College Cardiff, in Dresser (1985)	–	Wood charcoal, *Quercus* sp., *Corylus* & *Alnus* (oak, hazel, alder)	Tourist Cliff, W1 ore vein	Working at surface, junction of West 2 & North 1 veins, area around Vivian's Shaft	760–410
HAR-4845	2940 ± 80	–	–	–	Wood charcoal, *Alnus* (alder)	Loc.5	Underground working off Roman Shaft, 27m below surface	1390–925
OxA-14308	3344 ± 27	Brock, *et al.* (2010)	Pre-treatment (procedure AF) in Brock, *et al.* (2010)	-20.14	Human mandible (lower jawbone)	Tourist Cliff, W3/N2 ore veins	Working at surface, junction of West 3 & North 2 veins, north of Vivian's Shaft	1730–1535

Laboratory Number	Radiocarbon Age (BP)	Calculation Details	Laboratory Methods	δ13C(‰)	Sample ID	Sample Location	Location Detail	Calibrated Date Range cal BC, 95% probability
OxA-14309	3362 ± 28	Brock, *et al.* (2010)	Pre-treatment (procedure AF) in Brock, *et al.* (2010)	-20.57	Human clavicle (collarbone)	Loc.18	Large underground mined chamber (stope) beneath 'Lost Cavern', depth of 13m below surface off Vivian's Shaft	1740–1540
OxA-5397	3150 ± 50	Hedges, *et al.* (1996)	Pre-treatment, synthesis & measurement in Bronk & Hedges (1989)	-21.5	Wood charcoal, like *Prunus* (blackthorn)	Loc.7a	Underground working off 'Jackdaws Hole' level, eastern edge of Lost Cavern	1510–1285
SUERC-69412	3069 ± 29	Dunbar, *et al.* (2016)	Pre-treatment, synthesis & measurement in Dunbar, *et al.* (2016)	-25.2	Wood charcoal, *Quercus* sp. (oak)	Loc.36	North-west corner of Lost Cavern	1415–1235
SUERC-69760	2980 ± 30	Dunbar, *et al.* (2016)	Pre-treatment, synthesis & measurement in Dunbar, *et al.* (2016)	-26.8	Charcoal comprising thin section of branch wood, species n.d.	Loc.36	North-west corner of Lost Cavern	1375–1060
Wk-14192	3325 ± 41	Stuiver & Polach (1977), corrected for fractionation using reported value	–	–	Animal bone, species/number n.d.	North-West Corner	Rock outcrop at edge of Lost Cavern	1735–1505
Wk-14193	3118 ± 46	Stuiver & Polach (1977), corrected for fractionation using reported value	–	–	Wood charcoal, species/form n.d.	Loc.7c (West)	Narrow passage near bottom of deep West 8 vein, cliff overlooking south-west corner of Lost Cavern. Located off Loc.7c (Underground)	1495–1265
Wk-14194	3255 ± 84	Stuiver & Polach (1977), corrected for fractionation using reported value	–	–	Wood charcoal, species/form n.d.	Tourist Cliff, East 8 vein (Middle)	Small working at surface, East 8 ore vein, east of Vivian's Shaft	1745–1310
Wk-14197	3183 ± 57	Stuiver & Polach (1977), corrected for fractionation using reported value	–	–	Animal bone tool fragment, species n.d.	Tourist Cliff, E1 vein (South)	Narrow north-south working in bedrock floor, southern end of East 1 ore vein	1610–1295
Wk-14199	3127 ± 42	Stuiver & Polach (1977), corrected for fractionation using reported value	–	–	Wood charcoal, species/form n.d.	Tourist Cliff, E1 vein (South)	Working at surface, southern end of East 1 ore vein	1500–1285

Appendix 3: Sampled archaeological contexts and references for the radiocarbon dates from the Great Orme mine and its associated production sites. BM-2753 – calibrated date range is AD.

Laboratory Number	Sample Location	Sampled Archaeological Context	References
WET ORE PROCESSING SITES			
BETA-148793	Ffynnon Rufeinig	Context (312), a distinct layer comprising intermingled clay, yellow dolomitic sand and gravel with malachite/azurite fragments and 13 green-stained animal bone fragments. 312 is one of a complex succession of layers making up a tip interpreted as well-sorted wet ore processing (washing) waste. Documentary and archaeological evidence for ore reprocessing and other activities at this site from the 19th century AD onwards	Wager & Ottaway (2019, p. 29)
BM-2753	*Ffynnon Galchog*	*Tip of well-sorted and occasionally graded dolomitic silt, sand and gravel. Mineralogically resembled spoil sequences from the Great Orme mine*	*Lewis (1990a); Lewis (1996, p. 146); Ambers & Bowman (1998, p. 424)*
COPPER SMELTING SITE			
BETA-127076	Pentrwyn	Bulk sample from deposit (029) of charcoal-rich dark silt infilling a small, roughly conical hole (030), 4cm in diameter and 120cm in depth, cutting into an occupation layer. Contained fragments of slag and some vitrified material. Drawn section indicates possibility of small mammal burrowing. Dateable Mesolithic and (possibly) Medieval activity at the site, as well as natural erosion. Site truncated by road construction in late 19th century AD	Smith, Chapman, *et al.* (2015, pp. 68–9, table 9)
SUERC-39896	Pentrwyn	Grey, charcoal-rich occupation layer (107), up to 25cm thick, extending into a horizontal fissure under a rock underhang. Numerous finds including metal fragments, bone, shell, charcoal, stone and flint. (NB Smith, Chapman, *et al.* 2015, p. 68 refer to sampled context as the adjacent and apparently contemporary layer 108, but elsewhere throughout the text the sampled location is recorded as 107.) Dateable Mesolithic and (possibly) Medieval activity at the site, as well as evidence for animal burrowing and natural erosion. Site truncated by road construction in late 19th century AD	Smith, Chapman, *et al.* (2015, pp. 68–9, table 9)
SUERC-39897	Pentrwyn	Grey, charcoal-rich occupation layer (107), up to 25cm thick, extending into a horizontal fissure under a rock underhang. Numerous finds including metal fragments, bone, shell, charcoal, stone and flint. (NB Smith, Chapman, *et al.* 2015, p. 68 refer to sampled context as the adjacent and apparently contemporary layer 108, but elsewhere throughout the text the sampled location is recorded as 107.) Dateable Mesolithic and (possibly) Medieval activity at the site, as well as evidence for animal burrowing and natural erosion. Site truncated by road construction in late 19th century AD	Smith, Chapman, *et al.* (2015, pp. 68–9, table 9)
SUERC-44867	Pentrwyn	Fill (112) at the base of a small pit (feature 111) cut into layer 108. Feature 111 is interpreted as a simple reducing smelting furnace due to morphology and associated finds of copper prills and slag fragments. Dateable Mesolithic and (possibly) Medieval activity at the site, as well as evidence for animal burrowing and natural erosion. Site truncated by road construction in late 19th century AD	Smith, Chapman, *et al.* (2015, pp. 68–9, table 9)
GREAT ORME MINE			
BETA-65894	South of Loc.21	Small deposit of charcoal collected from the rock surface at the extreme end of a very narrow, sinuous, blind passage. No evidence for late modern disturbance	Lewis (1996, table 1)
BETA-65895	Loc.7c (Underground)	Bulk sample from two contexts: layer (5006) of large dolomite boulders, well-calcined together, and a hard-packed clayey layer (5007) with plentiful small stones. (5006) and (5007) both sealed by layer (5005) of dark-brown clayey silt with occasional dolomite pebbles, interpreted as natural sediment. No evidence for late modern disturbance	Lewis (1996, table 1; App. C, 4–5); Unpublished context record, Great Orme Mines
BETA-65896	Loc.7c (Underground)	Clayey silt horizon (5010) interpreted as a temporary working floor. Stratigraphically earlier than (5007), (5006) and (5005) within same sequence of deposits. No evidence for late modern disturbance	Lewis (1996, table 1; App. C, 4–5); Unpublished context record, Great Orme Mines
BM-2641	Loc.21	Large deposit of charcoal interpreted as the residue from a fire-set in a crevice in the sidewall of an 8m-long passage, at a similar depth and approximately 20m west of BM-2751. Evidence for late modern disturbance at far end of passage only	Lewis (1989); David (1992); Ambers & Bowman (1994, p. 105)
BM-2645	Tourist Cliff, Loc.007	Deposit of morphologically prehistoric spoil (context 002) containing numerous bone and charcoal fragments in the base of a short, 3m-long passage. Sealed by a layer of clayey silt interpreted as possible evidence of hiatus. No evidence for late modern disturbance	Lewis (1989); Ambers & Bowman (1994)
BM-2751	Loc.18	Context 006 in a stratified sequence of morphologically prehistoric spoil at the north-eastern boundary of the stope, where it connects to smaller, more restricted workings. Associated with stone and bone tools and fragmented charcoal from fire-setting. The sequence of prehistoric deposits was found sealed beneath spoil of characteristic late modern morphology	Dutton, *et al.* (1994, table 1); Lewis (1996, table 1; App. C, 10–11); Ambers & Bowman (1998); A. Lewis 2021, pers. comm., 10 August
BM-2752	Loc.18	Context 043 in a stratified sequence of morphologically prehistoric spoil at the south-eastern boundary of the stope, where it connects to smaller, more restricted workings. Associated with stone and bone tools and fragmented charcoal from fire-setting. The sequence of prehistoric deposits was found sealed beneath spoil of characteristic late modern morphology	Dutton, *et al.* (1994, table 1); Lewis (1996, table 1; App. C, 10–11); Ambers & Bowman (1998); A. Lewis 2021, pers. comm., 10 August

Laboratory Number	Sample Location	Sampled Archaeological Context	References
BM-2802	Loc.7c (Surface)	Small pocket of prehistoric spoil described as 'undisturbed' by the excavators, found within an overhang towards the base of the north-west side of the Lost Cavern	Dutton, *et al.* (1994, table 1); Ambers & Bowman (1998, p. 414)
BM-3117	Loc.35	Undisturbed deposit on a ledge in a passage in one of the deepest parts of the mine	Lewis (1996, table 1; App. C, 16–17); Ambers & Bowman (1999, p. 189); A. Lewis 2021, pers. comm., 10 August
BM-3118	Loc.11	Presumed fire-setting debris in a small passage backfilled with spoil to a depth of 0.8m. Cut by a 19th-century AD shaft that had holed through c.1.5m of rock into the earlier tunnel	Lewis (1996, table 1; App. C, 7); Ambers & Bowman (1999, p. 189); A. Lewis 2021, pers. comm., 10 August
CAR-1184	Tourist Cliff, Loc.007	Context 022 in a deposit of morphologically prehistoric spoil in the base of a confined, tubular surface channel. Sealed by a layer of dolomitic sand interpreted as possible evidence of hiatus. No evidence for late modern disturbance	Lewis (1989, p. 43, 1990d, p. 7); Dutton, *et al.* (1994, table 1)
CAR-1280	Tourist Cliff, N1 ore vein	Base of a context (026) comprising angular spalls of burnt rock, charcoal and grey clay-silt, interpreted as evidence of fire-setting. Sealed by a layer of dolomitic sand interpreted as possible evidence of hiatus. No record of late modern disturbance. The deposit of burnt rock was found immediately below red-pink concave wall-rock, indicating that this is in-situ evidence for Bronze Age fire-setting	Dutton, *et al.* (1994, p. 261, table 1)
CAR-1281	Tourist Cliff, W1 ore vein	Compacted layer (165) of mottled clay heavily flecked with charcoal, interpreted as a working floor, and directly associated with a small pocket of working (045) containing a cattle rib gouge. 165 was sealed by a 'localised deposit' (163). No evidence for late modern disturbance	Dutton, *et al.* (1994, p. 258, table 1)
HAR-4845	Loc.5	Revetted deposit of morphologically prehistoric spoil, of the same lithology as the surrounding country rock (clay wayboard), containing charcoal, and at least one bone tool and one cobblestone mining tool, on a ledge in the sidewall at the end of a passage. There is evidence for late modern partial reworking and enlargement of the passage	James (1990, p. 4); Lewis (1996, table 1)
OxA-14308	Tourist Cliff, W3/ N2 ore veins	Surface spoil in a cave-like opening on the west side of the Lost Cavern, close to an area excavated by Gwynedd Archaeological Trust from which Bronze Age dates have been recovered (reported in Dutton, *et al.* 1994)	Lewis (1996, pp. 96, 126); David (2005, p. 146); James (2011, table 1.2)
OxA-14309	Loc.18	Spoil cleared during preparation of the Visitors' Route, just off the stope	David (2005, p. 146); James (2011, table 1.2)
OxA-5397	Loc.7a	Charcoal embedded in the base of a calcite flowstone in a series of passages containing morphologically prehistoric spoil, charcoal, bone tools, bone tool marks and one cobblestone mining tool. There are very few signs of late modern activity in the main passage and the presence of large, very well-preserved stalactites at that location indicates that the impact of this later activity on the prehistoric residues was minimal	Hedges, *et al.* (1996, p. 195); Lewis (1996, table 1; App. C, 3)
SUERC-69412	Loc.36	Uppermost layer (004), rich in bronze metal fragments, charcoal and heavily fire-reddened rock fragments, c.10m into an underground passage. As there is no rock wall evidence for in-situ fire-setting, 004 is interpreted as redeposited fire-setting waste originating from another, unknown location further underground. Late modern disturbance is restricted to a different part of the working	Jowett (2017)
SUERC-69760	Loc.36	Compacted layer (005) with some bronze metal fragments and no fire-setting residues, underlying context 004. Compaction may have arisen as the tunnel was mined, partly filled in and then used for access. Late modern disturbance is restricted to a different part of the working	Jowett (2017)
Wk-14192	North-West Corner	Deposit in a short, 5m-long tunnel, 'Passage A'. Passage A had no evidence for late modern disturbance and contained two cobblestone mining tools and 20 animal bone fragments. Bone tool marks were identified on the rock walls next to its entrance	David (1998, p. 95); Unpublished database, Great Orme Mines (ref. no: VIV-98, S-40040)
Wk-14193	Loc.7c (West)	Charcoal fragments found in a deposit interpreted by the excavator as an undisturbed, primary context. 205 bone fragments (including 43 tools) and 5 cobblestone mining tools (including one showing, unusually, modification for hafting) excavated at the same location. No evidence for late modern disturbance	David (1999, p. 79); Unpublished database, Great Orme Mines (ref. no: VIV-99, S-41423)
Wk-14194	Tourist Cliff, East 8 vein (Middle)	Thin silty layer, containing 2 cobblestone mining tools and 28 bone fragments (including 5 tools). Charcoal is not listed as a find in the published excavation report and so its association with this context has to be inferred. Overlain by a nearly sterile, compacted clayey deposit of shale origin with a few modern finds. No evidence for late modern disturbance	David (2001, p. 118); Unpublished database, Great Orme Mine (reference number VIV-01, S-57017)
Wk-14197	Tourist Cliff E1 vein (South)	Deposit of morphologically prehistoric spoil in deep, narrow groove in bedrock floor. At least 84 other bone fragments, including 26 tools, as well as a scapula tool and an unusually long, complete rib tool, were excavated in the same deposit, plus one cobblestone mining tool	David (2002, p. 100); Unpublished database, Great Orme Mine (reference number VIV-02, S-61392)
Wk-14199	Tourist Cliff E1 vein (South)	Deposit containing more than 1,320 bone fragments (including 218 tools) and 15 cobblestone mining tools or tool spalls. No evidence for late modern disturbance and the position of several of the rib bone tools indicated they had been intentionally placed. Charcoal is not listed as a find in the published excavation report and so its association to this context has to be inferred	David (2002, pp. 99–100, 2003, p. 96); Unpublished database, Great Orme Mine (reference number VIV-03, S-61808)

Appendix 4: Radiocarbon dates from the Great Orme mine and the Pentrwyn copper smelting site used in Bayesian Chronological Models 1 and 1_v2 (sample details and interpretation courtesy of P. Marshall).

Laboratory Number	Radiocarbon Age (BP)	Sample	Sample Location	Interpretation
COPPER SMELTING SITE				
BETA-127076	3310 ± 80	Charcoal fragments	Pentrwyn	*TPQ:* unidentified sample
GREAT ORME MINE				
BETA-65894	3010 ± 60	Charcoal, *Quercus* sp.	South of Loc.21	*TPQ:* potential age-at-death offset
BETA-65895	3220 ± 70	Wood charcoal	Loc.7c (Underground)	*TPQ:* unidentified sample
BETA-65896	2970 ± 70	Wood charcoal	Loc.7c (Underground)	*TPQ:* unidentified sample
BM-2641	3000 ± 50	Charcoal, *Quercus* sp., immature	Loc.21	Short-lived sample provides date for mining activity at site
BM-2645	3290 ± 60	Bone	Tourist Cliff, Loc.007	Short-lived sample provides date for mining activity
BM-2751	3230 ± 50	Bone	Loc.18	Short-lived sample provides date for mining activity at site
BM-2752	3070 ± 50	Bone	Loc.18	Short-lived sample provides date for mining activity at site
BM-2802	3180 ± 80	Charcoal, *Quercus* sp. & *Alnus*	Loc.7c (Surface)	*TPQ:* potential age-at-death offset
BM-3117	3000 ± 45	Wood charcoal	Loc.35	*TPQ:* unidentified bulk sample
BM-3118	3000 ± 50	Wood charcoal	Loc.11	*TPQ:* unidentified bulk sample
CAR-1184	3370 ± 80	Wood charcoal, *Alnus glutinosa*	Tourist Cliff, Loc.007	Short-lived sample provides date for mining activity
CAR-1280	2970 ± 70	Charcoal, *Quercus* sp. & *Alnus*	Tourist Cliff, N1 ore vein	*TPQ:* potential age-at-death offset
CAR-1281	2450 ± 60	Charcoal, *Quercus* sp., *Corylus* & *Alnus*	Tourist Cliff, W1 ore vein	*TPQ:* potential age-at-death offset
HAR-4845	2940 ± 80	Charcoal, *Alnus*	Loc.5	Short-lived sample provides date for mining activity at site
OxA-14308	3344 ± 27	Human bone	Tourist Cliff, W3/N2 ore veins	*TPQ:* relationship to mining activity not clear
OxA-14309	3362 ± 28	Human bone	Loc.18	*TPQ:* relationship to mining activity not clear
OxA-5397	3150 ± 50	Charcoal, *Prunus*	Loc.7a	Short-lived sample provides date for mining activity at site

Appendix 5: Outcomes of the evaluation of radiocarbon sample 'quality' for the Great Orme mine and its associated production sites. Samples from single-entity, short-lived species with either an inferred or demonstrably secure archaeological association achieved the highest overall 'sample quality' score (3 or 4). Samples of wood charcoal of both unidentified species and entity from the mine, all with an inferred functional association with their recovery context, are indicated by a total sample quality score of '1id'. These samples were included as TPQ dates in Chronological Model 2. Samples with an archaeological association score of '0' were assessed as unsuitable for chronological modelling (indicated by 'n/a'). Archaeological association score: 0 – none; 1 – inferred; 2 – demonstrated; sample maturity score: 0 – definite or likely age-at-death offset, 1 – no offset, id – not known; sample entity score: 0 – bulk sample, 1 – single entity, id – not known.

			Archaeological Association		Sample Maturity		Sample Entity			
Lab. Number	Sample Id.	Sample Location	Assessment	Score	Assessment	Score	Assessment	Score	Total 'Sample Quality' Score	Model 2
WET ORE PROCESSING SITE										
BETA-148793	Animal bone fragments (2) with blue-green copper staining, species n.d.	Ffynnon Rufeinig	At the site level a functional relationship between the sample and ore processing can be inferred, by comparing the archaeological evidence with comparable evidence elsewhere for Bronze Age ore processing. At the deposit level, a functional relationship between the sample and the date of formation of its specific find deposit is less secure, given the documented late modern activity (including ore processing) at the site	1	Bone: short-lived sample	1	Bulk sample – could be from more than one animal. Possible that all the bone was brought to the site in the same processing event	0	2	No
GREAT ORME MINE										
BETA-65895	Wood charcoal, species/form n.d.	Loc.7c (Underground)	No reported evidence for late modern disturbance in the vicinity, hence the possibility that the sample was redeposited in this later period is low. Functional relationship between the sample and its find location can be inferred	1	Not known	id	Bulk sample from two contexts. Increased probability that it includes reworked material of earlier/later date	0	1	No
BM-2645	Animal bone fragments (3), species n.d.	Tourist Cliff, Loc.007	No reported evidence for late modern disturbance in the vicinity, hence the possibility that the sample was redeposited in this later period is low. Functional relationship between the sample and its find location can be inferred	1	Bone: short-lived sample	1	Bulk sample – could be from more than one animal. Cannot determine securely from the archaeological evidence whether all the bone relates to the same event	0	2	No
BM-2751	Animal bone, species/number n.d.	Loc.18	Despite the evidence for late modern activity, reasonable confidence that the sample was from a primary context. The charcoal fragments recovered from the sampled context were fairly large, indicating minimal post-depositional disturbance. The spoil lithology points to a functional relationship between the deposit and the working in which it was found	1	Bone: short-lived sample	1	Uncertain whether the dated sample included one or more fragments. If more than one, it is possible that each bone came from a different animal and relates to events of different date	id	2	No
BM-2752	Animal bone, species/number n.d.	Loc.18	Despite the evidence for late modern activity, reasonable confidence that the sample was from a primary context. The charcoal fragments recovered from the sampled context were fairly large, indicating minimal post-depositional disturbance. The spoil lithology points to a functional relationship between the deposit and the working in which it was found	1	Bone: short-lived sample	1	Uncertain whether the dated sample included one or more fragments. If more than one, it is possible that each bone came from a different animal and relates to events of different date	id	2	No

			Archaeological Association		Sample Maturity		Sample Entity			
CAR-1281	Wood charcoal, *Quercus* sp., *Corylus* & *Alnus* (oak, hazel, alder)	Tourist Cliff, W1 ore vein	No reported evidence for late modern disturbance in the vicinity, hence the possibility that the sample was redeposited in this later period is low. Functional relationship between the sample and its find location can be inferred (there is no mention of specific evidence for fire-setting at this location, other than the abundant charcoal)	1	Oak: potential old-wood-offset sample	0	Bulk sample from three different species. This could be due to wood procurement strategies, but the high degree of charcoal dissemination increases the possibility that the sample contained redeposited material, despite its secure archaeological context	0	1	No
OxA-14308	Human mandible (lower jawbone)	Tourist Cliff, W3/N2 ore veins	Sample was not retrieved from a distinct context (such as a grave pit). Despite the likelihood that the remains are those of an individual involved in mining at the site, no functional relationship can be demonstrated between the sample and its find location. This conclusion is reinforced by the recovery of OxA-14309, possibly from the same individual, in a different part of the mine	0	n/a	n/a	n/a	n/a	n/a	No
OxA-14309	Human clavicle (collarbone)	Loc.18	See OxA-14308	0	n/a	n/a	n/a	n/a	n/a	No
SUERC-69412	Wood charcoal, *Quercus* sp. (oak)	Loc.36	Sample was from a context comprising spoil redeposited in prehistory. No functional relationship can be demonstrated between the sample and its find location	0	n/a	n/a	n/a	n/a	n/a	No
SUERC-69760	Charcoal from thin section of branch wood, species n.d.	Loc.36	There is a probable functional relationship between the context and the passage in which it was found, but the presence of a few bronze fragments indicates inclusion of intrusive material from the overlying layer 004	0	n/a	n/a	n/a	n/a	n/a	No
Wk-14192	Animal bone, species/number n.d.	North-West Corner	Although no context details are available in the published literature, a functional relationship between the bone and its find location can be inferred from the presence of bone tool marks on the walls. The lack of late modern activity means the probability that the sample was redeposited in this later period is low	1	Bone: short-lived sample	1	Uncertain whether the dated sample included one or more fragments. If more than one fragment, it is possible that each bone came from a different animal and relates to events of different date	id	2	No
BM-2641	Branchwood charcoal and twigs, *Quercus* sp. (oak)	Loc.21	Reasonable confidence that this sample was from a primary context as the excavators were careful to point out the spatial distinction between the sampling site and the location of the late modern activity. Therefore, the possibility that this sample was redeposited in this later period is low. Function relationship between the sample and its find location can be inferred	1	Immature wood: short-lived sample	1	Single-entity sample	1	3	Yes
BM-2802	Branchwood charcoal, *Alnus* sp. (alder)	Loc.7c (Surface)	No reported evidence for late modern disturbance in the vicinity, hence the possibility that the sample was redeposited in this later period is low. Functional relationship between the sample and its find location can be inferred	1	Branch wood: short-lived sample	1	Single-entity sample	1	3	Yes
CAR-1184	Wood charcoal, *Alnus glutinosa* (alder)	Tourist Cliff, Loc.007	No reported evidence for late modern disturbance in the vicinity, hence the possibility that the sample was redeposited in this later period is low. Functional relationship between the sample and its find location can be inferred	1	*Alnus glutinosa*: short-lived sample (Barrett, Brown & Plunkett, 2019)	1	Single-entity sample	1	3	Yes

Lab code	Material	Location	Archaeological Association		Sample Maturity		Sample Entity				
HAR-4845	Wood charcoal, *Alnus* (alder)	Loc.5	Despite the evidence for late modern activity, reasonable confidence that the sample was from a primary context. The spoil lithology points to a functional relationship between the deposit and the working in which it was found. There is no evidence for later disturbance on the ledge itself and the retaining wall comprises predominately prehistoric pack. Functional relationship between the sample and its find location can be inferred	1	*Alnus*: short-lived sample (Barrett, Brown & Plunkett, 2019)	1	Single-entity sample	1	1	3	Yes
OxA-5397	Wood charcoal, like *Prunus* (blackthorn)	Loc.7a	Despite the evidence for late modern activity, the sample was from a sealed primary context. Although there was no other evidence for fire-setting at this location, the discovery of nests of charcoal point to its use for lighting. From this a functional relationship between the sample and its find location can be inferred	1	Blackthorn: twiggy, short-lived sample	1	Single-entity sample	1	1	3	Yes
Wk-14197	Animal bone tool fragment, species n.d.	Tourist Cliff, E1 vein (South)	No reported evidence for late modern disturbance in the vicinity, hence the possibility that the sample was redeposited in this later period is low. Functional relationship between the sample and its find location can be inferred	1	Bone: short-lived sample	1	Single-entity sample	1	1	3	Yes
BETA-65894	Wood charcoal, *Quercus* sp. (oak)	South of Loc.21	No reported evidence for late modern disturbance in the vicinity, hence the possibility that the sample was redeposited in this later period is low. Functional relationship between the sample and its find location can be inferred	1	Oak: potential old-wood-offset sample	0	Single-entity sample	0	1	2	Yes (TPQ)
BETA-65896	Wood charcoal, species/form n.d.	Loc.7c (Underground)	No reported evidence for late modern disturbance in the vicinity, hence the possibility that the sample was redeposited in this later period is low. Functional relationship between the sample and its find location can be inferred	1	Not known	id	Not known	id	id	1id	Yes (TPQ)
BM-3117	Wood charcoal, species/form n.d.	Loc.35	No reported evidence for late modern disturbance in the vicinity, hence the possibility that the sample was redeposited in this later period is low. Functional relationship between the sample and its find location can be inferred	1	Not known	id	Not known	id	id	1id	Yes (TPQ)
BM-3118	Wood charcoal, species/form n.d.	Loc.11	No reported evidence for late modern disturbance in the vicinity, hence the possibility that the sample was redeposited in this later period is low. Functional relationship between the sample and its find location can be inferred	1	Not known	id	Not known	id	id	1id	Yes (TPQ)
CAR-1280	Large pieces of wood charcoal, *Quercus* sp. & *Corylus* (oak, hazel)	Tourist Cliff, N1 ore vein	No evidence for late modern disturbance in the vicinity. Functional relationship between the sample and its find location can be demonstrated	2	Oak: potential old-wood-offset sample	0	Bulk sample of two different species. As the context is very secure, the charcoal was probably the residue from a single event (the mixed sample reflecting wood procurement strategies)	0	0	2	Yes (TPQ)
Wk-14193	Wood charcoal, species/form n.d.	Loc.7c (West)	No reported evidence for late modern disturbance in the vicinity, hence the possibility that the sample was redeposited in this later period is low. Functional relationship between the sample and its find location can be inferred	1	Not known	id	Not known	id	id	1id	Yes (TPQ)

			Archaeological Association		**Sample Maturity**		**Sample Entity**			
Wk-14194	Wood charcoal, species/form n.d.	Tourist Cliff, East 8 vein (Middle)	No reported evidence for late modern disturbance in the vicinity, hence the possibility that the sample was redeposited in this later period is low. Functional relationship between the sample and its find location can be inferred	1	Not known	id	Not known	id	1id	Yes (TPQ)
Wk-14199	Wood charcoal, species/form n.d.	Tourist Cliff, E1 vein (South)	No reported evidence for late modern disturbance in the vicinity, hence the possibility that the sample was redeposited in this later period is low. Functional relationship between the sample and its find location can be inferred	1	Not known	id	Not known	id	1id	Yes (TPQ)
COPPER SMELTING SITE										
BETA-127076	Wood charcoal, species/form n.d.	Pentrwyn	Sample could be residual due to animal disturbance. Functional relationship between the sample and the date of formation of the hole infill cannot be demonstrated	0	n/a	n/a	n/a	n/a	n/a	No
SUERC-44867	Twig wood charcoal, *Ilex* (holly)	Pentrwyn	Despite all the other activity at this site, a close functional relationship between the dated charcoal and pit 111 can be demonstrated	2	Twig: short-lived sample	1	Single-entity sample	1	4	Yes
SUERC-39896	Nutshell charcoal, single piece, *Corylus* (hazel)	Pentrwyn	Less secure archaeological association than for SUERC-44867 as the datable material was from the layer into which pit 111 was cut. Despite all the other activity at the site, a functional relationship between the charcoal and this smelting pit can be inferred. Represents a TPQ for the smelting pit	1	Nutshell: short-lived sample	1	Single-entity sample	1	3	Yes (TPQ)
SUERC-39897	Twig wood charcoal, single piece, *Corylus* (hazel)	Pentrwyn	Less secure archaeological association than for SUERC-44867 as the datable material was from the layer into which pit 111 was cut. Despite all the other activity at the site, a functional relationship between the charcoal and this smelting pit can be inferred. Represents a TPQ for the smelting pit	1	Twig: short-lived sample	1	Single-entity sample	1	3	Yes (TPQ)

Appendix 6: Finds of copper/bronze, tin and lead metal fragments, ingots and casting paraphernalia from hoards, isolated finds and burials in north Wales. Data sources: Portable Antiquities Scheme (PAS) online database; Historic Environment Record (HER) for Gwynedd and Clwyd-Powys; other (as specified in 'References'). Search criteria: any finds associated with the post-smelting stages of bronze production (i.e., remelting, refining, alloying, casting, smithing and finishing).

Locality	County	Object Details	Context	Context Details	Object Association	References
EARLIER BRONZE AGE						
Object Type – Metal						
Porth Dafarch	Anglesey	Fragment (bronze rivet?)	Burial	In one of three barrows/cairns on low-lying ground at inner end of inlet. Barrow 1 (now mostly destroyed) positioned on rocky knoll	Barrow 1 contained 2 urns: 1) bones of a woman and dog, 'bronze rivet', smaller accessory cup with incomplete child skeleton; 2) inverted urn containing accessory cup with ashes	HER PRN GAT1773 [Accessed online 12-03-20]; Smith (2003); Wiles (2007)
Ty'n-y-pwll	Anglesey	2 copper/bronze fragments	Burial	Barrow formerly containing seven cremation urns (three lost), some in cists	Razor, bracelet fragment	HER PRN GAT4356 [Accessed online 12-03-20]; Baynes (1909); Smith (2003)
Bryn Ellyllon, Mold ('Golden Barrow')	Flintshire	16 thin bronze fragments hammered to a wavy edge	Burial	Barrow (now destroyed)	Gold 'cape'; fragments of sheet gold; bronze fragment possibly from sword or dagger blade; large number of amber beads (only one surviving); human bone	HER PRN CPAT119803 [Accessed online 12-03-2020]; Morgan (1990)
Object Type – Stone Mould						
Bwlch-y-Maen	Conwy	Open one-piece mould for a flat axe (one matrix only). One side and upper face well-trimmed and smooth. Remaining faces are only roughly shaped	Isolated find	Recovered by ploughing of field on slope of steep-sided river valley	None when found	HER PRN GAT3741 [Accessed online 12-03-20]; Williams (1924); Britton (1963, fig. 8); Webley, Adams & Brück (2020, p. 238)
Bodwrdin	Anglesey	Bivalve mould (opposing matrices missing). Sandstone. Matrix on all four faces, two for end-looped socketed spearheads; one matrix for object with no known archaeological metal equivalents (double-looped socketed chisel)	Isolated find	Not recorded	None when found	HER PRN GAT2530 [Accessed online 12-03-20]; Britton (1963, fig. 16); Barber (2003, pp. 118–9); Webley, Adams & Brück (2020, p. 253)
New Mills, Newtown	Powys	Matrices for halberds only. Only known one-piece stone mould from Britain/Ireland with no flat axe matrices	Isolated find	Found at edge of grass field overlooking steep-sided river valley	None when found	Green (1985); Cowie & O'Connor (2009, p. 323); Webley, Adams & Brück (2020, p. 238)
Deansfield, Bangor	Gwynedd	Two bronze bivalve moulds: 1) both valves & exterior decoration for looped palstave; 2) both valves (one broken) & exterior decoration for unlooped palstave. Cast decoration of nested triangles formed from raised lines like that on examples from Norfolk	Hoard	Not recorded	Looped trident-patterned palstave; not made in either mould	HER PRN GAT2304 [Accessed online 12-03-20]; Lynch (2000, fig. 3.9); Webley and Adams (2016, fig. 1); Webley, Adams & Brück (2020, p. 256)
Conwy Community	Conwy	Two bronze valves (matching pair) for a palstave mould. Single mid-rib decoration, differing slightly on each valve	Isolated find	Found on rough pasture by metal detecting	None when found	Gwilt (2021) [Treasure Case 17.12]

Locality	County	Object Details	Context	Context Details	Object Association	References
Object Type – Mould Ingate/Casting Jet						
Flintshire	Flintshire	Copper/bronze object likely to be from the funnel/well of a composite two-part mould	Isolated find	Not recorded	None when found	Beeton (2019)
LATER BRONZE AGE						
Object Type – Bronze Mould						
Gwernymynydd, Hafod Mountain	Flintshire	Both valves of a single mould, fused together and with exterior decoration, for ribbed socketed axe	Hoard	Found on rough pasture by metal detecting, close to summit of low hill (excavation found no further information)	Four 'late' looped palstaves, two ribbed socketed axes	HER PRN CPAT17459 [Accessed on-line 13-03-20]; Webley and Adams (2016, App. 1)
Llwyn-mawr	Gwynedd	One valve, undecorated, for 'late' looped palstave	Isolated find	Not recorded	None when found	HER PRN GAT3909 [Accessed on-line 12-03-20]; Webley and Adams (2016, App. 1)
Object Type – Ingot						
Guilsfield, Rhuallt	Powys	Numerous (26) copper/bronze ingots	Hoard	Found on pasture while drain digging	More than 120 objects in total, including many spearheads (different types), spear ferrules, broken swords, chapes (scabbard covers), socketed axes, 'late' looped palstaves, two socketed gouges, faulty spear casting; other production evidence [itemised]	HER PRN CPAT96 [Accessed on-line 13-03-20]; Davies and Lynch (2000, p. 182, fig. 4.14)
Cwm Cadnant Community	Anglesey	Three copper ingot fragments (with small amounts of lead) of plano-convex or cake form	Hoard	Found dispersed and close to surface on rough pasture while metal detecting	Gold penannular ring with silvery gold inlaid decoration; probably personal ornament worn on ears/nose, not hair	Gwilt (2015) [Treasure Case 13.13]
Abergele Community	Conwy	Copper/bronze ingot or cake	Hoard	Found in a field pasture, carefully arranged in a small pit, with a probable marker stone adjacent	Two chapes; seven socketed axes; three palstaves	Gwilt, et al. (2019) [Treasure Case 17.07]
Object Type – Metal Fragment						
Guilsfield, Rhuallt	Powys	Copper/bronze fragments of different types, including 'plate scrap' (2); rivet fragment (1); sheet metal fragment (1)	Hoard	Found on pasture while drain digging	As above	HER PRN CPAT96 [Accessed on-line 13-03-20]; Davies and Lynch (2000, p. 182)
Llangollen Community	Denbighshire	Four lead alloy fragments	Hoard	Found within 2m of each other at depth of 12–15cm, on sloping grassland with slate bedrock near surface	Fragment of lead alloy (with Sn 10–20%, Cu 5–10%) socketed artefact (form unrecognised); socketed axe fragment of same composition	Oakden (2012b, 2012c) [Treasure Case 12.11]
Llangwyllog	Anglesey	Tin 'oval mount'	Hoard	Found while stream widening	Bronze leaf-shaped razor; bronze tweezers; bronze wire armlet/bracelet; small bronze rings; small bronze disks or buttons and hollow bronze ring (possible harness fittings?); hard green stone ring; wedge-shaped jet bead; 18 amber beads	HER PRN GAT2144 [Accessed online 13-03-20]

Locality	County	Object Details	Context	Context Details	Object Association	References
Object Type – Mould Ingate/Casting Jet						
Guilsfield, Rhuallt	Powys	One item	Hoard	Found on pasture while drain digging	As above	HER PRN CPAT96 [Accessed online 13-03-20]; Davies and Lynch (2000, p. 182, fig. 4.14)
POSSIBLY LATER BRONZE AGE?						
Object Type – Ingot						
Trefnant	Denbighshire	Cast copper/bronze ingot or cake	Isolated find	Not recorded	None when found	Oakden (2012a)
Object Type – Metal Fragment						
Guilsfield, Rhuallt	Powys	Copper/bronze 'blob', c.2cm wide, presumably waste or spillage during casting	Isolated find	Not recorded [but could be related to eponymous hoard]	None when found	HER PRN CPAT131069 [Accessed online 13-03-20]
Trefnant, Llanarch Hall	Denbighshire	Copper/bronze casting waste: possible casting jet	Isolated find	Not recorded	None when found	McIntosh (2008); HER PRN CPAT119230 [Accessed online 18-03-20]
Object Type – Mould Ingate/Casting Jet [?]						
Trearddur	Anglesey	Unidentified copper/bronze object: possible casting jet	Isolated find	Found while metal detecting; no further details	None when found	Allnatt (2017)